VISUALIZING MUSIC

MUSICAL MEANING AND INTERPRETATION

Robert S. Hatton, Editor

VISUALIZING MUSIC

Eric Isaacson

INDIANA UNIVERSITY PRESS
BLOOMINGTON AND INDIANAPOLIS

This book is a publication of

Indiana University Press
Office of Scholarly Publishing
Herman B Wells Library 350
1320 East 10th Street
Bloomington, Indiana 47405 USA

iupress.org

 The paper used in this publication meets the minimum requirements of the American National Standard for Information Sciences—Permanence of Paper for Printed Library Materials, ANSI Z39.48–1992.

Manufactured in the United States of America

First printing 2023

Library of Congress Cataloging-in-Publication Data

Names: Isaacson, Eric J., [date]
Title: Visualizing music / Eric Isaacson.
Other titles: Musical meaning and interpretation.
Description: Bloomington : Indiana University Press, 2023. | Series: Musical meaning and interpretation | Includes bibliographical references and index.
Identifiers: LCCN 2022059540 (print) | LCCN 2022059541 (ebook) | ISBN 9780253064721 (hardcover) | ISBN 9780253064738 (paperback) | ISBN 9780253064745 (adobe pdf)
Subjects: LCSH: Musical analysis. | Music—Charts, diagrams, etc. | Music—Mathematics | Musical scales and intervals. | Musical notation. | Visualization. | BISAC: MUSIC / Instruction & Study / Theory | MATHEMATICS / Graphic Methods
Classification: LCC MT90 .I83 2023 (print) | LCC MT90 (ebook) | DDC 780.1/4—dc23/eng/20230106
LC record available at https://lccn.loc.gov/2022059540
LC ebook record available at https://lccn.loc.gov/2022059541

Contents

Preface

My interest in viewer-centered information design grew out of an interest in user-centered computer interface design. That interest first developed in graduate school when Gary Wittlich invited me to serve as his graduate assistant to program a set of ear-training drill programs that became known as ETDrill. This was in the pre-Windows era, so the graphical interfaces needed to be developed from scratch, and it took a lot of careful planning to create a user interface that was clear and efficient—concerns that carried over when I directed development of the Windows version in the mid-1990s. In the late 1990s, I started brainstorming tools for teaching music theory. Some of these ideas were put to use when, in 1999, I received a grant from the US Department of Education that provided time (and staff) to design a self-contained, web-based course to be called Music Fundamentals Online. Because there was to be no live instructor, the interface had to work for students with only minimal explanation. This required moving from an intuitive understanding of interface design to a more research-oriented approach. In 2000, I joined a team of collaborators at Indiana University (IU) to secure a $3 million National Science Foundation grant to develop a successor to IU's groundbreaking Variations Digital Music Library Project. My roles there included using the system to help design teaching tools for faculty and exploring content visualization.

One of the outcomes was development of the Variations Audio Timeliner, which was later spun off as a separate tool, now maintained by Brent Yorgason. But it also inspired my initial interest in automated music visualization, which subsequently developed into a more general interest in the art of creating pictures that tell musical stories. By 2005, I had become aware of data visualization guru Edward Tufte. After reading a couple of his masterful books and then attending one of his eight-hour seminars, I started seeing the figures in the books, articles, and conference handouts in my field (music theory) with fresh, critical eyes. While what I saw was only sometimes terrible, it was also only sometimes great. I knew this wasn't because the field lacked brilliant, creative people—because it doesn't—so it seemed clear to me that we needed to find a way to raise awareness that design *matters* and to have access to design principles to guide the decisions that arise from that awareness. I taught my first seminar on the topic in 2007. Four seminars, three sabbaticals, twelve years of administration, and two conducting side gigs later, here we are.

Music scholars create a *lot* of images. Over the last fifteen years, I have looked critically at thousands of them, scanned more than 1,800, and from those have selected some 500 for inclusion in this volume. There are many truly remarkable images that I have not used because they escaped my notice until too late. My aim was not to curate only the best images, however, but rather to choose those that help illustrate larger principles. May you find the principles as eye-opening as I have, and may they shape how you communicate with others through images so that they enhance, rather than impede, understanding.

Acknowledgments

Many people have contributed to this project in important ways, and it gives me great pleasure to recognize their inspiration and their feedback.

Several early software development projects fueled my interest in human-centered design and laid the groundwork for the research that has culminated in this book. I want to acknowledge those who played key roles in those projects: Gary Wittlich, who invited me to serve as programmer for ETDrill while I was in graduate school; developers who worked with me on subsequent projects, including Jodi Graham (ETDrill for Windows), Art Samplaski (the Multimedia Music Theory Teaching Project), and Will Findlay and Brent Yorgason (Music Fundamentals Online); and leaders and key collaborators on the Variations2 Digital Music Library project, including Jon Dunn, Mark Notess, Don Byrd, and especially Brent Yorgason (again).

Students who have enrolled in seminars I have taught on the subject have provided valuable feedback and have inspired me with their creativity. These include Emily Barbosa, Timothy Best, Jack Bussert, Mark Chilla, Nicole DiPaolo, Nick Farmer, Loida Raquel Garza, Mitia Ganade D'Acol (who, along with Calvin Peck, scanned many of the final images), John Heilig, Melinda Leoce, Anne Liao, Tatiana Lokinha, Robin Kaleo Macheel, Martin McClellan, Michael McClimon, Stephen McFall, Ji-Yeon Min, Jack Nighan, Mitchell Ohriner, Despoina Panagiotidou, Anna Peloso, John Reef, Connor Reinman, Lev Roshal, Alexander Shannon, Emily Truell, Wade Voris, and Lizhou Wang. I am pleased to be able to include a few of their most exemplary projects here. Reid Merzbacher set some of the musical examples.

I have benefited from conversations with IU colleagues whose expertise touches on topics central to the book, especially Katy Börner, Douglas Hofstadter, and Marty Siegel. Jill T. Brasky and Julian Hook read early drafts and helped sharpen my focus. Two anonymous reviewers offered excellent suggestions that have markedly improved the book. I received valuable feedback at talks given at Ohio State University, McGill University, Sydney Conservatorium of Music, the IU Graduate Theory Association Symposium and Colloquium Series, the Society for Music Theory Music Informatics Group, and elsewhere. I am particularly grateful to Robert Hatten, whose enthusiasm for the project at a key juncture helped push it forward. The project has been supported by a subvention from the Society for Music Theory and from grants-in-aid from the Indiana University Office of the Vice President for Research, for which I am most grateful. This project could not have come to fruition without the hardworking team at Indiana University Press, particularly Allison Chaplin, Sophia Hebert, and the remarkable Nancy Lightfoot.

Finally, to my children, Anna, Sarah, and John, who all grew up and left the nest during this project, and especially to my beloved partner, Manju: your support has been invaluable.

Accessing Audiovisual Materials

Audiovisual materials are available for this book and can be viewed online at https://publish.iupress.indiana.edu/projects/visualizing-music.

VISUALIZING MUSIC

Introduction

The home-buying process typically starts by poring over photos that sellers and their agents have staged and curated to highlight a house's features. While these photos are valuable, they do not show all there is to know about the suitability of a house. Photos cannot convey proximity to amenities like public transportation and schools. For this, a local map is useful. Photos cannot show whether the position of a house on its lot will allow for the vegetable garden you want. This requires a drawing of the site plan. Photos cannot provide a clear sense of the sizes of the rooms. Hence, house listings sometimes provide room dimensions. While these dimensions can tell us whether the dining room is big enough for your grandparents' kitchen table, they don't convey how the rooms flow from one to another. Here, a drawing of the floor plan is valuable. Yet none of these image types provides the emotional punch that ultimately seals or sours a deal. That requires actually visiting the property, so you can scan the neighboring houses for over- or under-manicured lawns, assess traffic noise, walk the lot, wander slowly from room to room, take stock of room dimensions and connections, stare at the views out the windows, study the kitchen appliances, picture your car in the garage, imagine hosting family holiday gatherings or department parties, relate the home to the one you grew up in, picture your children and their friends playing in the backyard, and so on.

Nevertheless, as essential as the in-person house tour is in forming a holistic assessment and evoking the raw emotional response one needs when buying a house, the visual representations are often invaluable. Do you want to walk to a grocery store or allow your kids to ride their bikes to a park on their own? You'll want that map to note the distance and major cross streets. Do you want to know if your king-size bed and antique dresser will fit in the primary bedroom? You'll want the list of dimensions. Do you want to know if there is space to add a sunroom?

That site plan will be helpful. Do you want to compare the layouts of two houses you like? Floor plans that provide a synoptic view of each will facilitate that. And, after looking at twenty homes, you're going to appreciate the reminder the photos provide of their external and internal features. In short, the process of really *understanding* a house requires both direct sensory engagement with it and supplementary information—information that often takes the form of images: photos, drawings, or tables.

The same is true of music. To feel the emotional force of a work, one must experience it. There is no substitute for taking in a piece through the ears and, if possible, the eyes and the rest of the body. Yet music is ephemeral. We experience it in real time, after which it exists only in the traces of our memories. Although it is possible to communicate about music as it occurs (a classroom instructor playing an excerpt at the piano might ask students to listen for a particular feature ". . . here!"), many aspects of music cannot be easily conveyed this way.

Therefore, when we want to understand the architecture of a piece of music and, importantly, when we want to communicate with others about that understanding, we often draw pictures. The acoustically noisy nature of music and music's temporal essence make this more challenging than drawing a floor plan or taking photos of a house. We are aided, however, by the fact that many of our Western conceptions of music are metaphorically linked to the spatial realm. For example, we think of notes as objects and of time as flowing left to right. We therefore draw representations of physical objects, structures, and relationships as analogues to (what we conceive of as) musical objects, structures, and relationships. Musical drawings are surprisingly adept at fixing the ephemeral into something more concrete, allowing us to ponder music in *our* time, not *its* time.

We have been creating musical images for millennia. From Euclid's description of the monochord there have been neverending developments in musical notation; from Schenker's analytic system to the animated illustrations sometimes found in electronic journals and online video sites, musicians and especially music scholars have continually sought ways to convey aspects of music visually. A word commonly used for the intentionally communicative use of images is *visualization*.

This is a book about the art of music visualization. In it, I explore what makes musical visualizations effective or ineffective, with the aim of enabling those who want to communicate about music through images to do so as well as possible. I do this from the perspective of the human visual apparatus (both low-level perceptual and higher-level cognitive aspects) and current thinking on information visualization. The book and accompanying online supplement are inspired above all by the work of Edward Tufte (1995, 2001, 2003, 2006, 2020), whose fingerprints are evident throughout. I examine hundreds of images drawn from both historical sources and more recent books and articles. Although the bias is toward Western art music and theories relating to it, many images in the book pertain to popular and world musics, and the principles discussed can apply to any music. The images include symbolic musical representations (including Western music notation); representations of pitch, time, and other individual musical parameters; conceptual representations, including the depiction of musical spaces and musical structure; visual representations of music as it sounds; the presentation of music analyses; metamusical visualizations of information relating to historical trends and stylistic influences; and visualizations involving "big data."

I do not explore musicians' mental pictures of music, which are sometimes described as "musical imagery" (see, for example, the book with that title by Godøy and Jørgensen [2001]). I do not discuss purely artistic renderings of musicians or musical instruments, though there are countless beautiful examples of them. Nor do I discuss what particular images might tell us about musical life and practice at the time they were created. Those interested in iconography and related topics might see Judd's (2006) inspiring monograph on the subject.

This book has two components. The printed monograph in your hands describes core principles and illustrates them with nearly three hundred images. An extensive online supplement contains nearly two hundred additional images, chosen to further illustrate the principles or show alternative approaches to visualizing a topic. That supplement contains the majority of the many color images, although twelve colorplates are included in this printed volume. You may find it helpful to read near a computer or tablet so you can examine images in the online supplement and their descriptions when referenced.

This book is in six parts, each consisting of a series of short vignettes that focus on a particular visualization principle or musical topic. Part 1 describes the human visual system, explores and illustrates some of the principles summarized above, and concludes with a case study describing why I consider Western music notation to be a system that is highly optimized to take advantage of these capabilities and to adhere to these principles. The next three parts explore the treatment of musical spaces (part 2), musical time (part 3), and an assortment of topics, such as pitch, texture, timbre, formal models, and such metamusical topics as pitch-class set tables, schematic and procedural representations, instrument ranges, and translations (part 4). Part 5 turns to music analytical visualizations, of the sort employed by music scholars. It also looks at visualizations of information *about* music, including information tracing musical styles and their influences, and the visualization of "big data" such as that resulting from corpus studies. Part 6 provides practical advice for creating conference handouts and slideshows, placing musical examples in books and articles, and creating posters for interactive conference sessions. It concludes with a concise overview of the tools and techniques that can be employed to produce high-quality visualizations.

In discussing individual images, I explore the decisions made about visual representations—what is represented, what information has been omitted (and at what cost)—and the technical strengths in their design. I explore the state of the art in music visualization, particularly in the many color images included as colorplates and as part of the online supplement. Some of the pictures have been chosen because they represent models of good or even exceptional visual design. Others have been chosen because they represent different approaches to a particular issue. Often, I suggest improvements or provide alternatives to illustrate the principles at hand.

In assessing musical images, I am generally interested in the visualizations themselves, not the theories underlying them. With rare exceptions, I have chosen to exclude from the prose itself the names of the scholars who produced the images, though they are always credited in the accompanying captions. Each of the hundreds of scholars whose visualizations I have looked at has produced both praiseworthy images and images that could be improved. Often, the images for which I suggest improvements are those of which I am particularly fond. I consider everyone who finds themselves listed in these pages to be partners in the quest for effective visual communication about music. I hope they will accept my comments in the constructive spirit with which they are intended.

It is likewise important to acknowledge that the technical features of any image are constrained by available technology. In several instances, for example, I offer alternative drawings involving the use of color, which would have been impractical, if not impossible, when an image was originally produced. The use of shades of gray—or in some historical periods, even the drawing of straight lines—was simply not possible. I raise the issues entirely with future images in mind.

The book's primary audience includes music scholars, particularly music theorists, musicologists, music educators, and ethnomusicologists, along with composers, editors and editorial board members, publishers, and anyone whose business it is to communicate about music through visual means. For those interested in learning more, in addition to Tufte's books, website (https://www.edwardtufte.com), and seminars, the writings of physicist Richard Feynman and the pages of the *New York Times* are consistently full of information visualizations of exceptional quality.

You may not agree with all of my comments or suggestions, and I am certain many readers will have better ideas about how to depict some of the topics discussed here. That would make me very happy.

Part 1

Preliminaries

What makes a visualization compelling? In his beautiful books *The Visual Display of Quantitative Information* (2001) and *Beautiful Evidence* (2006), Edward Tufte outlines many principles of effective information design that are also appropriate for the display of musical information. I summarize the most central of these here. Many will be explored in more detail in part 1, and they will reappear throughout the book.

"Show causality, mechanism, explanation, systematic structure" (2006, 128) Effective images might tell a story, convey a sense of narrative, reveal cause-and-effect relationships, or demonstrate how something is structured—or, better, why. Western tonal music hinges on the dynamic interplay between expectations and their realization or avoidance. Theories of music are therefore nothing if not attempts to explain and define mechanism and systematic structure. Images can be powerful allies in describing our theories.

"Show comparisons, contrasts, differences" (Tufte 2006, 127) We understand things better when they are contextualized. A score of 297 means nothing without more information, such as the scale of possible values (relatively poor on a 200–800 SAT test), the norm (very high LDL cholesterol, where 100–130 is considered desirable), or a measure of human achievement (in pounds, it would tie the world record in the women's weight-lifting clean and jerk in the sixty-three-kilogram division). Likewise, the meaning of the musical note D will vary if it is followed by an E♭ (it was $\hat{7}$) or a G (it was $\hat{5}$) or an F♯ (it was $\hat{1}$). The same applies to the display of complex information. Putting information alongside other information helps give it meaning.

"Graphical excellence is nearly always multivariate" (Tufte 2001, 51; see also 2006, 130) A three-dimensional object almost always conveys more information than does a two-dimensional image of it, such as a photograph. The additional dimension provides a sense of depth and allows for the ability to look at the object from different angles. We can generalize the idea of *dimension* to that of *variable*, or some characteristic of the thing we are studying, which might take one of several values. Visualizations that express more variables are more informative than those showing fewer variables and therefore enable a richer understanding of what is being studied. Music is richly multivariate, with pitch and time being only the most obvious dimensions. Their compounds, harmony and form, are often discussed as separate dimensions, and many other dimensions, including dynamics, texture, and instrumentation (or timbre more generally), can play a role in telling a rich musical story.

"Graphical excellence requires telling the truth about the data" (Tufte 2001, 51) Outright fabrications are rare in music graphics, but sometimes the choice of representation will necessarily distort a dimension when providing another perspective, in a way analogous to how a two-dimensional projection of the Earth necessarily distorts the scale and shape of the three-dimensional globe's features. In music, this occurs regularly in the display of pitch and time information. For example, diatonic, chromatic, and frequency pitch spaces look different when mapped onto a physical dimension. Compare, for instance, staff lines, which favor diatonic distance; piano roll notation, which favors chromatic distance; and the vertical-axis mapping of pitch frequency, which features an exponential spacing rather than a linear one. Likewise, a proportional representation of musical time will be quite different depending on whether the underlying grid is based on beats, measures, phrases, or seconds.

"Graphical excellence consists of complex ideas communicated with clarity, precision, and efficiency" (Tufte 2001, 51) Effective images don't just illustrate; they illuminate. They produce responses such as "Aha, now I understand," "That makes more sense," or "I never thought of it that way." Clarity comes from making sure an image's principal aim is at the forefront. Precision involves excising ambiguities of meaning. Efficiency is achieved by using only as much information as is needed to make a point. Designers can apply this principle iteratively while creating a visualization. They can step back and ask, "Is this clear? Is it precise? Is it efficient?" Then, after making adjustments, they can repeat the process.

"Graphical excellence is that which gives to the viewer the greatest number of ideas in the shortest time with the least ink in the smallest space" (Tufte 2001, 51) This principle blends a couple of previous ones. The best images are often information rich, but to the maximum extent possible, every part of an image should be essential—that is, the information in it should pertain to the point of the image. A good application of this principle is to take out a (metaphorical) eraser and remove any ink (also potentially metaphorical) whose removal does not adversely affect the image.

"Completely integrate words, numbers, images, diagrams" (Tufte 2006, 131) Stronger images are self-documenting, inviting the viewer to linger because of the richness of the information they contain. While it is tempting to force viewers to read our carefully crafted prose to understand what we are trying to say about an unadorned musical example, the score excerpt is often a more effective place to make our essential points clear. Arrows, labels, text, and other annotations can enliven a musical example and assert our central claims, while the accompanying prose can then walk the viewer through the narrative the image conveys, to reinforce, convey subtlety and nuance, and contextualize.

Many of these principles seem self-evident, yet they are ignored surprisingly often, particularly when graphic designers sacrifice clarity, depth, and accuracy for cleverness, cuteness, or deception. Fortunately, egregious violations in musical contexts are rare compared to what one finds in many newspapers and magazines. Nevertheless, many images we find gracing our literature would benefit from an assessment based on these principles.

Part 1 expands on these principles to provide a theoretical framework for the topics of the remainder of the book. Chapter 1 gives an overview of low-level human visual processing, beginning with the eye, while chapter 2 explores the other end of the cognitive system, the role of metaphor in shaping human understanding. Chapters 3 through 12 expand on and illustrate the principles outlined above, with numerous examples from music scholarship. Chapter 13 makes the case that Western music notation has evolved into a highly effective system for communication from composer to performer.

Chapter 1 Leveraging the Power of the Brain

Visualizations should allow viewers to quickly locate the information we want them to focus on and to appropriately prioritize that information. To create visualizations that do so, it helps to understand how the human vision system works and then to leverage that system to improve the effectiveness of our visual communication. That is, when we design musical graphics, we should make these innate perceptual abilities work for us rather than against us. Descriptions of the visual system can be found in many places. Much of this chapter draws from the excellent and accessible introduction to the topic found in Ware (2008).

Light enters our eyes through the pupil and falls on an area at the back of the eye called the *fovea centralis* (or *fovea*). Our ability to resolve detail is exponentially greater at the center of the fovea than at its edges. Our eyes continually dart from point to point in the visual field to keep the object of our interest centered on the fovea. This allows us to see that object in the finest detail possible, while nearby information is available in our peripheral vision

Information that hits the fovea is sent to the primary visual cortex, an area at the back of the brain also known as V1. This area of the brain is a rapid feature detector. It acts preconsciously, almost instantaneously detecting boundaries in the visual field, allowing us to identify basic shapes and some of their key characteristics, including relative size, orientation, motion, and stereo depth, as well as color along three channels: red–green, blue–yellow, and luminance (light–dark). Our ability to detect differences in these features is automatic. If we are looking at a scene in which one object is markedly distinct in one of those features relative

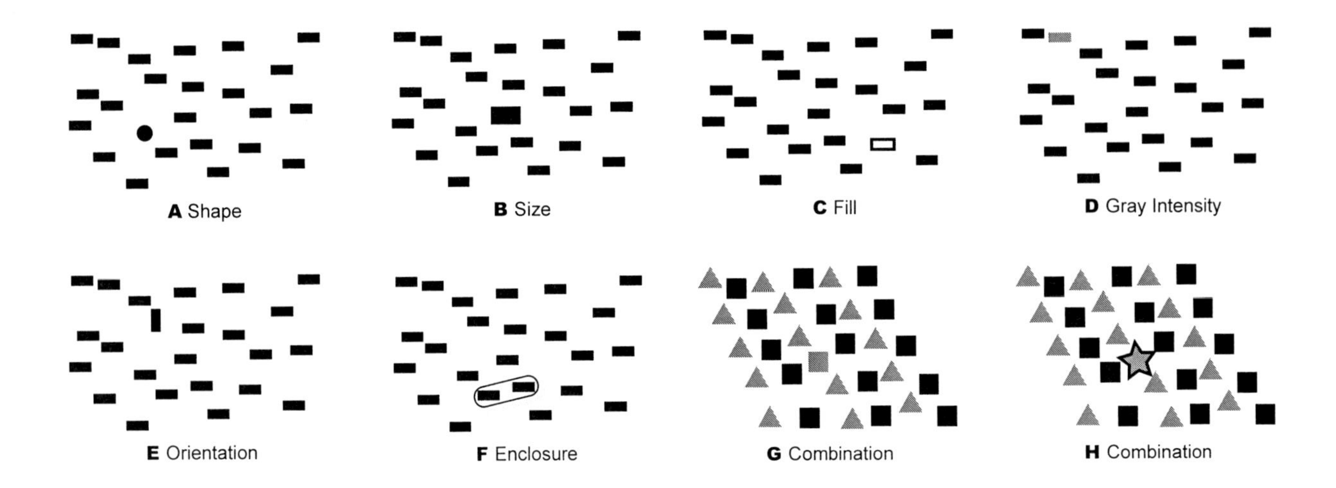

1.1 Various "pop-out" features. Based on similar images in Ware (2008).

to other objects, it will pop out at us without any cognitive effort. In figure 1.1, the pop-out effect is apparent in images A through F, where the eye immediately identifies the unique element. The effect is strongest when the background is relatively homogeneous with respect to the visual channel along which the single object is popping out. In C, for example, the lone open rectangle pops out because all the other rectangles are solid. When two or more channels compete with each other, however, the pop-out effect is greatly reduced, and we must then actively scan the visual field for the object we're looking for. In G, finding the lighter square requires an active search because it does not stand out as obviously. Its shape, distinct among the light-colored objects, is not distinct among the dark ones, and its color, distinct among the squares, is not distinct among the triangles. If we increase the number of ways the object contrasts, as in H, however, the object again pops out of the visual texture. Color and relative motion, not shown here, also produce pop-out effects.

It is easy to imagine the evolutionary survival benefits of visual pop-out effects. They helped our ancestors to avoid obstacles on the ground when running, to quickly distinguish fruit among a tree's leaves and branches and to tell good fruit from bad, to recognize when a potentially deadly snake was slithering in the grass ahead of them, and so on. In modern times, it comes in handy when sitting down in a chair, crossing the street, or facing a one-hundred-mile-per-hour fastball.

In visual design, we can take advantage of this ability to instantaneously detect visual features to ensure that the information that is most important stands out, while less important information recedes into the background. In figure 1.2, the curved segmentation lines pop out against the music, which involves largely straight lines. No effort is required to discern them.

1.2 Guerino Mazzola, Stefan Göller, and Stefan Müller, *The Topos of Music Geometric Logic of Concepts, Theory, and Performance* (Basel: Birkhauser, 2002), 119.

Our ability to quickly apprehend simple shapes would be sufficient if the world around us consisted of simple shapes, but it doesn't. Real-world objects are much more complex, and the rest of the visual system enables us to move from the almost instantaneous perception of basic shapes to engaging with the richer visual world we inhabit. Other areas of the brain involved in vision allow us to detect patterns, and then patterns of patterns, extracted from the features detected by the V1 layer and to link these patterns with information and concepts in our memory, to interpret those patterns, and to determine how to respond—for example, whether to focus more intently on one region of the field of vision, to turn our attention toward another area, or to take some action.

Effectively conveying the hierarchy of the information that we are presenting (background, middleground, foreground), separating the distinct layers of information, and allowing the viewer to quickly navigate these layers are among the central tasks in the design of visual information. When it is not immediately apparent how the information is structured, we can help the visual system by offering stronger clues. In figure 1.3, viewers will more quickly recognize familiar constellations within a seemingly random jumble of points when those points are grouped with contrasting boxes.

1.3 Creating visual hierarchy. Orion, Ursa Major, and Cassiopeia.

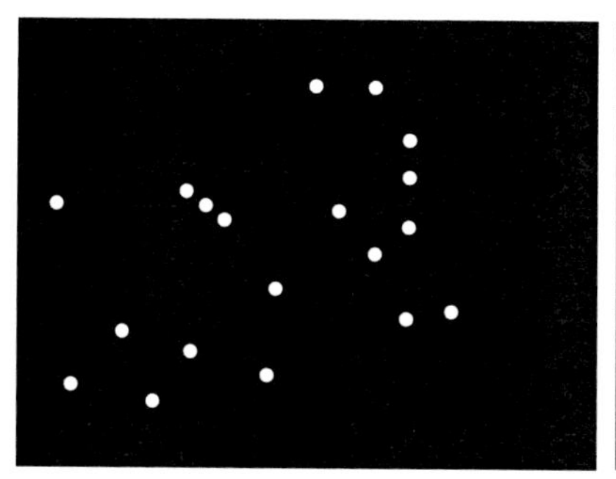

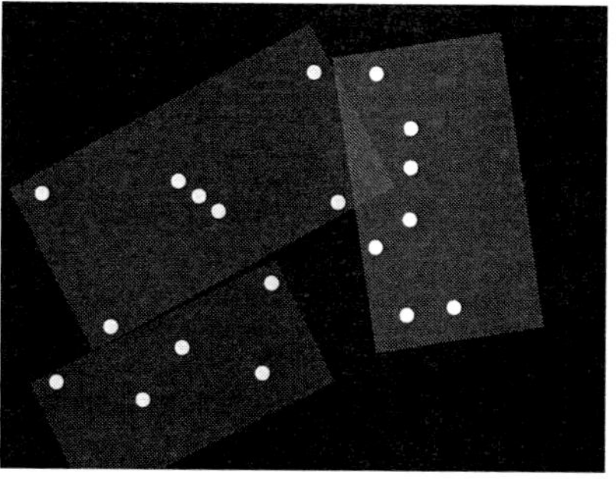

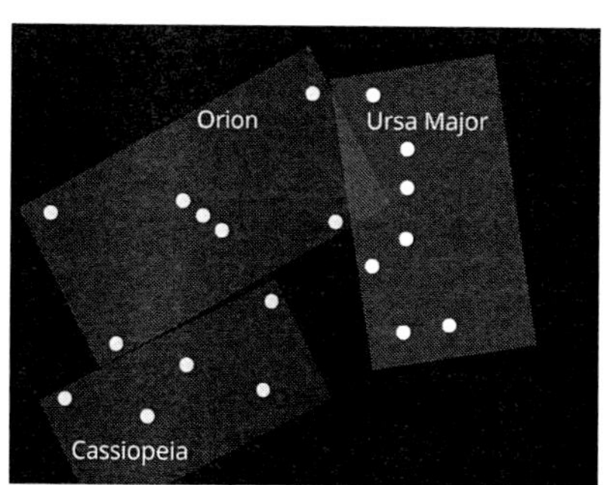

The ways in which the visual processing system can inform information design will be a steady undercurrent throughout the book, explicitly in much of part 1 and implicitly through much of the rest. When we harness the pop-out effects of the visual system and the human tendency to recognize patterns and hierarchy, we can help the eyes better appreciate what the ears already know so well—that music can be beautiful indeed.

Chapter 2 The Role of Metaphor

Since our images can only *be about* music, rather than *being* music, choosing a representation that conveys "musical truth" is essential. Sometimes a representation will simply involve labels, such as instrument names or measure numbers. Often, however, it will involve some kind of metaphor. Lakoff and Johnson's (1980) claim that metaphor is central to human cognition is widely accepted, and their work has been extended by music scholars, including Zbikowski (2002) and Larson (2012). As Spitzer (2004) details, metaphorical musical language is found throughout history. A commonly used metaphor in Western music likens notes to physical objects that exist within a spatial framework. As Zbikowski (2002) notes, the use of Pythagoras's discovery of ratios to define musical intervals involves a mapping between *size* and *pitch*, based on the relationship between the sizes of anvils and the production of various pitches (larger object = lower pitch; see fig. 2.1). This metaphysical mapping still exists to an extent through our association of larger instruments (such as the tuba) with what we describe as lower pitches and smaller instruments (such as the piccolo) with higher pitches. Nevertheless, since Pythagoras, Westerners have come to think of pitch more often using the concepts high and low. Replacing a metaphor involving size with one involving space invites all manner of extensions: ascending and descending lines, intervals as distances that can be measured, and so on. Our quantized conception of pitch also allows for the concept of scale, from the Latin word for ladder, and the distinction between steps and leaps. In visual contexts, pitch height is almost invariably rendered on a vertical axis—toward the top of the page, that is, the end that is farther from our bodies, which is how height is often drawn.

2.1 Lawrence Zbikowski, *Conceptualizing Music: Cognitive Structure, Theory, and Analysis* (Oxford: Oxford University Press, 2002), 9.

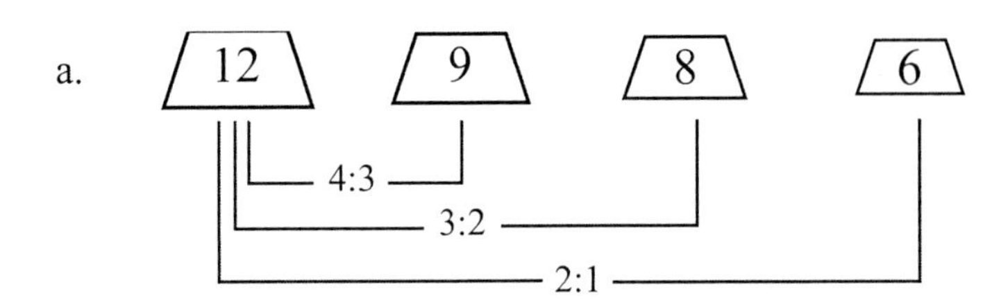

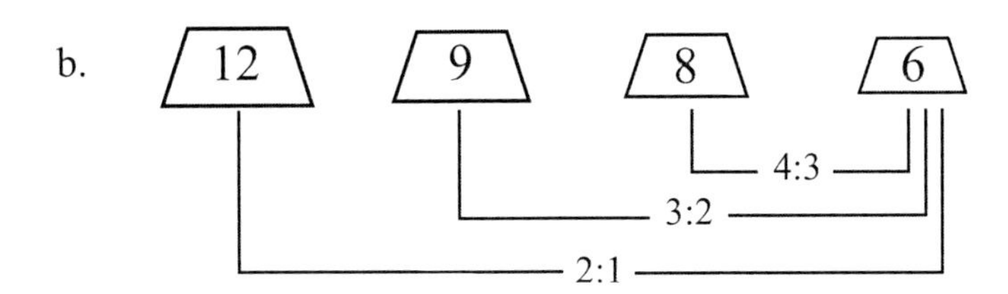

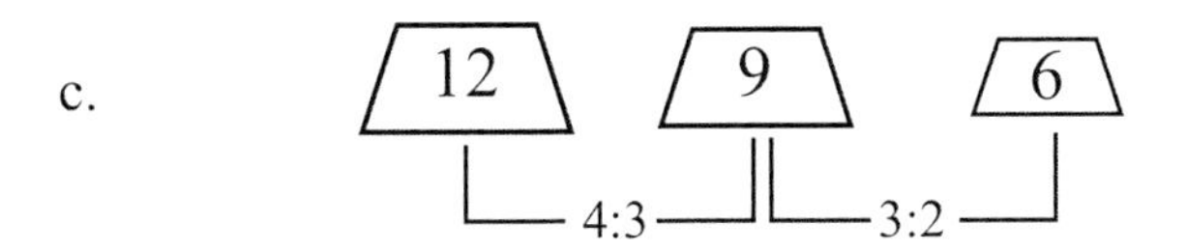

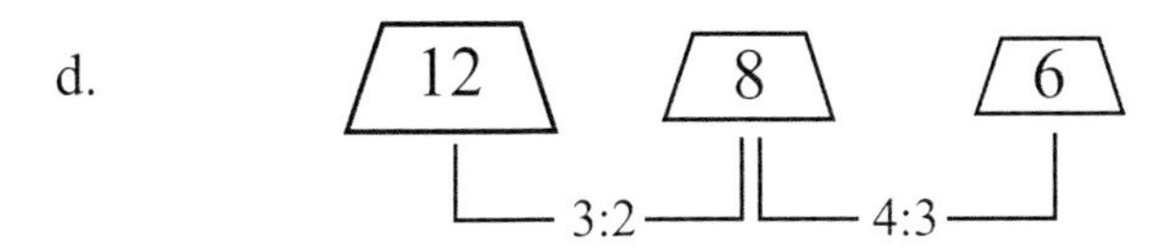

The metaphors for pitch then extend to things made up of pitches. For instance, we describe how a melody rises and falls as though charting the flight of a bird, with each pitch representing a snapshot of its location at a point in time. Larson (2012, chap. 3; see fig. 2.2 for example) explores spatial metaphors for music, detailing a theory of "musical forces," including analogues of the physical forces, gravity, magnetism, and inertia.

Visualizations of musical time frequently invoke the common metaphor of time as a line moving from left (past) to right (future). This metaphor appears in general historical timelines and even in our writing, which in many languages involves motion from left to right and then top to bottom. In music, spatial distance can represent measurable temporal spans (as suggested in online fig. 2.3). Part 3 explores this and other ways scholars measure time in musical images.

The Moving Music Metaphor

Source (Physical Motion)		Target (Music)
Physical Object	——>	Musical Event
Physical Motion	——>	Musical Motion
Speed of Motion	——>	Tempo
Location of Observer	——>	Present Musical Event
Objects in Front of Observer	——>	Future Musical Events
Objects Behind Observer	——>	Past Musical Events
Path of Motion	——>	Musical Passage
Starting/Ending Point of Motion	——>	Beginning/End of Passage
Temporary Cessation of Motion	——>	Rest, Caesura
Motion over Same Path Again	——>	Recapitulation, Repeat
Physical Forces	——>	"Musical Forces"

2.2 Steve Larson, *Musical Forces: Motion, Metaphor, and Meaning in Music* (Bloomington: Indiana University Press, 2012), 68.

Other metaphors we use to describe music include the following:

- Aspects of our physical experience, such as pace, cadence, tension, relaxation, and climax
- Aspects of the physical world, such as growth, evolution, development, trees, color (colorful music, chromatic harmony, blue notes), and light (dark vs. bright timbre, brilliance)
- References to the physical and visual arts, including texture; architecture; foreground, middleground, and background; and styles such as impressionism, expressionism, and minimalism
- References to other musical styles, particularly involving musical topics, such as galant, pastoral, and military
- Aspects of language, including grammar (online fig. 2.4), poetic meter (online fig. 2.5), and rhetoric
- States of life or moods, such as vivace, morendo, brooding, happy

Not all of these are amenable to visual representation, of course, but when designing a visualization, one must select the best metaphor for the task. It should be a metaphor that the reader will understand, one in which the relationships between the musical and nonmusical domains are clear.

Although conventional representations are often the most appropriate, there is value in considering other possibilities. Can the information be represented as ordered in some way? If so, then one could consider plotting it along a line, such as a horizontal or vertical axis. Information that is modular requires an extra dimension to plot, like two-dimensional clockface representations of the one-dimensional circle of fifths (fig. 2.6). But ordered information might also be plotted

2.6 Nicolas Slonimsky, *The Road to Music*, rev. ed. (New York: Dodd, Mead, 1960), 41.

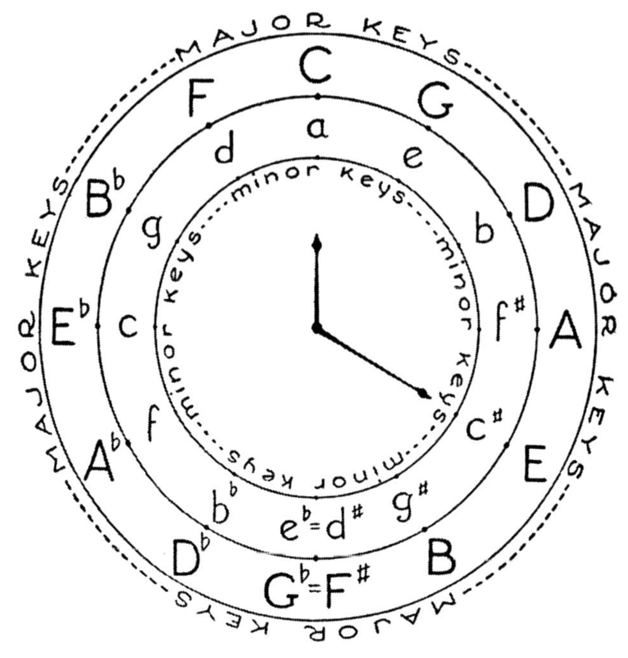

using a dark-to-light scale. Is mapping to another kind of physical space suggested? Schenker's influential analytic system of tonal hierarchy (discussed more fully in chap. 48) might evoke a sense of depth perception such as that associated with perspective drawing (online fig. 2.7). Or it might suggest architectural metaphors with an inner core, the skin, and the finishing touches. It is important to remember, however, that for commonly employed metaphors, such as the high–low spatial metaphor for pitch, the more entrenched a convention, the greater the justification needed to defy it.

Chapter 3 Multivariate Images

The spatial dimensions of height, width, and depth are of course familiar. Information can be dimensional as well. For instance, airline data includes departure date and time, arrival date and time, departure and arrival (and layover) airports and gates, airline, flight number, plane type, meal options, fares, mileage, and seat availability. Fantasy baseball typically operates on ten statistical dimensions (including home runs and earned run average), plus categorical dimensions such as players' names and positions. Since these informational dimensions are often not orthogonal to one another—that is, we don't think of them as sitting at right angles to one another in some multidimensional space—it is easier to think of such information as tracking multiple variables rather than multiple dimensions.

The ten-day weather forecast in figure 3.1 (colorplate 1) is intensely multivariate. The image plots over 2,300 data points, spanning eighteen variables that are displayed in a variety of ways. At the top, both date and day of week are listed for the next ten days (1). The forecast daily high (2) and low (3) temperatures are listed using conventional colors for hot (red) and cold (blue). Next is an overall daily forecast, rendered in both graphical and textual form (4). Below this is the predicted type (5) and amount (6) of precipitation. The former is shown in blue when precipitation is anticipated and gray when not. The distinct icons for snow (❄) and rain (💧) serve the function of a currency symbol, which indicates how to interpret the number of inches (snow has different liquid content than rain). Next come daily sunrise and sunset times (7). Note how the hours of darkness extend downward through the remainder of the graphs using subtle shading.

Below this information, the graphic shows hourly predictions of temperature (8) and dew point (9). These lines use strongly contrasting colors of similar saturation, though the colors have no particular meaning in this case. Since the gridlines do not actually convey any information, it is appropriate that they are shown in gray, except that the current time is highlighted by a heavier gray line, and the freezing mark (32 degrees Fahrenheit) is indicated with a pale-blue line.

Next come hourly projections of humidity (10) and barometric pressure (11), again in contrasting colors. The lines have different scales. The one for humidity is on the left, and the one for barometric pressure is on the right. This segment of the graph also shows the projected amount of cloud cover, shaded gray (12), and the chance of precipitation, shaded blue when the precipitation is expected to be rain (13) and violet when it is expected to be snow (14). These three measures are cleverly nested—it cannot snow unless there is precipitation, and there can be no precipitation without cloud cover. The shading is transparent, which allows the underlying grid to remain visible. The next segment of the display shows projected precipitation, measured in total accumulation (15) and hourly liquid content (16). Finally, at the bottom are projected wind speed (17) and direction (18) at four-hour intervals. The wind speed uses a standard line graph, while the wind direction is indicated by the orientation of the icon.

Multivariate images such as this allow us to compare variables and see their interdependence. For instance, we can see the often-inverse relationship between barometric pressure and cloud cover as the black line moves in roughly contrary motion with the gray shaded area. In short, this is a masterful assembly of disparate information mapped through time, sensibly aggregated by category. All that is missing is confidence that the predictions will actually come to pass.

Few topics lend themselves to the display of eighteen variables, but the best visualizations are indeed multivariate—they "escape flatland," as Tufte says (2006, 137). When creating a visualization, we should always ask ourselves whether additional variables could enhance the informational story being displayed. Conveying disparate dimensions while keeping them separate and yet appropriately integrated is one of the challenges in the design of information visualization. (The suggestion to add informational layers might seem to contradict the admonition to conserve ink outlined in the introduction to part 1 and explored further in chapter 8. But the act of eliminating wasteful ink and that of adding useful ink are both in service of enhancing visual communication.)

Music is richly multivariate. All the familiar dimensions of music—pitch, time, timbre, texture, dynamics, articulation, form, and so on, along with their subcategories, combinations, and interactions—can vary in value. Music visualizations can therefore also be multivariate. Four examples follow.

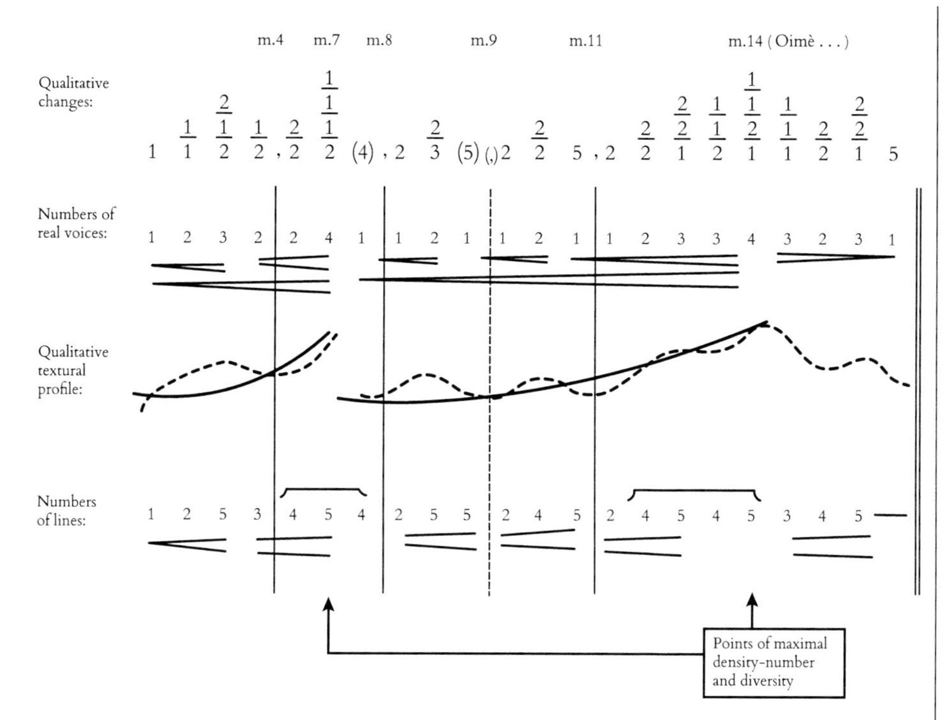

3.2 Wallace Berry, *Structural Functions in Music* (New York: Dover, 1987), 260.

In figure 3.2, several variables drawn from a Gesualdo madrigal flow left to right through time. The image combines discrete data with graphical interpretations of them that show processes of growth and decay. The numbers that count "real voices" and "lines" are underlaid with hairpin dynamic markings that make the growth processes clearer. The density of the "qualitative changes" in the top row are shown graphically in the third: a dashed line marks the foreground contour, while a thicker trend line maps out a deeper-level pattern of textural growth.

Although many musical images feature time as one dimension, not all do. Figure 3.3 provides eight kinds of information about hexachords, arranged in tabular form. (Chapter 9 explores the display of tabular data in general, while chapter 37 explores pitch-class set tables in particular.) See online figure 3.4 for another multivariate image.

Figure 3.5 (colorplate 2) is a model of excellence in multivariate visual design and will serve to preview design principles explored in more detail later. The image shows the characteristics of standard orchestral instruments. It contains at least seven dimensions of information. It provides instrument names in English, with translations into Italian, French, and German. It groups instruments into families using color (reds for woodwinds, greens for brass, blues for strings) and brackets. For each instrument, it displays ranges for elementary school, high school, and professional musicians, coded by both color shade and line width. Grayscale bars

Name	ic-vector	Z	TC (Transpositional Combination)	CUP (Complement Union Property)	Inclu-sion Family	Property Family	Invariance Vector
6-1[01235]	[543210]		3-1[012] @ 3			All-comb	11001100
6-2[012346]	[443211]		3-3[014] @ 2	6 & 4-1[0123]		P2-6	10000100
6-3[012356]	[433221]	6-36[012347]	3-1[012] @ 4	Z: 6 & 4-2[0124]	m	PZZ2-6	10000000
6-4[012456]	[432321]	6-37[012348]	3-1[012] @ 4	Z: 6 & 4-3[0134]	N	PZZ2-6	11000000
6-5[012367]	[422232]		3-8[026] & 1		t		10000100
6-6[012567]	[421242]	6-38[123478]	3-1[102] @ 5	Z: 3 & 4-9[0167]	t	MZ PZZ4-9	11000011
6-7[012678]	[420243]		3-1[012] @ 6 3-4[015] @ 6 3-9[027] @ 6		t	All-comb	22222222
6-8[023457]	[343230]		3-6[024] @ 3			All-comb	11111111
6-9[012357]	[342231]		3-4[015] @ 2	6 & 4-10[0235]		P2-6	10100101
6-10[013457]	[333321]	6-39[023458]	3-2[013] @ 4	Z: 6 &4-4[0125]	N	PZ3-12 PZZ2-6	10000000
6-11[012457]	[333231]	6-40[012358]		Z: 6 & 4-11[0135]		MZ PZZ2-6	10000010
6-12[012467]	[332232]	6-41[012368]		2 & 4-9[0167]	m	PZ4-9	10010000
6-13[013467]	[324222]	6-42[012369]	3-10[036] @ 1		O	PZ4-28	11000000
6-14[013458]	[323430]		3-4[015] @ 1	3-6[024] & 3-12[048]	N		10101010
6-15[012458]	[323421]			6 & 4-7[0145]	N	P2-6	10000100
6-16[322431]	[322431]			6 & 4-17[0347]		P2-6	10100101
6-17[012478]	[322332]	6-43[012568]		3-5[016] & 3-12[048] 4 & 4-9[0167]		PZZ3-12 PZ 4-9	10010000
6-18[012578]	[322242]		3-8[026] @ 5		t		10000100

just to the right of the wind instrument ranges show the relative loudness of each register. Other nice details include qualitative descriptions of the various registers of the woodwind and brass instruments, the location of pedal tones in trumpet and trombone, the available natural harmonics for the string instruments, and octave designations for ranges that extend an octave or more beyond the staff.

When designing visualizations of musical information, the first task is to determine which features in particular will make the intended point and to ask what additional information would further enrich the display, provide useful additional context, or enhance understanding. The next is to choose an appropriate representation for each dimension. Finally, the representations must be integrated in a way that is visually harmonious, attractive, and truthful.

Many factors go into deciding how to represent the various dimensions: Is the information in a dimension that is ordered in some way? If so, is the information continuous or discrete? Is it measured internally (though our perceptions), externally (through some abstract standard), or a combination of the two? Is it represented in a modular space of some kind? Is the information sharp or fuzzy? Is it categorical? Is it nested or hierarchical? Is it temporal? Is an understanding of the dimension informed by some kind of metaphor? How do the parameters interact? Are the dimensions independent, parallel, or orthogonal? For purposes of the visualization task, which dimensions are primary, and which are secondary? Is the primary interest in complex musical objects such as chords, pitch-class sets, melodies, voices, sections, or whole works?

Different kinds of information invite different kinds of visual presentation. The decisions made in the design stage are critical in facilitating understanding of the completed product. Information that is ordered, whether lower to higher, earlier to later, softer to louder, or smaller to larger, can be represented in a number of ways. Variations of the horizontal and vertical axes are most common—they are often the easiest to represent visually, and they align well with the spatial metaphors through which we understand much about music. Standard music notation is based on just such a metaphor, as we will detail in chapter 13. But such characteristics as size, a color scale (light-to-dark, red-to-green), descriptive words (such as the standard dynamic and tempo labels), or simply numbers can also represent dimensional data. Information that is not dimensional, such as categorical information, can be conveyed using words, colors, images, or complex diagrams.

***Facing*, 3.3** Robert Morris, *Class Notes for Advanced Atonal Music Theory*, vol. 2 (Lebanon, NH: Frog Peak, 2001), appendix A.

CHAPTER 4 Telling a Story

Peter Westergaard concluded his keynote address to the 1994 meeting of the Society for Music Theory with these literally poetic words:

> Tones are driven,
> drawn, deflected by the energies
> of pitch-time space. . . .
> At any given
> level, pitches describe trajectories
> through fields whose forces are defined by pitches
> just one level back. Time trips are filled
> with unexpected bumps and lurches—which is
> why we have to find a way to build
> a geometry of sounds in time that shows
> not just how music is, but how it goes. (Westergaard 1996, 21)

In that lecture, crediting Lewis Rowell, Westergaard observes that, when discussing music, the ancient Greeks often used spatial language to describe pitch and time, an orientation that Westergaard traces through the history of Western music theory. The Greek conception of music, however, was geometrical, "all state and no process," to again quote Rowell (1996, 5). Westergaard argues, however, that treating music as fixed ignores the *experience* of music, its pacing, its ebb and flow. Westergaard admonishes music scholars to devote more attention to the dynamic aspects of music, not just to the properties of musical events but to their behaviors and their complex interactions with one another and their environment.

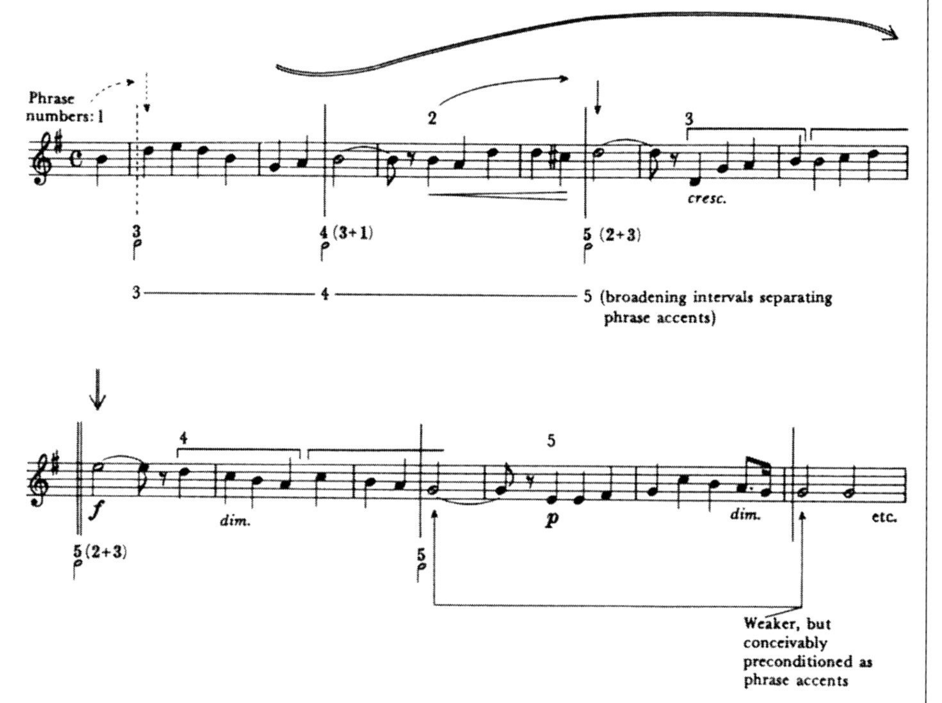

4.1 Wallace Berry, *Structural Functions in Music* (New York: Dover, 1987), 334.

An interest in "how music goes" leads one to explore things like cause and effect, expectation and fulfillment, continuity and discontinuity, growth and decay, stability and instability—to ask not just what and where but also how and why. Westergaard's plea aligns with Tufte's (2006, 128) admonition to "show causality, mechanism, explanation, systematic structure."

While we typically associate storytelling with prose, there are many ways to convey a sense of narrative graphically. Indeed, some of the most compelling images are those that convey cause and effect, antecedent and consequent, and a sense of process—in short, those that tell a story. Here are some images that do this admirably.

Figure 4.1 uses a dashed, then a solid, and finally a sweeping double arrow to direct us to hear a melody not just as a sequence of notes but as a narrative of flowing, directed motion. The arrows guide our listening and express an interpretation of the melody's shape that we can easily reproduce. (Meanwhile, long extra bar lines and meter signatures invite us to consider a metrical reading at odds with the notation.)

The analysis in figure 4.2 offers a compelling narrative of the cello part from "God-Music" in George Crumb's *Black Angels*. It conveys the dramatic shape of the

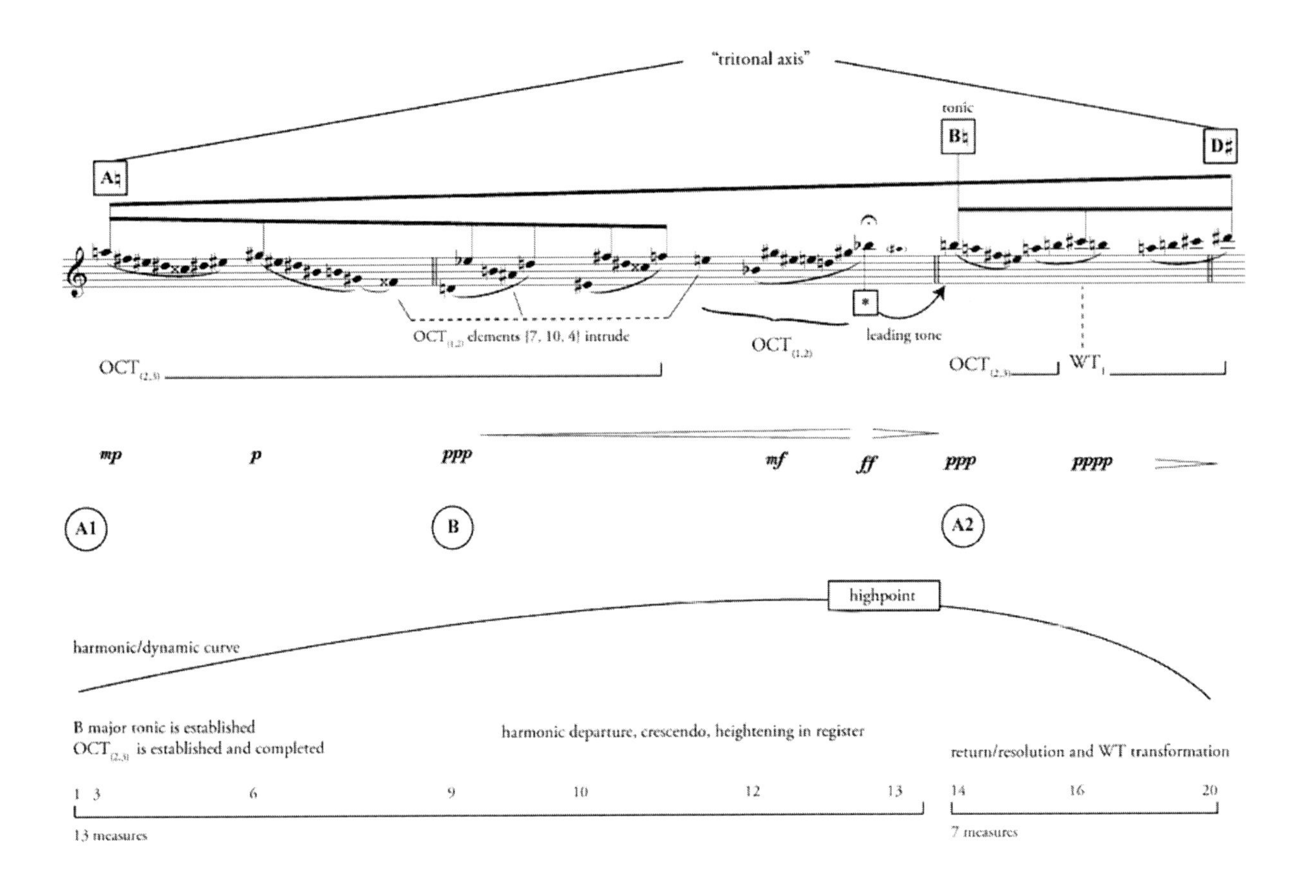

4.2 Blair Johnston, "Between Romanticism and Modernism and Postmodernism: George Crumb's *Black Angels*," *Music Theory Online* 18, no. 2 (June 2012): ex. 4, http://www.mtosmt.org/issues/mto.12.18.2/mto.12.18.2.johnston.html.

passage through the arc that leads us to the word *highpoint* in the lower part of the example, the hairpin crescendo in the middle that invites us to hear the dramatic shape, and the angled lines at the top of the image that direct our attention to the "tritonal axis." My choice of the words *lead*, *invite*, and *direct* highlight the active function of these annotations. The image guides, but the viewer is an active participant in its narrative.

Figure 4.3 chronicles the gradual revelation of the phrase "O Martin Luther King" in Luciano Berio's *Sinfonia*, movement 2. With time flowing left to right and top to bottom, we see the introduction of first the vowels, then the consonants, then full words, and finally the full phrase. See also online figures 4.4 and 4.5.

Note how successfully figure 4.6 conveys a graphical narrative of tonal progression. The upper part of the image expresses the probability of transitions (given in tabular form in the lower part) between chord roots in a collection of Baroque music. The thickness of the lines indicates the relative commonness of various progressions, making it easy to see which successions are most common. The layout illustrates the asymmetry in most chord pairs—V to I is more common than I to V, I to IV is more common than the plagal motion IV to I, and so on. A survey of the

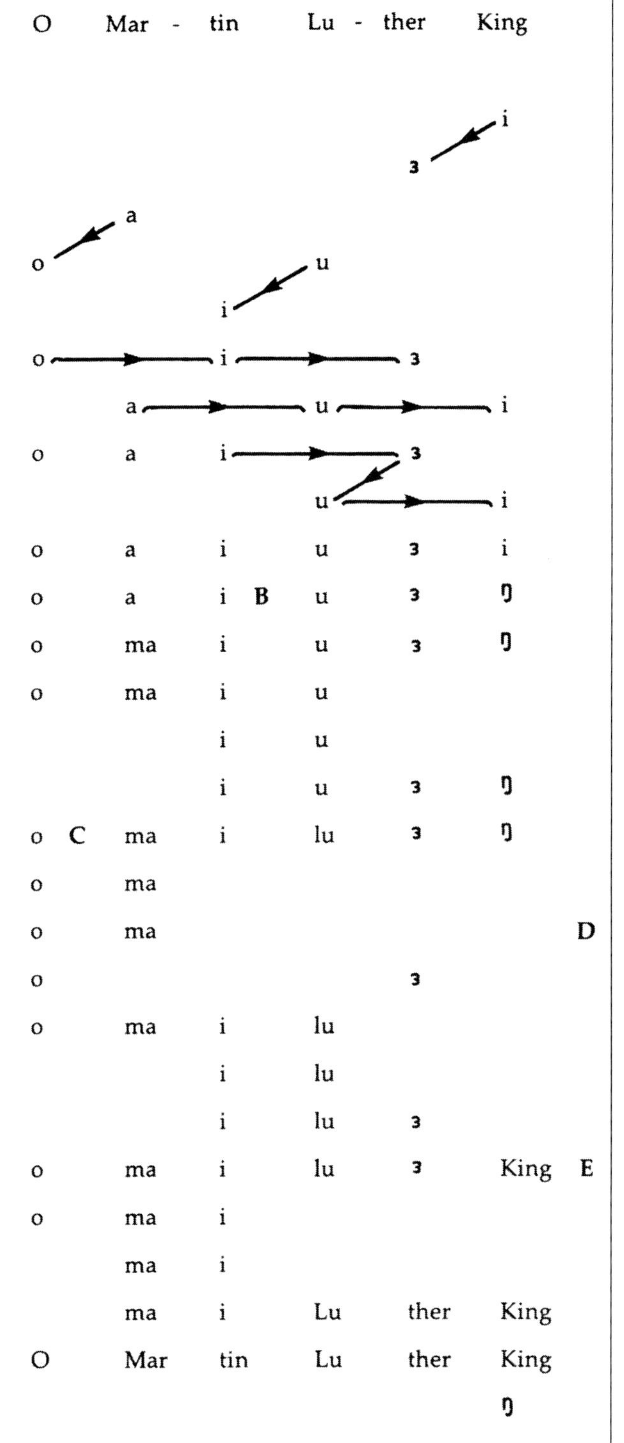

4.3 David Osmond-Smith, *Playing on Words: A Guide to Luciano Berio's "Sinfonia"* (London: Royal Musical Association, 1985), 38.

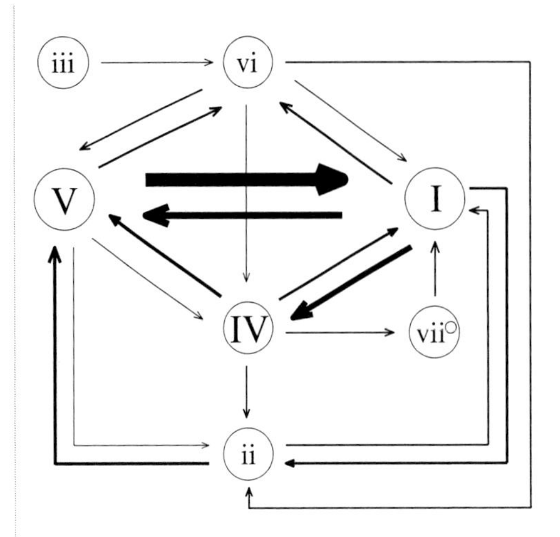

4.6 David Huron, *Sweet Anticipation: Music and the Psychology of Expectation* (Cambridge, MA: MIT Press, 2006), 251.

	→I	→ii	→iii	→IV	→V	→vi	→vii°
I→		0.143	0.025	0.309	0.376	0.120	0.025
ii→	0.274		0.040	0.032	0.564	0.040	0.048
iii→	0.063	0.127		0.340	0.063	0.383	0.021
IV→	0.304	0.132	0.025		0.380	0.025	0.132
V→	0.793	0.021	0.021	0.039		0.119	0.005
vi→	0.155	0.232	0.069	0.139	0.286		0.116
vii°→	0.583	0.000	0.027	0.138	0.194	0.055	

lines leaving a particular chord will indicate which other chords are most likely to follow it. The addition of numerical probabilities next to the lines would improve the image, but the story it tells is clear. Online figures 4.7 and 4.8 detail more abstract narrative readings.

While most Western tonal music is goal oriented and thus readily lends itself to narrative visual representations, as several of the images in this section attest, surface-level teleology is hardly a prerequisite for a storytelling image. Rather, an image tells the story that its designer, not the music, wishes to tell.

CHAPTER 5 Facilitating Comparison

We understand things better when we have something to compare them to. The comparison of something unknown to something known helps us understand the unknown better. When photographing an artifact, archaeologists place a ruler next to it to provide a sense of perspective. In music, the staff serves a similar function, allowing us to see the relative (diatonic) distance between pitches. (We will look at the staff from a cognitive perspective in chap. 13.) Graph paper can serve the same function for measuring distance in chromatic space, as in online figure 5.1. In each of these cases, primary objects (archaeological objects, pitches) gain meaning from their placement on a background grid.

Objects can also serve as references to one another. Figure 5.2 shows a rhythmic pattern first in standard notation and then as grouped by rests in the original. The image illustrates the difference between two modes of listening (formal vs. figural). Each makes more sense when put in the context of the other: each helps explains the other. Likewise, the spectrograms of figure 5.3 allow us to understand four vowel colors better because they are shown in the context of the others. The relative

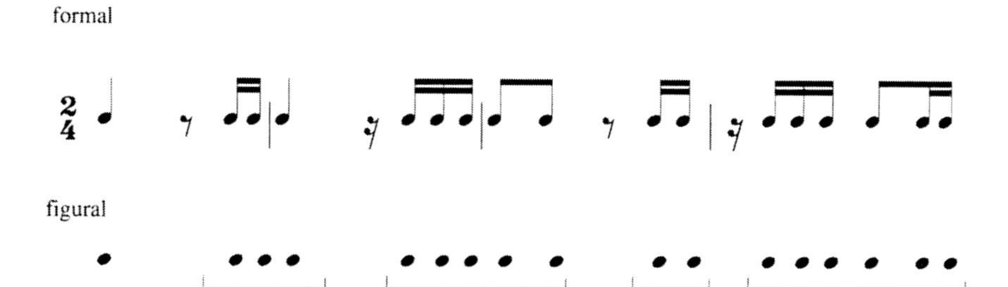

5.2 J. Kent Williams, *Theories and Analyses of Twentieth-Century Music* (Fort Worth: Harcourt Brace, 1997), 119.

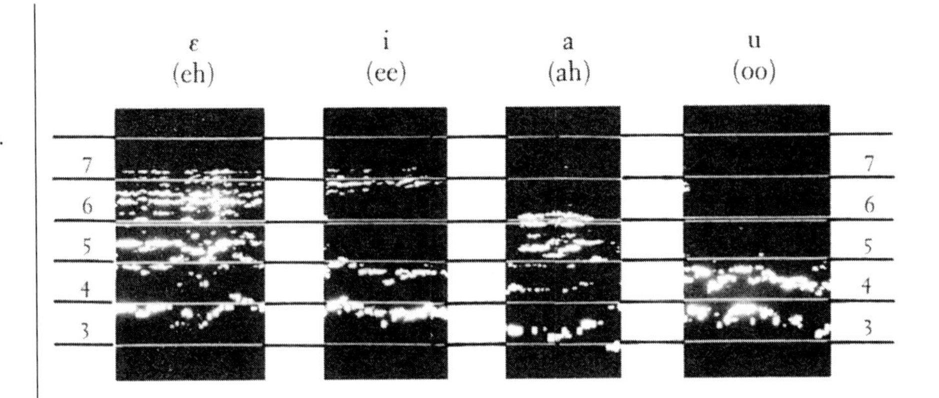

5.3 Robert Cogan, *New Images of Musical Sound* (Cambridge, MA: Harvard University Press, 1984), 10.

purity of *u* is more evident when we can see it alongside the roughness of *eh*. The layout encourages us to try out the sounds ourselves.

Tables are another technique for making comparisons. Figure 5.4 lists the characteristics of six variation genres. Reading left to right, we can see how a particular characteristic manifests in each genre. The columns, ordered by the typical length of the repeated thematic information (from "one or two measures" through "16 to 32 measures"), situate each variation type in the context of the others and, as a result, the genre as a whole. We will discuss the utility of tables in more detail in chapter 9. (See also online figs. 5.5 and 5.6.)

Comparative music notation is also an effective strategy. Through rhythmic alignment, figure 5.7 allows us to consider the similarities and differences among the many variations on a primary theme in Steve Reich's *The Desert Music*.

Showing comparisons is the first of Tufte's (2006, 126) principles of analytic design. When designing an image, it's worth taking time to ask whether providing a broader context, using a measuring stick of some kind, or adding another image would enhance or clarify the meaning.

COMPARATIVE TABLE OF DIVISION I.

<table>
<tr><th colspan="3">Basso ostinato</th><th colspan="3">Variation-forms</th></tr>
<tr><th>Ground-motive.</th><th>Ground-bass, or Basso ostinato proper.</th><th>Passacaglia.</th><th>Chaconne.</th><th>Small Variation-form.</th><th>Large Variation-form.</th></tr>
<tr><td colspan="6">Thematic basis</td></tr>
<tr><td rowspan="2">Motive of one or two measures, chiefly in bass.</td><td rowspan="2">Phrase of two or four measures, chiefly in bass.</td><td colspan="2">Period (or repeated Phrase) of 8 measures, of which the burden is</td><td rowspan="2">Double-period or 2-Part Song-form, 16 measures. Melody (chords, or bass, or formal design).</td><td rowspan="2">Usually 2- or 3-Part Song-form, 16 to 32 measures. Melody (chords, etc.)</td></tr>
<tr><td>usually the bass-line.</td><td>usually the chords (incidentally the melody or the bass).</td></tr>
<tr><td colspan="6">Distinctive traits</td></tr>
<tr><td rowspan="3">None.</td><td rowspan="3">None.</td><td>Minor mode.</td><td>Minor or major mode.</td><td rowspan="3">None.</td><td rowspan="3">None.</td></tr>
<tr><td colspan="2">Triple measure</td></tr>
<tr><td>($\frac{3}{4}$, $\frac{3}{8}$ or $\frac{6}{8}$, $\frac{3}{2}$).</td><td>($\frac{3}{4}$).</td></tr>
<tr><td colspan="6">Treatment</td></tr>
<tr><td>Homophonic; changing melodic, rhythmic and harmonic forms in upper (added) parts.</td><td>Homophonic; changing forms, phrase-group design.</td><td>Preponderantly polyphonic; thematic accompaniment of bass-theme.</td><td rowspan="3">Preponderantly homophonic; varying patterns of (chiefly) harmonic figuration, with approximate retention of Melody. Partly continuous, partly separated, variations.</td><td colspan="2">Chiefly homophonic, occasionally polyphonic; the variations completely separated, as a rule.</td></tr>
<tr><td colspan="3" rowspan="2">Structure continuous, with ordinary (transient) cadence interruptions.</td><td rowspan="2">Form of Theme retained, with unessential extensions.</td><td rowspan="2">Form of Theme treated with greater freedom, and transformed by Insertions and extensions. Elaboration, as well as Variation.</td></tr>
<tr></tr>
</table>

5.4 Percy Goetschius, *The Larger Forms of Musical Composition: An Exhaustive Explanation of the Variations, Rondos, and Sonata Designs, for the General Student of Musical Analysis, and for the Special Student of Structural Composition* (New York: G. Schirmer, 1915), 2.

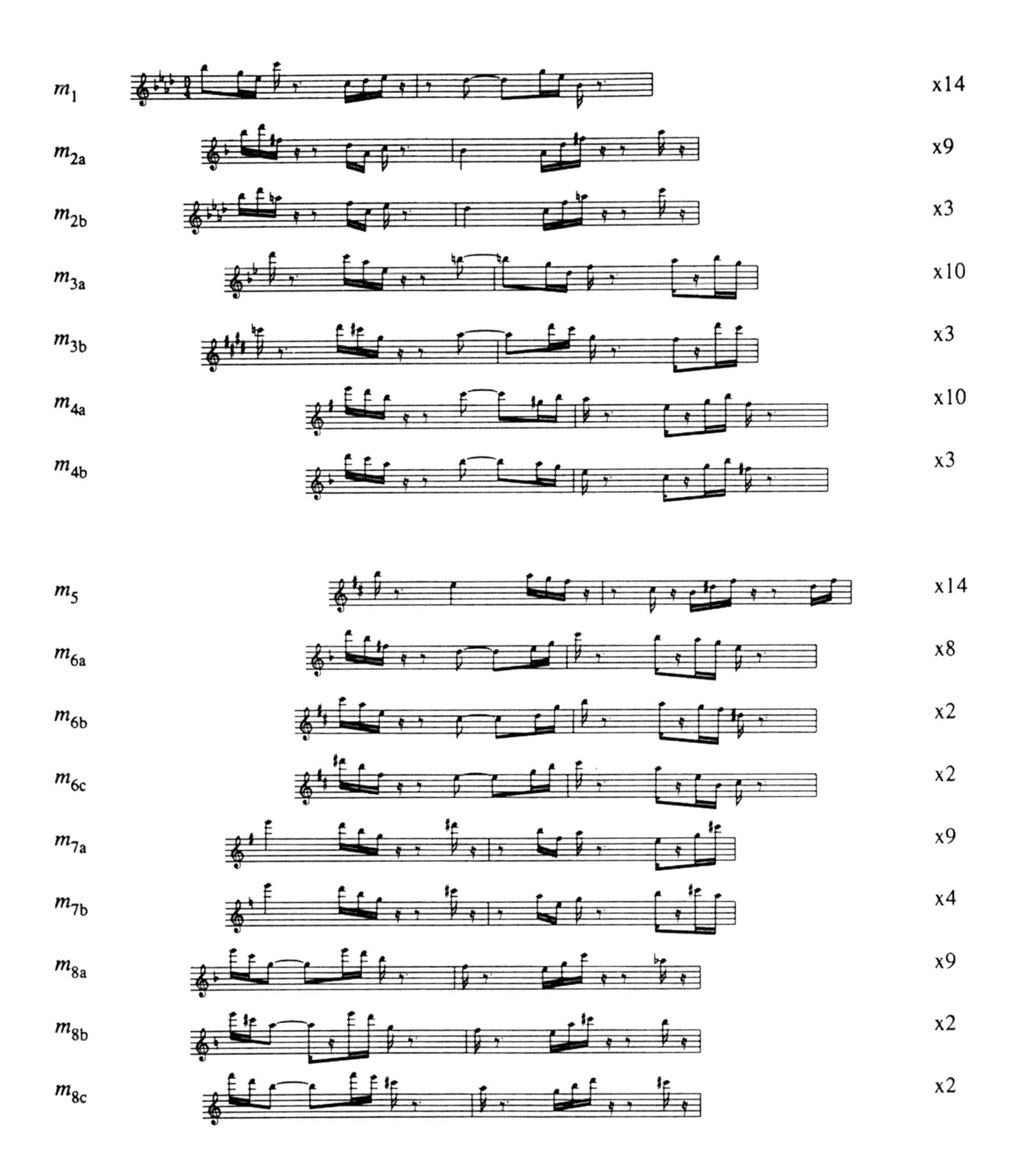

5.7 Ian Quinn, "Fuzzy Extensions to the Theory of Contour," *Music Theory Spectrum* 19, no. 2 (1997): 234.

CHAPTER 6 Information Layers

Often, visualized information can be thought of as occupying multiple conceptual layers, with each layer serving a different role and the layers differing in importance. A key part of the visual design process is determining how to differentiate these layers of information so the viewer understands their hierarchy.

Figure 6.1 contains at least three distinct layers: The first, the staff, serves as a grid but does not itself carry information. A second layer involves the musical notation—the clefs, noteheads, and accidentals—which do represent information.

6.1 Allen Forte, *The Structure of Atonal Music* (New Haven, CT: Yale University Press, 1973), 140.

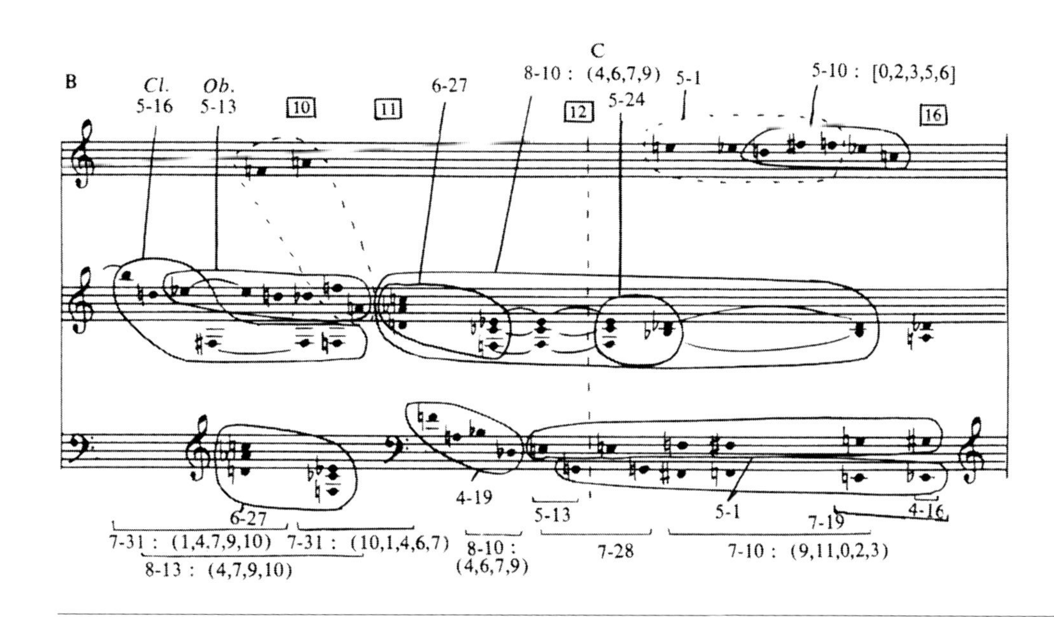

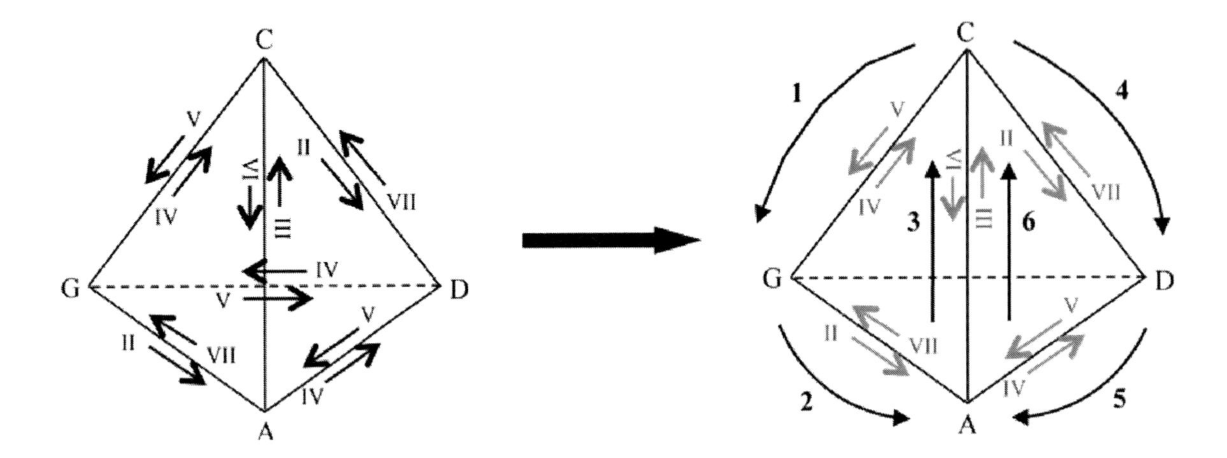

6.2 René Rusch, "Crossing Over with Brad Mehldau's Cover of Radiohead's 'Paranoid Android': The Role of Jazz Improvisation in the Transformation of an Intertext," *Music Theory Online* 19, no. 4 (December 2013): ex. 5c, http://www.mtosmt.org/issues/mto.13.19.4/mto.13.19.4.rusch.html.

Their characteristic shapes are visually distinct from the staff (see chap. 13). The final and most important layer contains the analytical apparatus overlaid on the notation, including the circles and brackets marking segments, the pitch-class set labels, and the lines that sometimes connect the segments and labels. Even though all the elements are the same color and are similar in thickness, their distinct shapes and orientations help keep the layers visually separated.

The left image in figure 6.2 contains essentially two layers: the first contains the prism and its vertices (chord symbols C, D, G, A), and the more important second layer includes the roman numerals and arrows that represent the functional harmonic relationships among these vertices. This second layer does not stand out from the first, and its significance is thus weakened. However, when a third layer is added in the image on the right (notating a path through the harmonic space), the second layer changes from black to gray, which makes it easier to distinguish the first two layers from each other. The curved lines marking several of the arrows and the straight lines in the center (labeled 3 and 6), which stand out from the gray symbols surrounding them, make the new third layer visually distinct. Online figure 4.4 (referenced earlier) employs the same strategy, separating foreground from background even more effectively by placing the background in gray.

The schematic representation in figure 6.3 also provides a good model for such images. It brings together multiple layers of textual information and, through its basic layout and the use of arrows, tells the story of the sonata form exposition. (The C box, because it has considerably more text than the others, is wider, unintentionally implying that the section is longer than other sections. Proportion is discussed in chapter 26.)

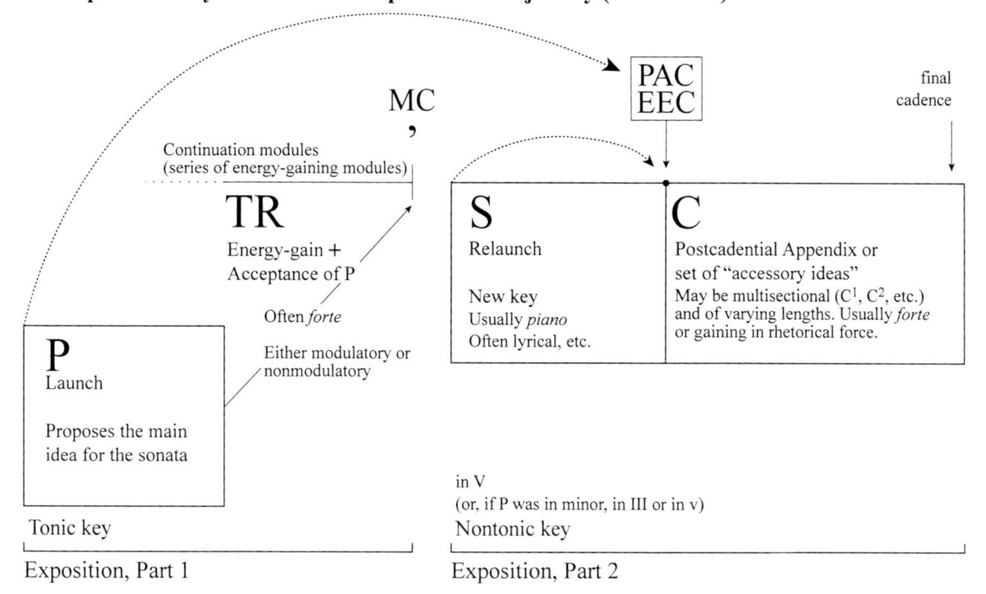

6.3 James Hepokoski and Warren Darcy, *Elements of Sonata Theory: Norms, Types, and Deformations in the Late-Eighteenth-Century Sonata* (New York: Oxford University Press, 2011), 17.

The theme of thoughtful information layering recurs throughout the book. Above all, information in the same layer should be internally connected, and layers should be clearly distinct from one another. To aid in this effort, designers can use an array of tools:

- Spatial layout (keeping layers physically separated)
- Shape (weight, style, curved vs. straight)
- Font (face, size, weight)
- Color (including shades of gray)
- Elements from music notation (such as in Schenkerian analysis)

Successful images employ characteristics that create pop-out effects for elements in the foreground and avoid inadvertent pop-out effects for items in background layers.

Chapter 7 Information Integration

Tufte (2006) advocates for the integration of words, numbers, images, and diagrams. Music scholarship is filled with images like figure 7.1. Although it shows a rich tapestry of relational information, the image falls short of its potential. The viewer must invest considerable effort to interpret the image because it contains only symbols, brackets, and lines, and the symbols used do not have conventional meaning, as they would, for instance, in a Schenkerian analysis (see chap. 48). To understand the image, the viewer must jump back and forth between prose and image. As an alternative, the version in figure 7.2 puts information from the prose directly onto the image:

1. The image pertains to Anton Webern's *Drei Lieder*, op. 25, no. 3, This information has been added to the caption.
2. The caption now also contains the purpose of the image: to show "the close registral integration of voice and piano in the 'developmental' part of the song" (Wintle 1996, 258).
3. Stemmed notes belong to a "deep line" referenced in an earlier example. A short text annotation referencing the previous figure near the first of these notes clarifies the point. The redrawing also increases the lengths of the stems and connects them with a beam to make these background pitches stand out more.
4. The letter *X* labels a particular voicing of (014) trichords (a 6th + a 3rd) that arise from notes 5–7 of the P and I forms of the piece's twelve-tone row. The redrawing takes the (014) label from the accompanying prose and explicitly shows it on each set. It also makes explicit both the row form in use and the order numbers.

5. In the original, lines connect the bracketed dyads and trichords when they appear in the same octave. The redrawing adds the text "Recurrences in register" to the first of these to clarify the meaning of the lines. Making the lines curved allows them to pop out against the image's mainly straight lines, and their contrasting thickness (thinner) and color (gray) allow them to stand out from the visual texture.
6. As the prose notes, "the three encircled arpeggios mark the declamatory beginning of each system (2, 3 and 4), with the boundary notes G♯/G in the first, G/G♯ in the second and F♯/G in the third making a circumlocution of the 'tonic'" (Wintle 1996, 258). The redrawing adds this essential information, making the purpose of the circles clear.

The redrawing has not changed the original conception in any way, but it allows the image to better stand on its own without the need to reference a prose description.

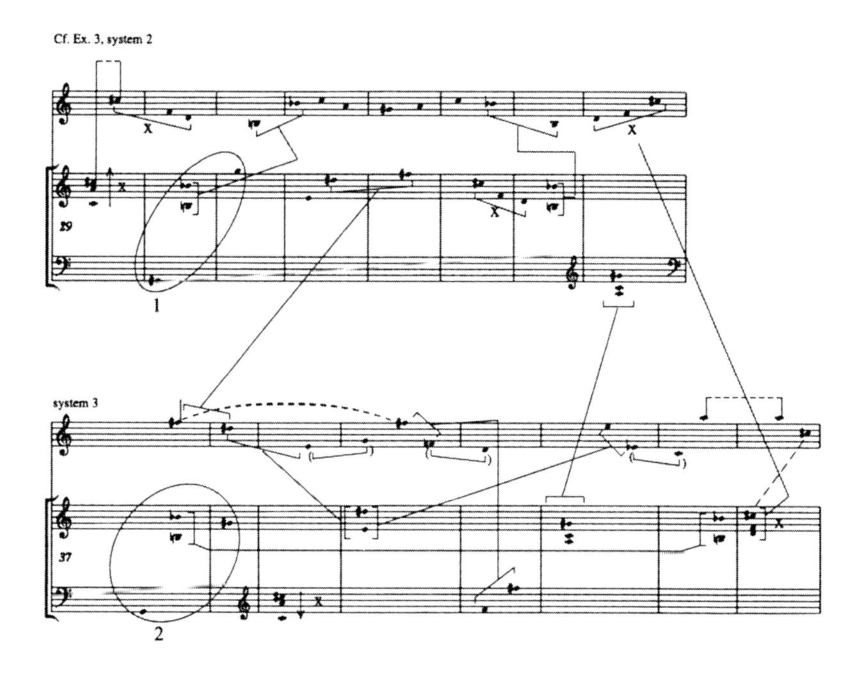

7.1 Christopher Wintle, "Webern's Lyric Character," in *Webern Studies*, ed. Kathryn Bailey, 229–63 (Cambridge: Cambridge University Press, 1996), 259.

7.2 Redrawing of fig. 7.1. Suggested new caption: "The close integration of voice and piano in the 'developmental' section of Anton Webern's *Drei Lieder*, op. 25, no. 3."

In addition to images such as figure 7.1, one also sometimes finds musical examples in which an author writing about a passage simply provides the excerpt, without adding commentary on the example itself. Why do images so often omit information that would enhance their effectiveness? Perhaps there is a fear of clutter. This can be addressed through effective graphical design. Perhaps the author lacks technical skills, but such work is easily outsourced. Perhaps there is an element of professional pride, in which the author does not want to give away too much in pictures, lest readers be tempted to skip the prose altogether. Yet consider how much more effective it is to put the information where it is most useful, where it enhances the excerpt directly. A severe-weather radar image is much more effective when it is overlaid on a map, which in turn is more effective when the street names are shown directly on the image. Information-rich images enhance and reinforce the prose. They invite viewers to linger, to see connections that are often

easier to comprehend visually than to simply read about, and even to find things the author didn't. Most importantly, they allow the author to focus on more substantive matters in the prose.

Figure 7.3 is a good model. The image includes pertinent analytic labels (harmonies throughout and, in mm. 74–75, an octave line), analytic interpretation ("cadential module"), and a narrative ("yielding to"). The annotations remain out of the area occupied by the music itself, keeping the information layers separated. Online figure 7.4 merits careful study as well.

This is not to suggest that piling more and more information into an image will make it better and better, only that many images would benefit from enhancement, annotation, and enrichment. The next chapter argues the opposite.

7.3 Andrew Davis, "Chopin and the Romantic Sonata: The First Movement of op. 58," *Music Theory Spectrum* 36, no. 2 (December 1, 2014): 284.

Chapter 8 Making Every Part of an Image Count

It can be tempting to spruce up an image with fancy stuff. Fortunately, few of us have the skill needed to produce infographics of the sort in figure 8.1. Whatever one's aesthetic reaction to the image, as information visualization, it is shallow. All the pictures, colors, and changes in font mask the fact that the entire image consists of about twelve pieces of data. In an informational graphic, the pertinent information should always be the most prominent, with as much of the image's effort devoted to information as possible.

I will start by picking on myself. Figure 8.2 represents entries in interval difference vectors using three-dimensional blocks, when in fact they contain just one dimension of information. (Interval difference vectors are formed by subtracting corresponding elements of two interval-class vectors from each other.) The ridiculous fill pattern on the fronts of the blocks further detracts. The redrawing in figure 8.3 addresses these concerns and makes some additional improvements. It replaces redundant verbiage with column headers, arranges the graphs vertically so parallel information is aligned in a single dimension, and sorts the images from most to least similar according to the similarity function being illustrated. It renders the dividing lines (which are useful in this instance to separate the segments of the image) in gray so they do not compete with the primary information. A sans serif font improves legibility.

In figure 8.4, the fills certainly contrast with one another, which is helpful in a stacked bar graph. Unfortunately, they also shimmer with unintended visual

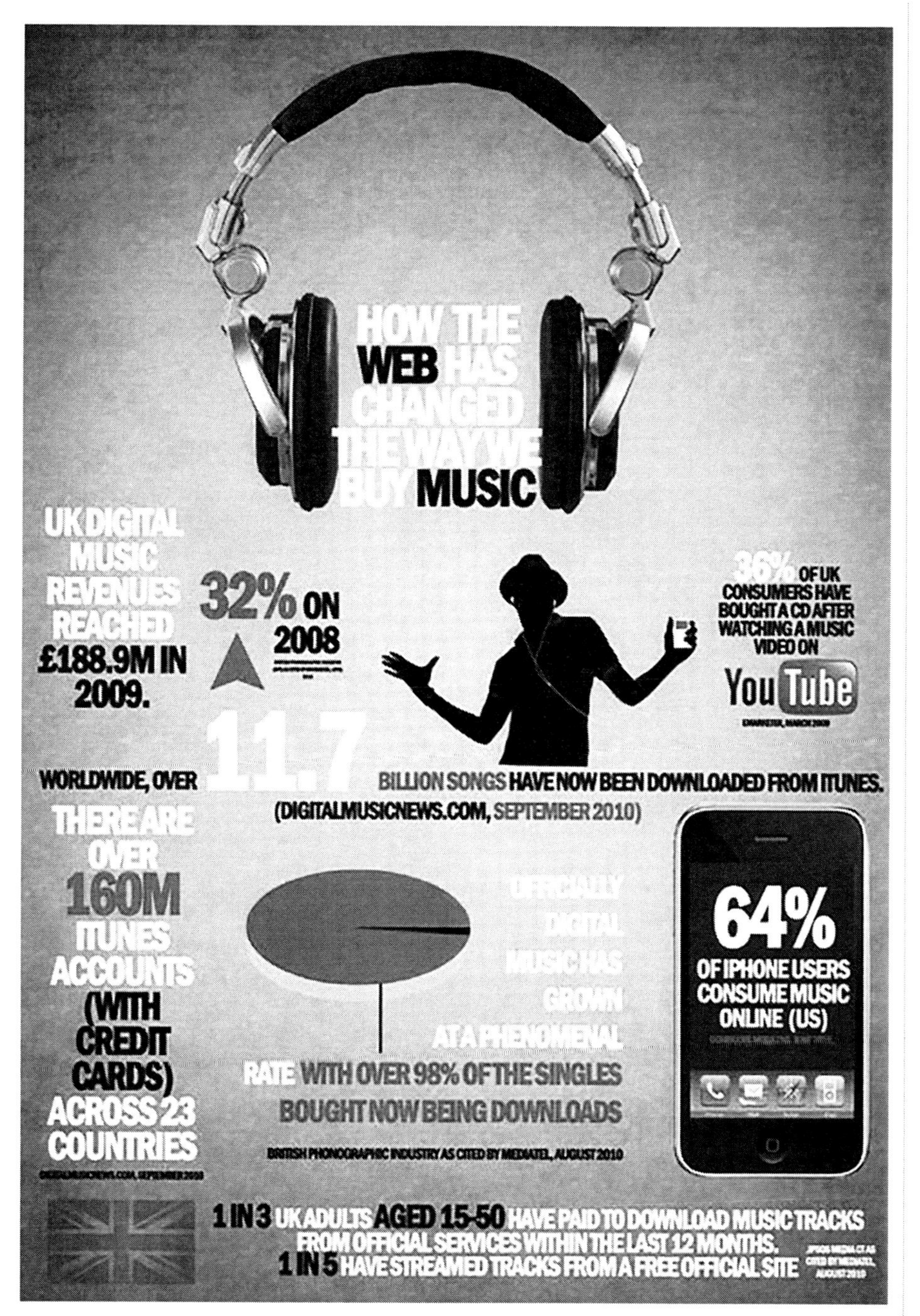

8.1 Catalin Zorzini, "How the Web Has Changed the Way We Buy Music [Infographic]," last modified September 27, 2010, http://inspiredm.com/how-the-web-has-changed-the-way-we-buy-music-infographic/ (site discontinued).

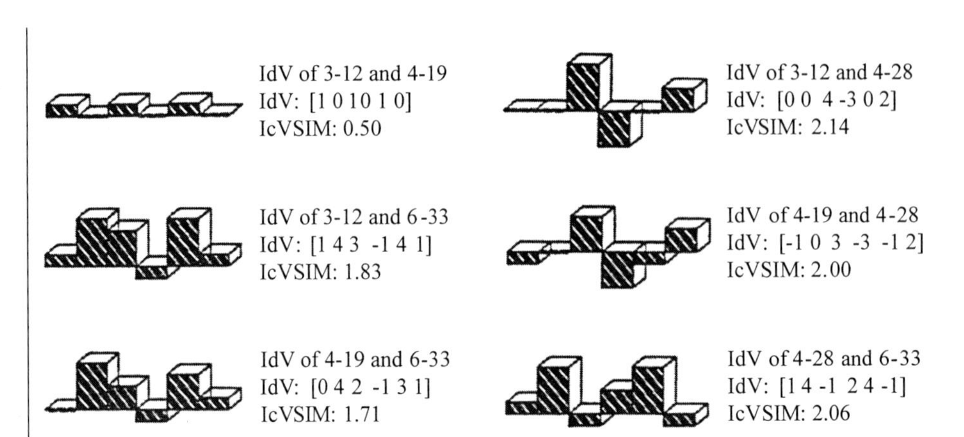

8.2 Eric J. Isaacson, "Similarity of Interval-Class Content between Pitch-Class Sets: The IcVSIM Relation," *Journal of Music Theory* 34, no. 1 (1990): 24.

8.3 Redrawing of fig. 8.2.

Set Classes	Difference Vector	IcVSIM
3-12 4-19	[1 0 1 0 1 0]	0.50
4–19 6–33	[0 4 2 -1 3 1]	1.71
3–12 6–33	[1 4 3 -1 4 1]	1.83
4–19 4-28	[-1 0 3 -3 -1 2]	2.00
4-28 6–33	[1 4 -1 2 4 -1]	2.06
3–12 4-28	[0 0 4 -3 0 2]	2.14

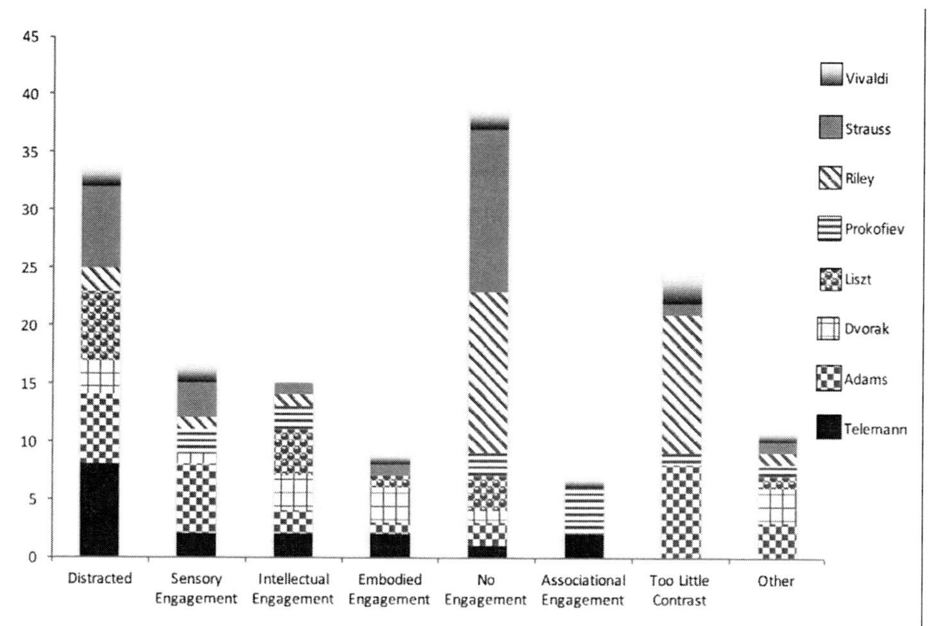

8.4 Elizabeth Hellmuth Margulis, "An Exploratory Study of Narrative Experiences of Music," *Music Perception* 35, no. 2 (2017): 242.

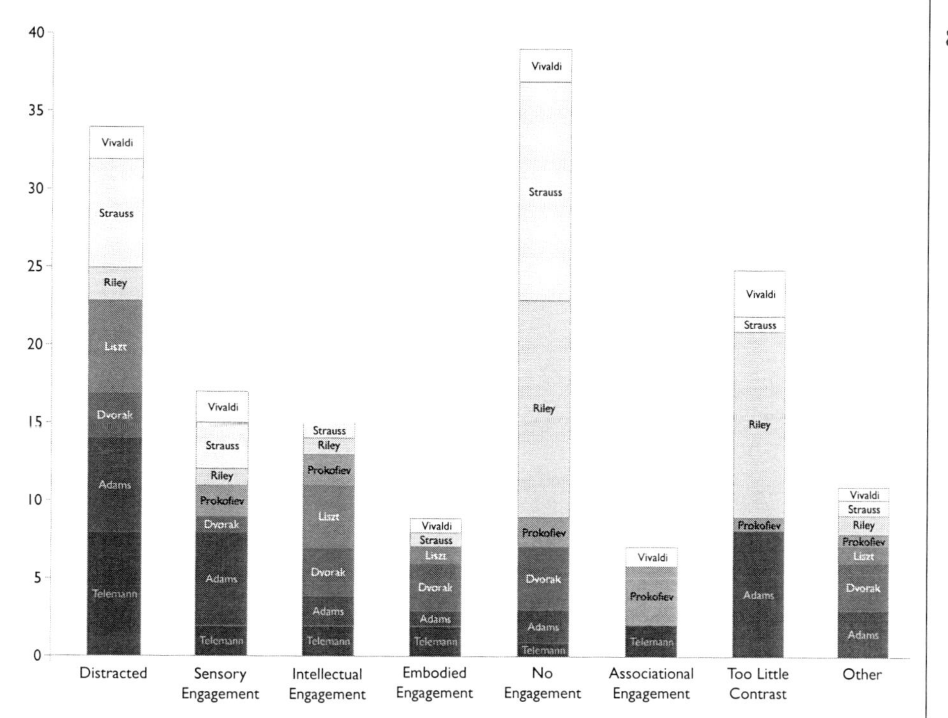

8.5 Redrawing of fig. 8.4.

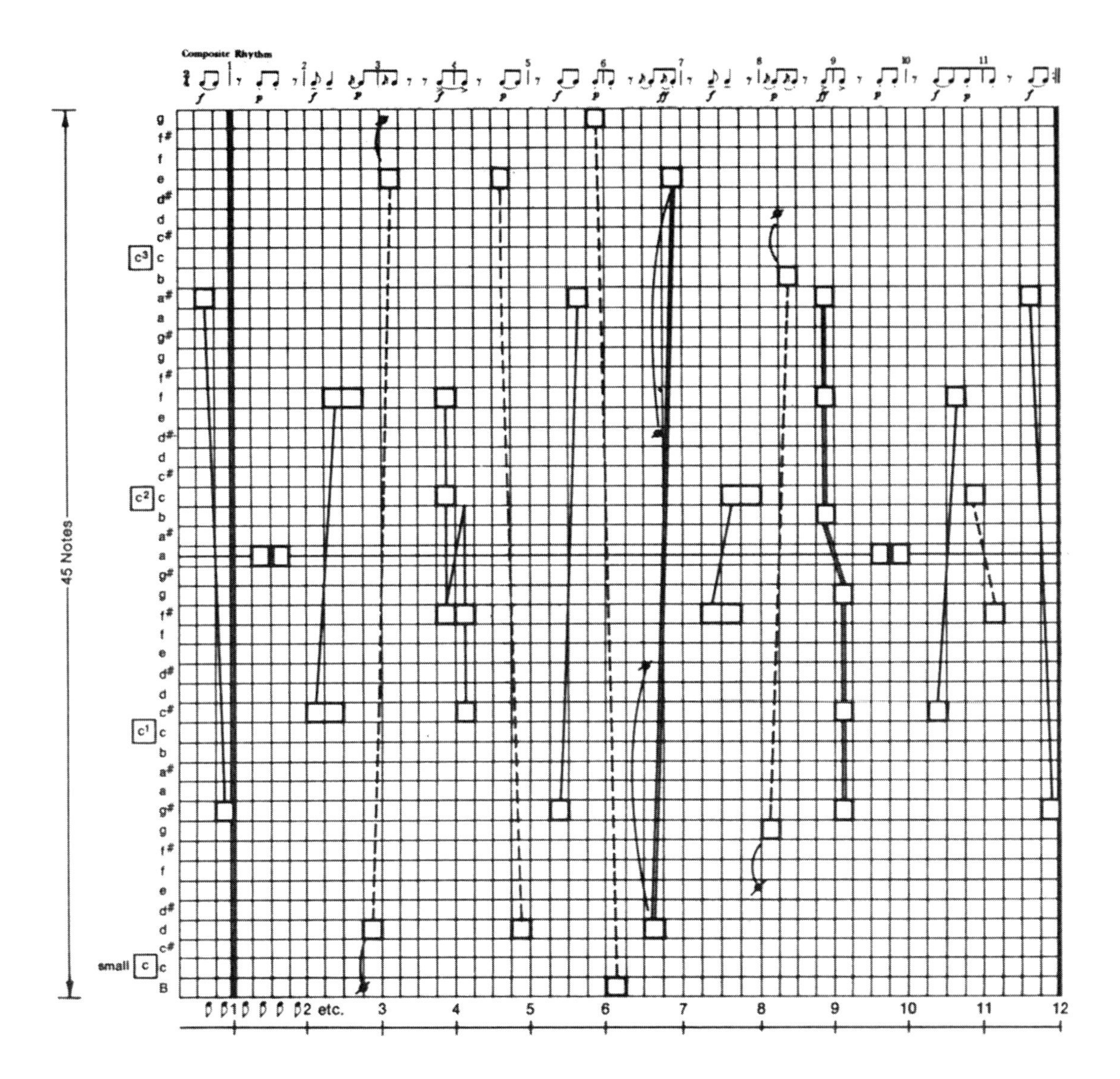

8.7 Gary E. Wittlich, "Sets and Ordering Procedures in Twentieth-Century Music," in *Aspects of Twentieth-Century Music*, ed. Gary E. Wittlich, 388–476 (Englewood-Cliffs, NJ: Prentice-Hall, 1975), 442.

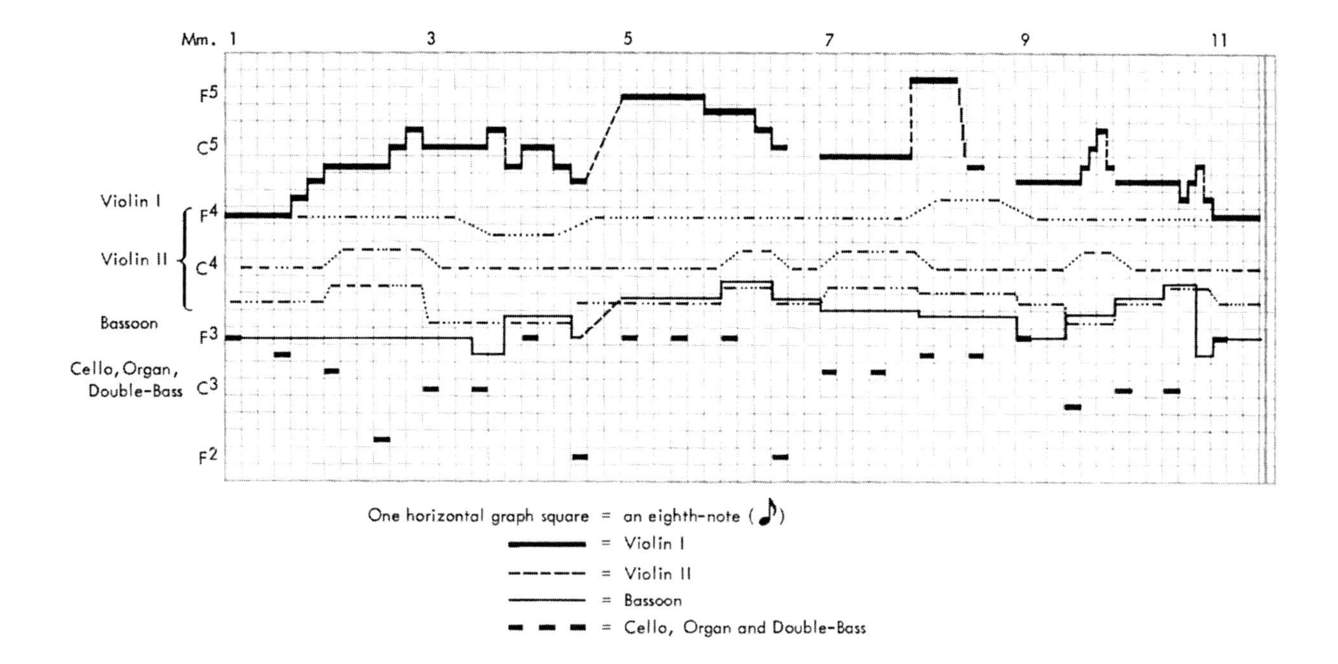

8.8 Robert Cogan and Pozzi Escot, *Sonic Design: The Nature of Sound and Music* (Englewood Cliffs, NJ: Prentice-Hall, 1976), 35.

artifacts and unintentionally evoke stereotypically garish clown outfits, a common side effect of using standard features of popular software. Also unfortunately, some of the fills are imprecise, including the upper end of "Prokofiev," which might be white or black, and, most seriously, "Vivaldi," whose gradient fill at the top of each bar fades from black to white, which happens to be the background color. The drawing in figure 8.5 replaces the fills with solid shades of gray. A thin gray border helps define the edges of the fills, particularly useful as they approach the background white. The redrawing also replaces the legend with labels directly on the bar segments they represent, always a better design choice. See also online figure 8.6.

Sometimes a graph is a good choice for presenting quantitative information that involves two dimensions. The gridlines that characterize graph paper inherently waste ink, however, since they do not themselves convey information. Moreover, physical graph paper is typically printed with cyan or light-gray ink, on which dark pencil lines stand out effectively. This contrast is often missing altogether in printed images. In figure 8.7, the gridlines have the same thickness and darkness as many of the content lines, and most of the content lines are nearly vertical, so they blend into the background, almost like camouflage. As a result, the image's content struggles for attention.

The grid in figure 8.8 works better. First, the gridlines are relatively thin and gray. The information-bearing lines also contrast more strikingly with the style of the gridlines, so they pop out. Still, reducing the number of gridlines, or eliminating them altogether, would improve a lot of wasted ink, with little loss of clarity.

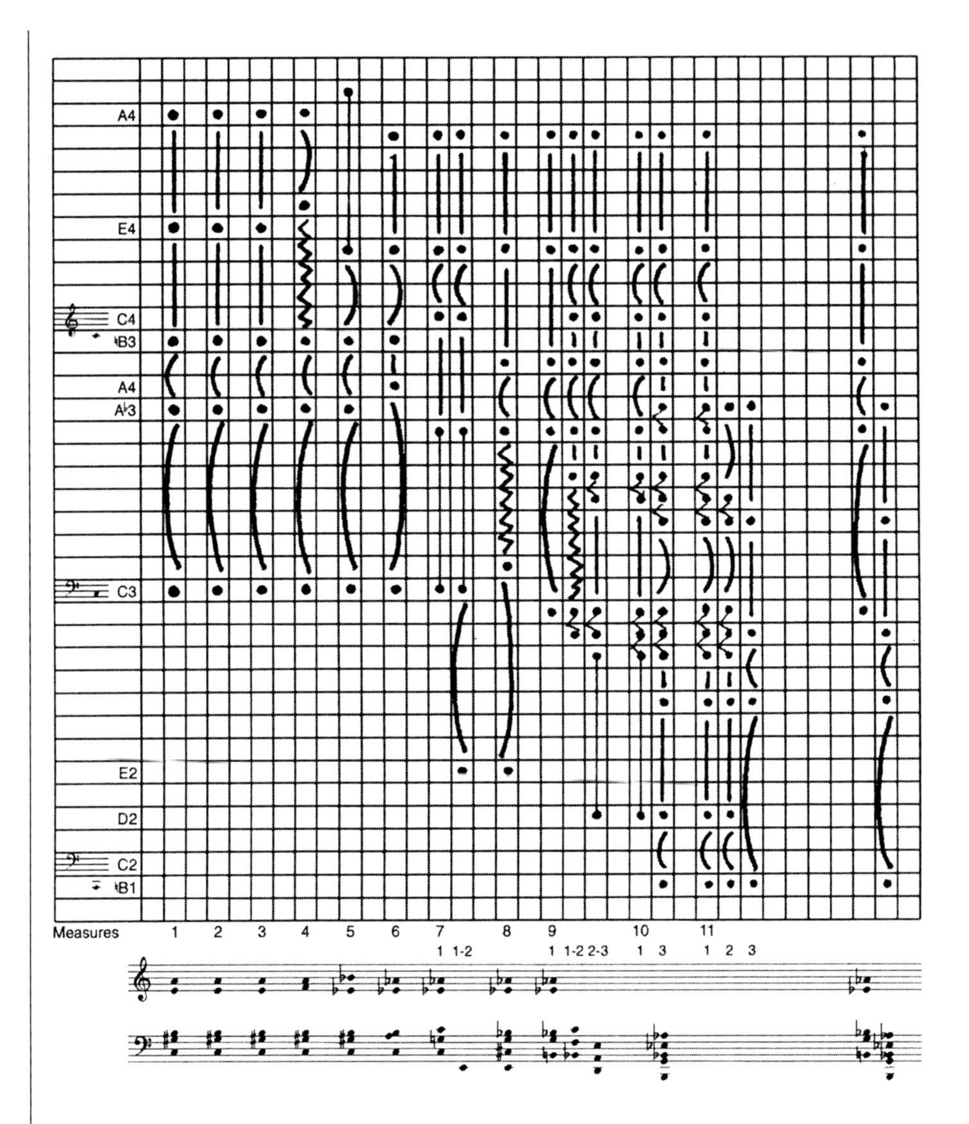

8.9 Erik Christensen, *The Musical Timespace: A Theory of Music Listening*, vol. 2 (Aalborg: Aalborg University Press, 1996), 29.

The grid in figure 8.9 is overbearing. The symbols representing intervals contrast in thickness and shape with the gridlines, however, so they stand about a bit more than in the previous figure. Nevertheless, foreground and background vie for attention. A redrawing in online figure 8.10 rethinks the task.

Figure 8.11 uses gridlines effectively. The image depicts part of a performance of a North Indian raga. It eliminates gridlines for the horizontal axis altogether, with minimal loss of precision. Time points are available for reference but do not demand attention. Pitch mapping on the vertical axis is more important. A dark solid line traces the performance's focal pitch, within an overall melodic range indicated by the shaded region, an approach that works well. Rather than using a grid

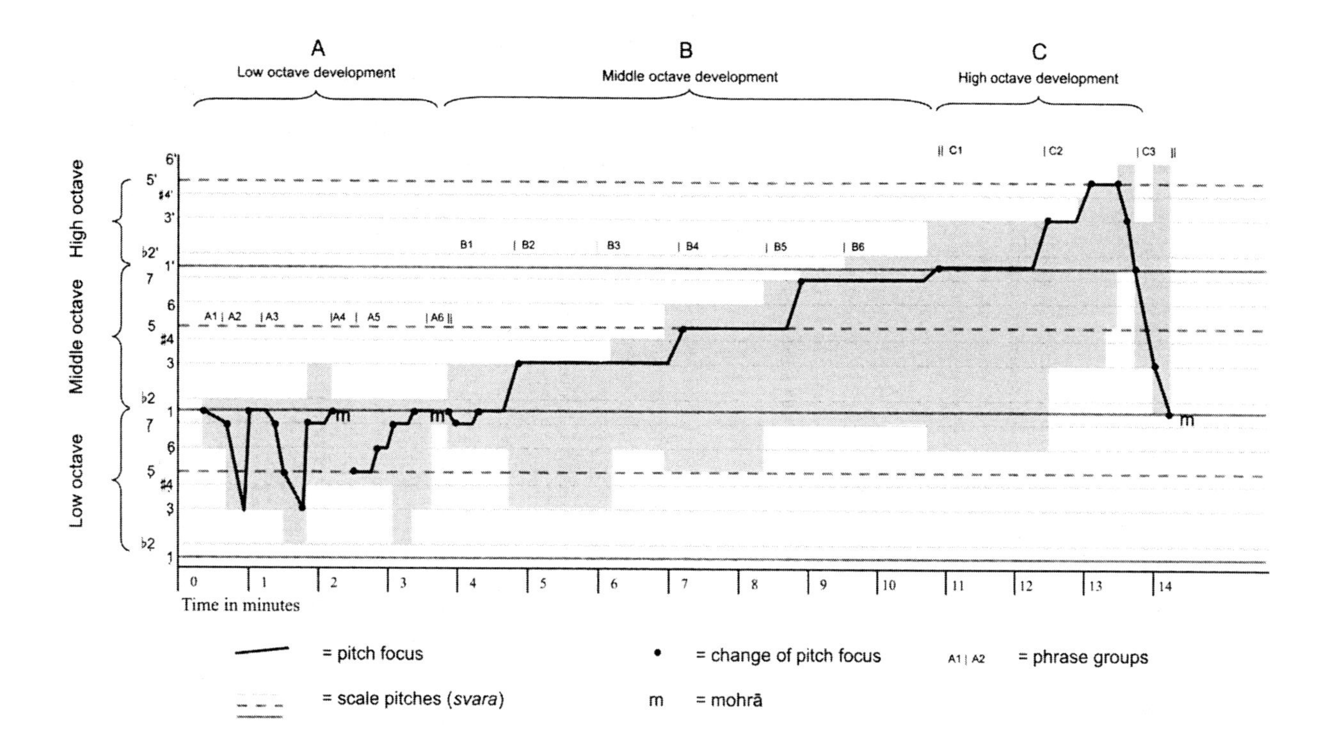

8.11 Richard Widdess, "Dynamics of Melodic Discourse in Indian Music: Budhaditya Mukherjee's Ālāp in Rāg Pūriyā-Kalyān," in *Analytical and Cross-Cultural Studies in World Music*, ed. Michael Tenzer and John Roeder (New York: Oxford University Press, 2011), 204.

of fixed weight, the image uses gray for the remaining lines. None of the gridlines competes with the image's information.

Tufte (2001, 96) describes the aim of giving primacy to the essential information, minimizing the visual impact of essential but unimportant information, and altogether eliminating unneeded elements as "maximizing the data-ink ratio." Removing any ink whose removal does not adversely affect the image is always a good idea.

Chapter 9 Presenting Tabular Data

Most tables present structured information within a two-dimensional grid that organizes and contextualizes the information. Tables are inherently multivariate, typically showing how systematic changes in two variables affect values in a third. They facilitate and encourage comparison.

Tables structured like figure 9.1 are quite common. Here, one can quickly look up tablature symbols and names for fret positions on the four strings of a biwa. The most common graphical improvement that one can make for tables is to reduce or eliminate the grid. (This will be a recurring theme in this chapter and throughout the book.) Figure 9.2 eliminates the grid, leaving the information itself fully intact but easier to see. The white space between rows and columns adequately organizes the information. Online figure 9.3 adds some subtlety to this generalization.

Across rows and down columns, tables by nature show differences. In cases such as figure 9.4, in which there are relatively few differences and those differences

	Open	Fret 1	Fret 2	Fret 3	Fret 4
IV	丄 *jô*	八 *hachi*	丨 *boku*	ム *sen*	也 *ya*
III	ク *gyô*	七 *shichi*	ヒ *hi*	ζ *gon*	之 *shi*
II	乚 *otsu*	下 *ge*	十 *jû*	乙 *bi*	コ *ko*
I	一 *ichi*	工 *ku*	几母 *bo*	フ *shû*	斗 *to*

9.1 Naoko Terauchi, "Surface and Deep Structure in the Tôgaku Ensemble of Japanese Course Music (Gagaku)," in *Analytical and Cross-Cultural Studies in World Music*, ed. Michael Tenzer and John Roeder (New York: Oxford University Press, 2011), 30.

	Open	Fret 1	Fret 2	Fret 3	Fret 4
IV	丄 *jô*	八 *hachi*	丨 *boku*	厶 *sen*	也 *ya*
III	ク *gyô*	七 *shichi*	ヒ *hi*	ㄛ *gon*	之 *shi*
II	乚 *otsu*	下 *ge*	十 *jû*	乙 *bi*	コ *ko*
I	一 *ichi*	工 *ku*	几母 *bo*	フ *shû*	斗 *to*

9.2 Redrawing of fig. 9.1.

are the point of the image, pop-out effects can guide the viewer to this critical information. Downplaying the grid would enhance the image further.

Although tables typically place categories along the left and top sides, other schemes are possible, as we have already seen in figure 9.5 (first encountered in chap. 5 as fig. 5.4), which describes the distinguishing characteristics of a number of variation genres. The image places headers (for example, "Thematic basis") above each row, rather than to its left. The grouping of three basso ostinato genres and three variation forms adds yet another dimension of information.

Creative layout can enhance a table's effectiveness. The image in figure 9.6 tabulates the chords in a key, showing (in modern parlance) which scale steps support chords of each quality (major/minor), in each position (root position, first and second inversions), in both major and minor modes. The design effectively answers questions like "Which chords in major keys have the fifth scale step in the bass?"

Album Version		**Extended Remix**	
Formal Section	*Number of bars*[12]	*Formal Section*	*Number of bars*
Introduction	9 (7 beat solo drum + 4+4)	Introduction	24 (4+4+1 bar of $\frac{2}{4}$ + 4+4+1 bar of $\frac{2}{4}$+2+4)
Verse ("Last night…")	8 (4+4)	Verse ("Last night…")	8 (4+4)
Chorus ("Tropical…")	8 (4+4)	Chorus ("Tropical…")	8 (4+4)
Guitar solo	4	Guitar solo	4
Verse ("I've been in…")	8 (4+4)	Verse ("I've been in…")	8 (4+4)
Chorus	8 (4+4)	Chorus	8 (4+4)
Guitar solo	4	Guitar solo	4
Bridge	9 (2+2+2+3)	Bridge + vamp	13 (2+2+2+7)
Synthesizer solo	4	Synthesizer solo	4
Verse ("Last night…")	4	Verse	4 ("Last night…")
Chorus	9 (4+4+1 bar of $\frac{2}{4}$)	Chorus	9 (4+4+1 bar of $\frac{2}{4}$)
Chorus	9 (4+4+1 bar of $\frac{2}{4}$)	Chorus	9 (4+4+1 bar of $\frac{2}{4}$)
		Chorus	9 (4+4+1 bar of $\frac{2}{4}$)
Outro	17 (4+4+1 bar of $\frac{2}{4}$+4+4)	Outro	26 (4+4+1 bar of $\frac{2}{4}$ +4+4+1 bar of $\frac{2}{4}$ +4+4)

9.4 Christine Boone, "Mashing: Toward a Typology of Recycled Music," *Music Theory Online* 19, no. 3 (2013): fig. 1, https://mtosmt.org/issues/mto.13.19.3/mto.13.19.3.boone.html.

9.5 Percy Goetschius, *The Larger Forms of Musical Composition* (New York: G. Schirmer, 1915), 2.

COMPARATIVE TABLE OF DIVISION I.

Basso ostinato			Variation-forms		
Ground-motive.	Ground-bass, or Basso ostinato proper.	Passacaglia.	Chaconne.	Small Variation-form.	Large Variation-form.
Thematic basis					
Motive of one or two measures, chiefly in bass.	*Phrase* of two or four measures, chiefly in bass.	*Period* (or repeated Phrase) of 8 measures, of which the burden is usually the *bass*-line.	usually the *chords* (incidentally the melody or the bass).	Double-period or 2-Part *Song-form*, 16 measures. *Melody* (chords, or bass, or formal design).	Usually 2- or 3-Part *Song-form*, 16 to 32 measures. *Melody* (chords, etc.)
Distinctive traits					
None.	None.	Minor mode. Triple measure ($\frac{3}{4}$, $\frac{3}{8}$ or $\frac{6}{8}$, $\frac{3}{2}$).	Minor or major mode. Triple measure ($\frac{3}{4}$).	None.	None.
Treatment					
Homophonic; changing melodic, rhythmic and harmonic forms in upper (added) parts.	Homophonic; changing forms, phrase-group design.	Preponderantly *polyphonic;* thematic accompaniment of *bass-theme.*	Preponderantly *homophonic;* varying patterns of (chiefly) harmonic figuration, with approximate retention of *Melody.*	Chiefly homophonic, occasionally polyphonic; the variations completely separated, as a rule. Form of Theme retained, with unessential extensions.	Chiefly homophonic, occasionally polyphonic; the variations completely separated, as a rule. Form of Theme treated with greater freedom, and transformed by Insertions and extensions. Elaboration, as well as Variation.
Structure continuous, with ordinary (transient) cadence interruptions.			Partly continuous, partly separated, variations.		

In der harten Tonart sind

Die Stufen der Tonleiter	wesentliche Dreyklänge	wesentliche Sextenaccorde	wesentliche Sextquartenacc.	zufällige Dreyklänge	zufällige Sextenaccorde	zufällige Sextquartenacc.
auf dem Grundtone	g e c		a f c		a e c	
auf der zweyten Klangstufe			h g d	a f d		
auf der dritten Klangstufe		c g e		h g e		c a e
auf der vierten Klangstufe	c a f				d a f	
auf der fünften Klangstufe	d h g		e c g		e h g	
auf der sechsten Klangstufe		f c a		e c a		f d a
auf der siebenten Klangstufe		g d h				g e h

9.6 Heinrich Christoph Koch, *Versuch einer Anleitung zur Composition* (Leipzig: Bey A. F. Böhme, 1782), 66.

and "Which first inversion major chords occur in a minor key?" Empty cells are not wasted space, as they reveal where in a scale the chord forms appear. The rows are ordered according to their ordinal position (ground tones, second step, third step, etc.), so the starting pitch goes up as one moves down the table. It would not be inappropriate to reverse the vertical ordering. Other examples of tabular data are found elsewhere in the book, as well as in online figure 9.7.

Figure 9.8 and online figure 9.9 summarize information across multiple works by Anton Webern and Tomás Luis de Victoria, respectively. The table in figure 9.8 documents the locations in Webern's sketchbooks for the rows used in his serial works, while online figure 9.9 lists modes used in Victoria's masses. In both, the tabular format supports a search for information about a particular work but also facilitates comparison among multiple works. Both appropriately eschew

	Row tables	Sketchbooks		
		dates	*material*	*location*
Op.19	1925/26 (top righthand corner, in red pencil: probably added later)			
Op.20	1926/27 (top righthand margin, below instrumentation and in same red pencil; both probably added later)		no sketches extant	
Op.21	1928 (top righthand margin, below instrumentation and in same red pencil; both probably added later)	Nov. Dec. 1927	sketches of mvt 2 opening, a row	I/15
		January 1928	another row, with its four permutations	I/16
			the Op.21 row, Thema of mvt 2	I/16
Op.22	1929/30	14•IX•1928	'Concert für Geige, Klarinette, Horn, Klavier u Streichorchester'	I/54
	(top right hand margin, below instrumentation; both in lead pencil and probably added later)	6•V,10•5	several preliminary rows	I/54
		[n.d.]	several more rows	I/53,56
		27•V•l929	another row, altered to become Op.22 row	I/56
		28·V•l929	first sketch of mvt 2 opening	I/58
Op.23	1933 (at end, in same red pencil ued for titles: probably written at the same time)	I•II•33	experiments with row and melody	II/52
			correct row written out with other forms in the following order: IX, X (P_9–R_9);[8] I, II (P_0-R_0), III, IV (I_0–RI_0), V, VI (P_6–R_6), VII, VIII (I_6-RI_6)	II/52
		4•IV•33	sketch of opening of Op.23/iii	II/51-2
Op.24	7•VII•1931 (at end, in lead pencil)	16•I•1931	outline of new work and row fragments	II/38
		19•I	fragments of music, not using row; SATORAREPO; *gilt* row and its retrograde, almost certainly added later	II/39
		4•Febr•l931	sketches of music, evolution of row	II/39-40
Op.25	1934 (at end, in same red pencil as titles, but different from red used for row nos; probably added later)	4•VII•34	experiments with (inv.) row; correct row with its permutations	II/75
			sketch of no.1 opening	II/75
Op.26	2-III•35 (at end, in green – same pencil used for last group of rows – badly faded)	19•II•35	a trial row, a melody, then opening melody of Op.26	III/22
		24•II	correct row and its permutations, labelled *gilt*	III/22
		14•III•35	short sketch of opening	III/22
Op.27	25•XI•35 (at end, in black ink)	14•X•35	numerous experiments with row	III/43
		16•X	page of sketches of mvt 3 opening, row changing	III/43
		undated	correct row and its permutations	III/44
			sketches of Op.27/iii	III/44
Op.28	23•I•37 (at end of p.2 (point at which he realised that he needn't write out any more rows),I n lead pencil)	17•XI•36	title, outline of movements added later	III/57
		undated	P_0 and R_0 written out, symmetrical relationships noted	III/57
		21•XI•36	first sketch of mvt 3 opening	III/57
Op.29	4•VIII•38 (at end of p.3, in ink)	l•VII•38	first work on row, sketches with 'wrong' versions	IV/2
		3•VIII	*gilt* row boxed and labelled	IV/2
		5•IX	first sketch for no.2 opening	IV/4
Op.30	7•V•40 (at end, in ink)	15•IV•40	sketch of opening cello figure	IV/48
		16•IV•	sketches of row	IV/48
		17•IV	many sketches of opening figures	IV/48
Op.31	14·VI·41 (at end, in ink) 3•XI•43 (in right margin at end, written vertically, in red pencil)	7•V•41	sketches of no.4 opening; sketch of row	IV/82

[8] *The appearance of these two rows is particularly interesting as they were not used in Op.23 and do not appear on the row table for that work*

the gridlines found more often than necessary in tables. The slightly greater white space between works used in online figure 9.9 guides the eye across the table.

Tables are an efficient way to present multidimensional information, particularly when the categories involved are clearly defined. Those creating tables must make two key design decisions: how to represent the data within the grid and how to minimize the grid itself so the data is primary.

***Facing*, 9.8** Kathryn Bailey, *Webern Studies* (New York: Cambridge University Press, 1996), 220–21.

Chapter 10 Small Multiples

A particularly useful visualization technique is the "small multiple." Tufte (2001, 270) describes the small multiple as resembling "the frames of a movie: a series of graphics, showing the same combination of variables, indexed by changes in another variable. . . . The design remains constant through all the frames, so that attention is devoted entirely to shifts in the data."

Figure 10.1, a study on segmentation in post-tonal music, uses the small multiple to show six potential segmentations of the first six notes of a Stefan Wolpe string quartet. The layout allows the eye to scan among the various possibilities, with the features that support each segmentation (+), and in some cases those that work against it (–), directly below each. Readers can jump from one segmentation to another and test the corresponding theory for themselves. Two other features make this image effective. First, most of the segmentation feature lists contain the specific values of those features. For instance, in segmentation 6, the circled segments share set class 2–4, registral interval 4, the dynamic pattern ***pf***, and so on. The pertinent information appears where it is most useful, saving the viewer from having to make an extra analytical step to interpret and verify the data. Second, the lines showing the segmentations are curved, rather than straight. The circles pop out, so we don't need to scan to find the segments; rather, we can see them at a glance. Online figure 10.2 shows small multiples at work pedagogically.

Small multiples can be especially useful to show how something changes over time. From an article describing an artificial intelligence system that "learns" stylistic tendencies in Wolfgang Mozart's juvenilia (K. 1–6), figure 10.3 shows the system's memory state at eight moments as it "listens" to a four-bar segment from

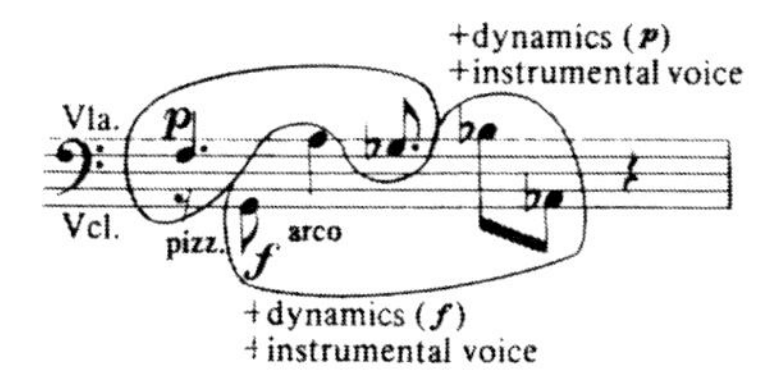

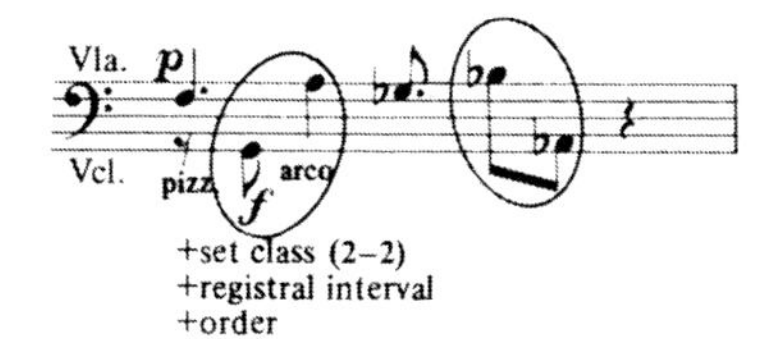

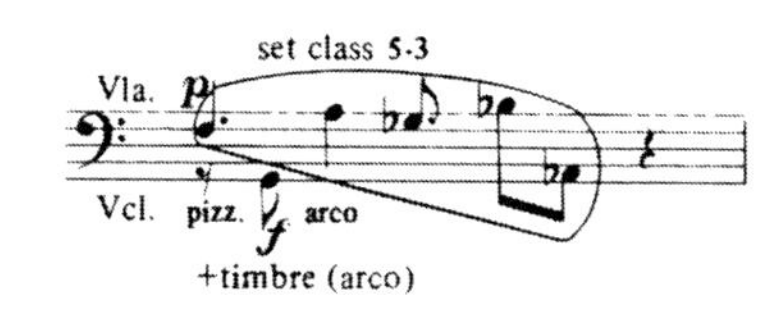

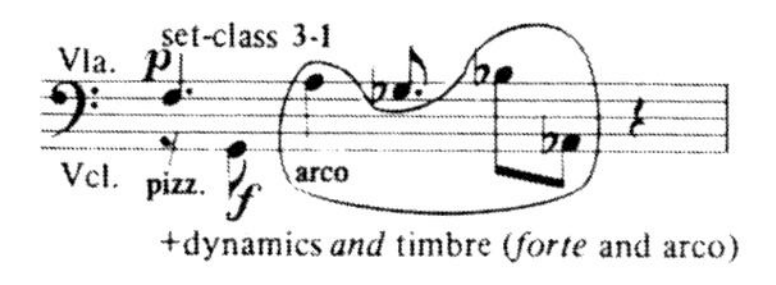

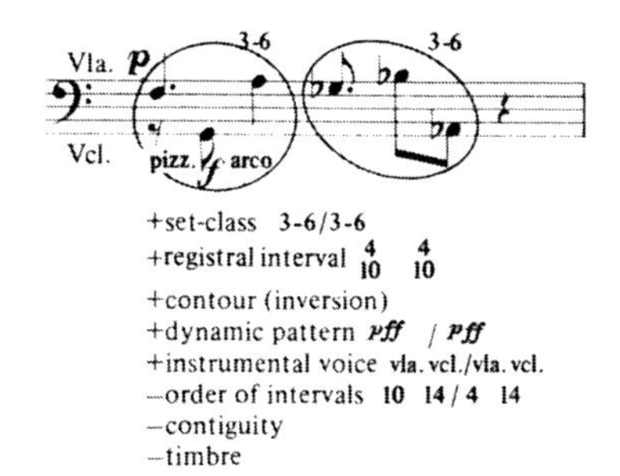

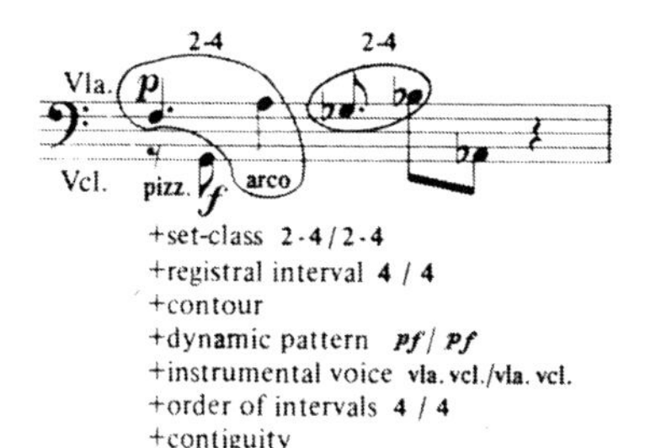

10.1 Christopher Hasty, "Segmentation and Process in Post-tonal Music," *Music Theory Spectrum* 3 (1981): 57.

the *Minuet*, K. 2, in the context of what it has learned previously from the other works. The sizes of the graphic elements (notes, melodic direction, and dissonance, indicated by *d*) indicate the strength of the components of each memory, which are triggered by schematic memory traces. In each trace, we can see the collective feature set evoked in the system's memory by each event. By comparing a trace to the ones that precede and follow it, we can see the effects of both fading memory and anticipation. To align the memory traces with the corresponding measures of music, the layout alternates the traces between two lines. This makes it harder to compare successive traces, however. The traces are aligned in online figure 10.4, which makes it easier to listen along with the childlike memory.

The strength of the small-multiple format is that it facilitates comparisons across multiple states, whether those are analytic interpretations, physical gestures, or points in time. It is exceptionally versatile. Additional examples appear elsewhere in the book, involving areas such as music cognition, performance timing, and corpus studies.

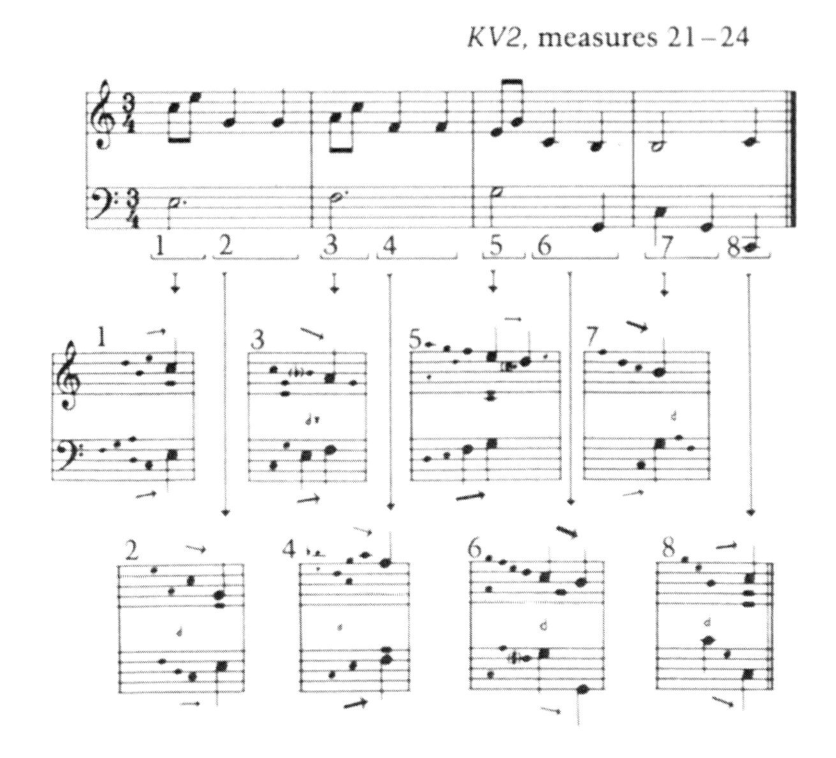

10.3 Robert O. Gjerdingen, "Using Connectionist Models to Explore Complex Musical Patterns," in *Music and Connectionism*, ed. Peter M. Todd and D. Gareth Loy (Cambridge, MA: MIT Press, 1991), 145.

Chapter 11 Using Color

Historically, color has been used relatively rarely in academic music scholarship owing to the prohibitive cost of color printing. Used appropriately, however, color can transform the display of information. Color can serve many purposes, including (1) to highlight, (2) to promote information layering, (3) to label, and (4) to represent scales of continuous values. Online figure 11.1 illustrates how effectively colored lines pop out of the visual texture and stay out of the way of the information they annotate.

See how the addition of color transforms an anodyne monochrome form diagram into an image with visual richness and a sense of structural hierarchy (fig. 11.2, colorplate 3). Contrasting hues serve as labels to differentiate the two major divisions, while different saturations of a hue label contrasting sections within each.

Colors serve as labels in the upper image in figure 11.3 (colorplate 4), in which the same color highlights rhyming words or phrases in a rap lyric. The effective design can be subtly improved, however, by tweaking a couple of features. The image requires eleven colors to label all the rhymes. It has been shown, however, that we can identify colors more quickly when they have common names associated with them, such as red, blue, or purple (Ware 2008, 77). Those looking at the *-ot* rhymes in this image might very well disagree on what name to use to describe the color. Also, some of the darker fills interfere with the legibility of the text. The redrawing in the lower image addresses both issues by using lighter and darker versions of five colors (blue, green, red, yellow, and orange), plus violet. In addition, it uses a more open font to improve legibility given the small font size, renders the horizontal

Facing, 11.4 C. Catherine Losada, "Between Modernism and Postmodernism: Strands of Continuity in Collage Compositions by Rochberg, Berio, and Zimmermann," *Music Theory Spectrum* 31, no. 1 (2009): 70.

gridlines in gray, and eliminates the vertical gridlines, except for those signifying the start of a beat. The effect is cleaner and keeps the focus on the text.

In figure 11.4, various hard-to-differentiate fill patterns mask the effectiveness of an otherwise excellent visual analysis of musical quotations in Luciano Berio's *Sinfonia*, movement 3. Each pattern represents a different quotation source. The patterns do not pop out, however, requiring viewers to actively scan the image to find each one. (Use of these patterns was necessitated by publication in a monochrome print format.) When colors replace textures as in figure 11.5 (colorplate 5), the sources of the quotations pop off the page. The highly contrasting colors eliminate the risk of confusing the references—each color (for example, green, brown, pink) pops out separately. Rendering the gridlines in a light color, cyan, mutes the otherwise intense grid and helps keep attention on the content, where it should be. The redrawing takes the liberty of reattaching the two halves of the image (they faced each other in the original), employs a sans serif font for improved legibility at this small size, and replaces the legend at the bottom of the image with direct labels, which are generally superior because they put the information at the place they are needed by the viewer.

Color can also be used to measure quantities. The luminance (light-to-dark) scale is useful for this. Figure 11.6 shows results of a probe tone experiment that recorded subjects' key judgments as they listened to a nine-chord progression that modulated from C major to D minor. The model's mathematics need not concern us here. Using a small-multiple format, the image maps subjects' responses following each chord (chords 1–3 in the top row, then 4–6, then 7–9) onto a tonal space in which keys related by perfect fifths are arranged in a roughly southwest-to-northeast axis, while the northwest-to-southeast axis alternates parallel and relative motions (see chap. 17). Shades of gray indicate the degree to which subjects judged each probe tone to be appropriate. The luminance scale ranges from light (stronger judgment) to dark (weaker judgment). Gray contour lines help distinguish the gray fills from one another, though the lines are thicker and darker than necessary and thereby add clutter to the image and obscure some of the pitch names. The pitch names face even greater competition from the darker regions, which almost entirely envelop them.

Color scales that do not rely on luminance are less effective at measuring quantities, as figure 11.7 (colorplate 6) illustrates. This "Tonal Landscape" shows the judgments of a key-finding algorithm that estimates the overall key within a "window" as it passes across a work from beginning to end. At the top of the triangle, the window spans the entire work, so the algorithm is estimating the work's overall key. The window gets progressively narrower the further down the triangle it goes. By the bottom, it is estimating the key very locally. There, the frequent color changes reflect the algorithm's short attention span. This particular image

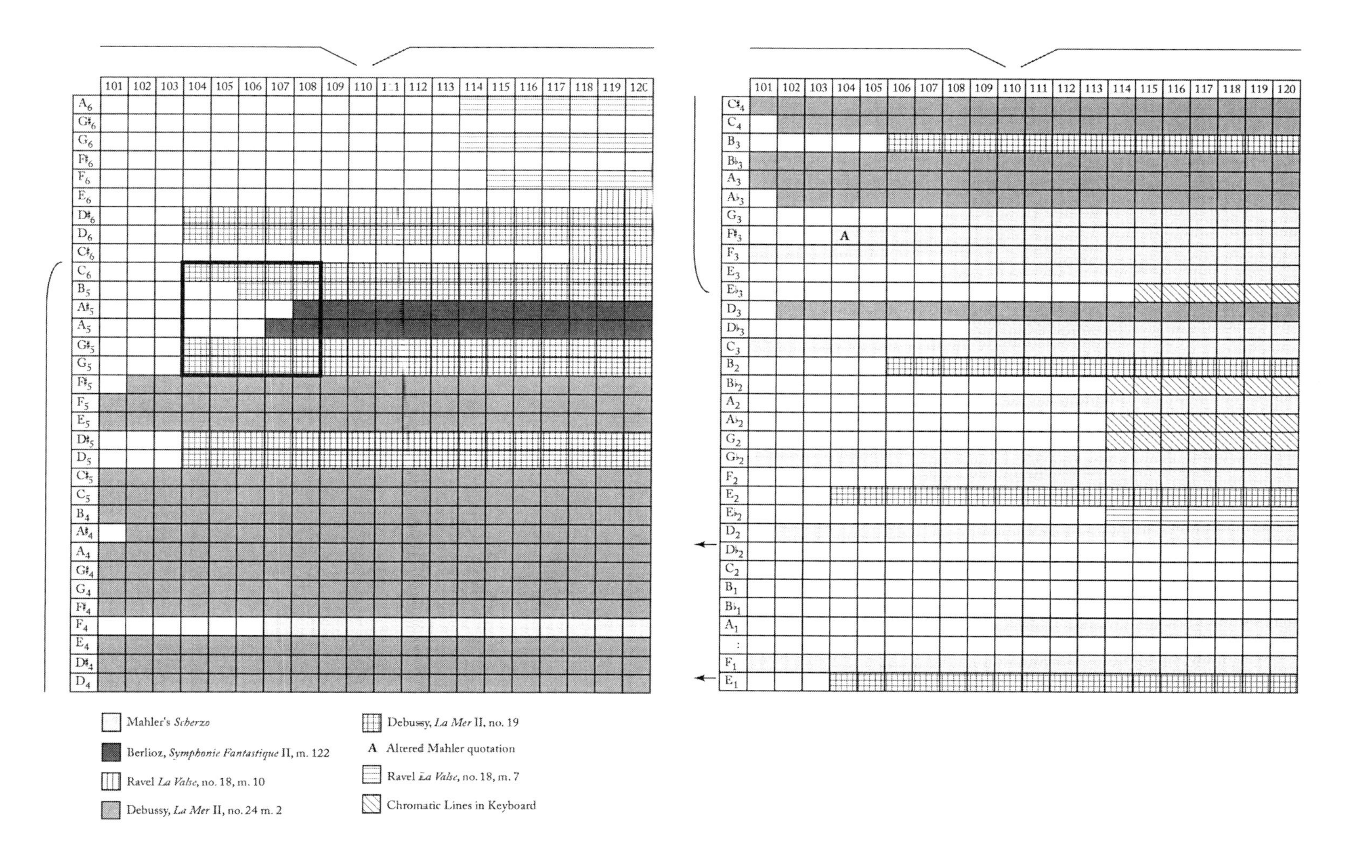
101 102 103 104 105 106 107 108 109 110 111 112 113 114 115 116 117 118 119 120
A6 G♯6 G6 F♯6 F6 E6 D♯6 D6 C♯6 C6 B5 A♯5 A5 G♯5 G5 F♯5 F5 E5 D♯5 D5 C♯5 C5 B4 A♯4 A4 G♯4 G4 F♯4 F4 E4 D♯4 D4
C♯4 C4 B3 B♭3 A3 A♭3 G3 F♯3 F3 E3 E♭3 D3 D♭3 C3 B2 B♭2 A2 A♭2 G2 G♭2 F2 E2 E♭2 D2 D♭2 C2 B1 B♭1 A1 : F1 E1
A
Mahler's Scherzo
Berlioz, Symphonie Fantastique II, m. 122
Ravel La Valse, no. 18, m. 10
Debussy, La Mer II, no. 24 m. 2
Debussy, La Mer II, no. 19
A Altered Mahler quotation
Ravel La Valse, no. 18, m. 7
Chromatic Lines in Keyboard

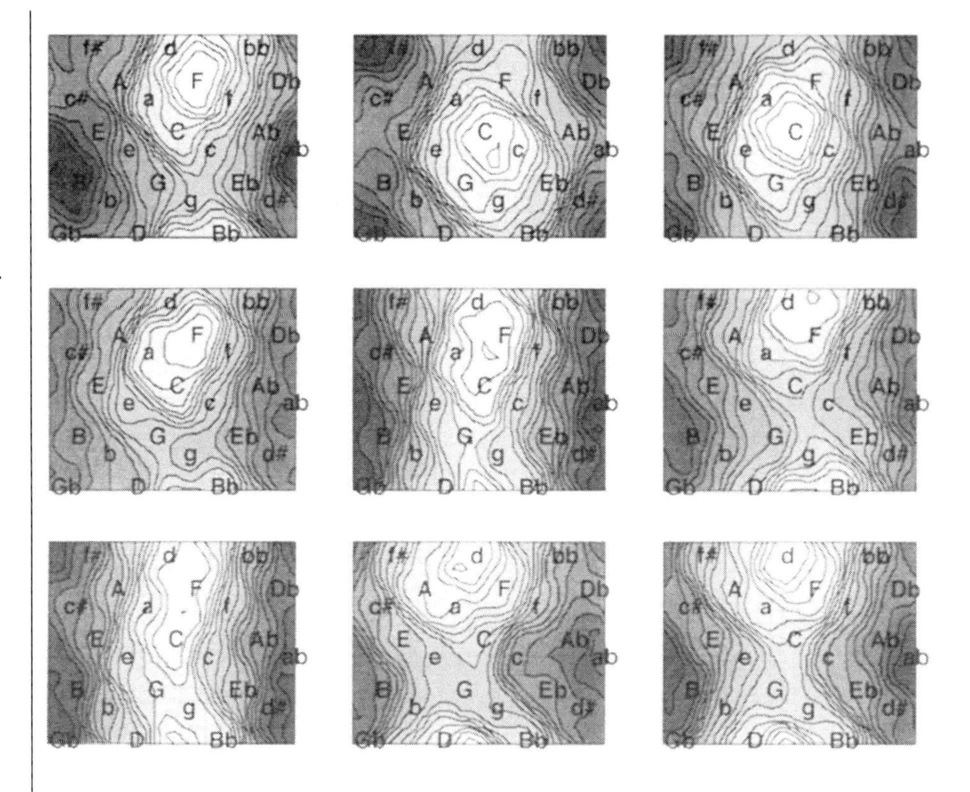

11.6 Carol L. Krumhansl and Petri Toiviainen, "Tonal Cognition," in *The Cognitive Neuroscience of Music*, ed. Isabelle Peretz and Robert J. Zatorre, 95–108 (New York: Oxford University Press, 2003), 98.

represents Frédéric Chopin's *Mazurka*, op. 24, no. 2. The color scale used appears below the image. The major keys associated with the seven diatonic steps are arranged in diatonic fifths. These are assigned respectively to the traditional ("ROY G. BIV") colors of the rainbow. Chromatic inflections use gradations within those basic colors—notice the halo of accidentals around the note C. Darker versions of the same color scale represent the parallel minor keys.

The color spectrum used *appears* to be ordered, yet it creates a curious musical effect. Visually proximate colors move in the direction of purple from C♮ to C♯ to C𝄪 to . . . G𝄫? The problem, of course, is that we do not consider chromatically adjacent keys to be at all closely related in the way the image implies by the use of chromatically adjacent colors. Instead, we more typically think of keys related by perfect fifth and relative keys to be most closely related. If we map the colors from figure 11.7 (colorplate 6) onto a circle of fifths, with relative keys in an inner ring (fig. 11.8, colorplate 7, left image), the nice color sequence vanishes. The three keys moving sharp-wise from C major seem harmonious and close to one another, but the same cannot be said going the other direction. The jump between red and purple is particularly jarring (to me). The five "shades" of C (green), which are visually similar, seem to be distributed nearly evenly. While the color scheme could be

11.9 Two-color gradient.

adjusted (see the right image in fig. 11.8, colorplate 7, for an effective rendering), it can never be fully fixed. We do not perceive a linear relationship between colors of the sort that would map onto a spatially oriented concept like tonal distance.

We can, however, judge relative darkness, which equates to distance between points on the luminance scale (that is, from black to white). Therefore, grayscale can effectively convey values (fig. 11.9). If we want to express both positive and negative values, we can use a color scheme like that in online figure 11.10, which uses increasing color saturation (that is, darkness) to convey increasing distance from a midpoint (such as zero) and uses different hues to represent negative (red) and positive (blue) values. Note that it is the light-dark continuum that varies, not the hues.

When using the color channels for emphasis, we need to remember that a nontrivial percentage of people, primarily males, suffer from some kind of color blindness, most commonly along the red-green color channel. It is therefore wise to consult a website that assesses color combinations for problems for those with different types of color blindness. It is also important to realize that the number of colors considered to be "common," and the boundaries between those colors, sometimes vary from culture to culture.

Color is one of the most powerful ways to enhance an image. It is useful for annotating an otherwise monochrome image, representing different information layers, serving as labels, or measuring quantities. Many additional examples are sprinkled throughout the book.

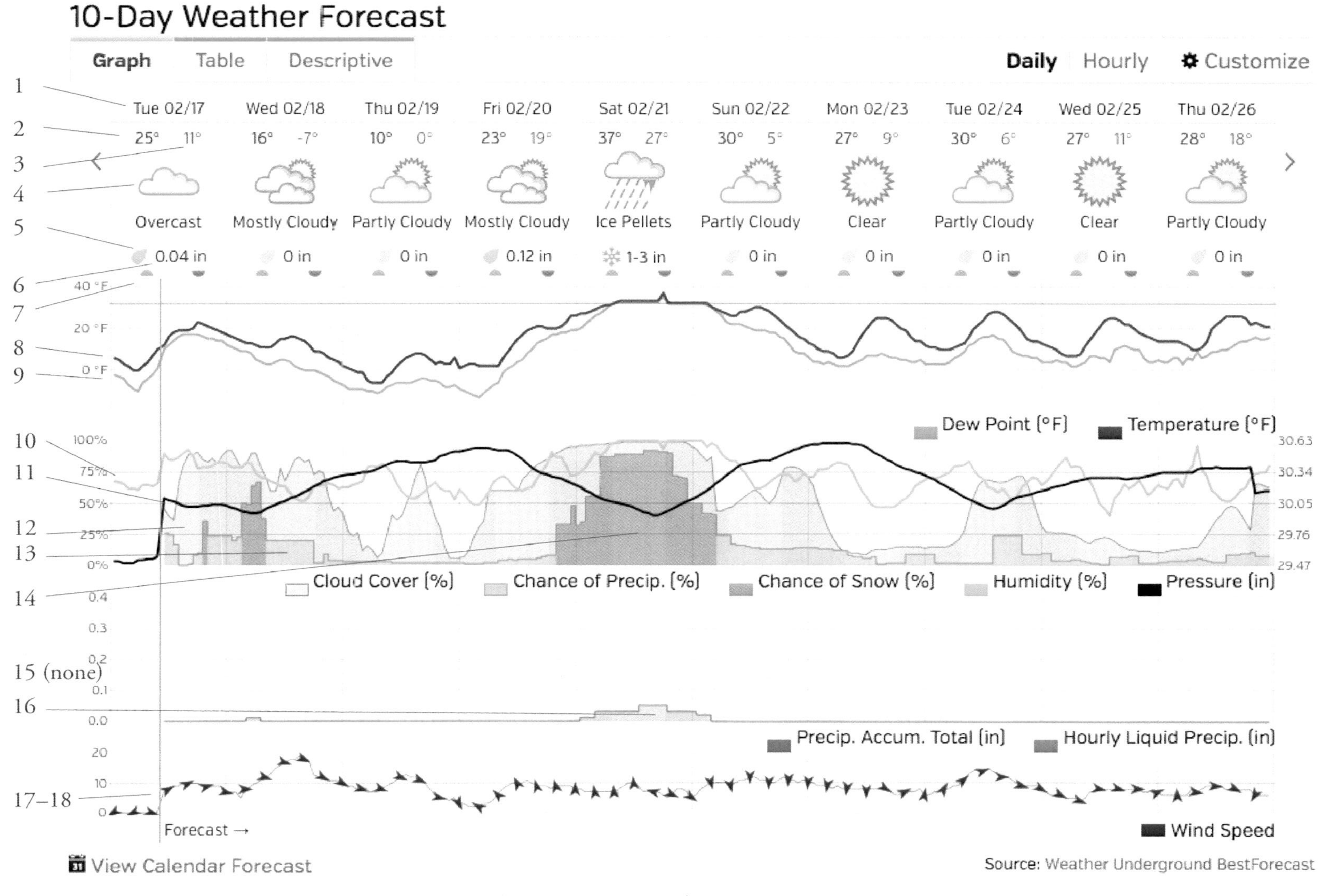

Colorplate 1 (Figure 3.1) Weather Underground, "Forecast for Bloomington, Indiana," accessed February 17, 2015, http://www.wunderground.com/cgi-bin/fir dweather/getForecast?query=Bloomington%2C+IN. Annotations added.

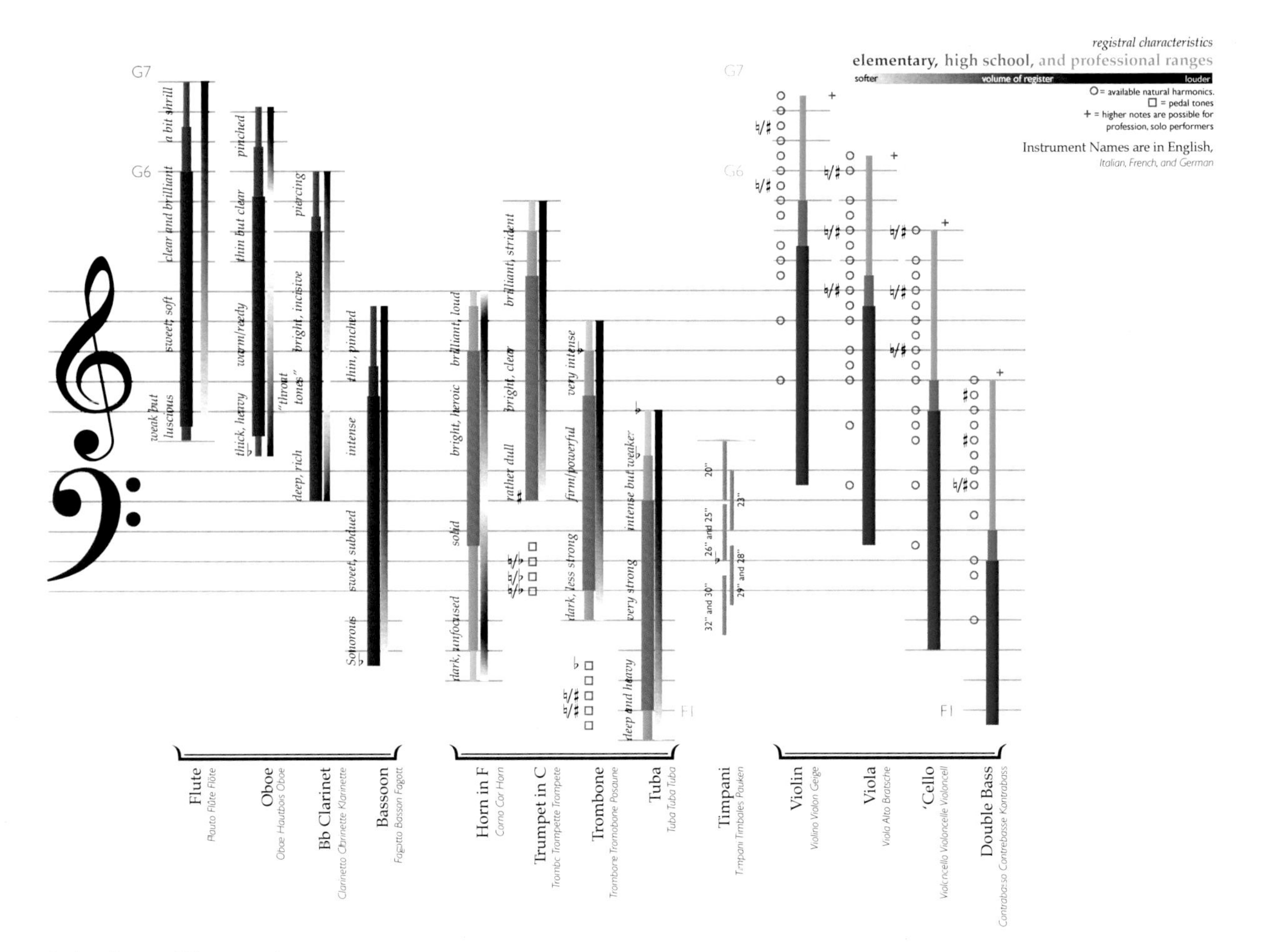

Colorplate 2 (Figure 3.5)
Mitchell Ohriner, 2007, unpublished.

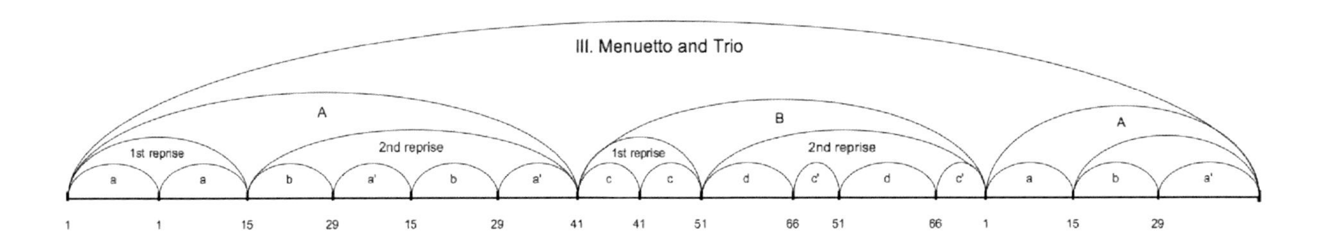

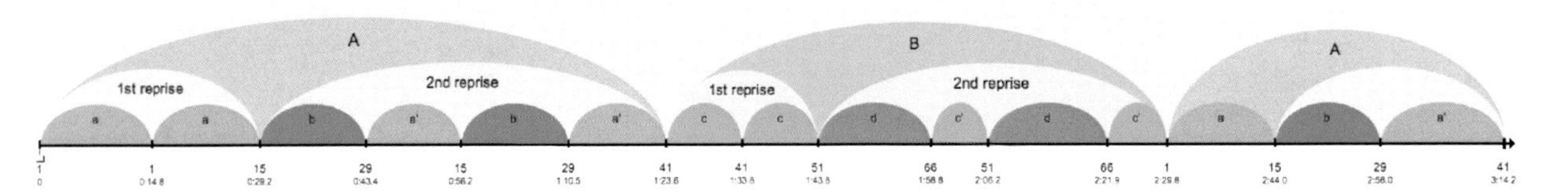

Colorplate 3 (Figure 11.2)
Two versions of a form diagram of Beethoven, *Sonata*, op. 2., no. 1, created with Variations Timeliner.

	1	x	y	z	2	x	y	z	3	x	y	z	4	x	y	z
1	So		nas-		ty	that	it's	prob'-	ly	some-	what	of a	tra-	ves-	ty	hav-
2	ing	me	dai-	ly	to-	tal	peo-	ple	you	can	call	me	your		ma-	je-
3	sty			keep	your	batt-	er-	y	charged				you	know	it	won't
4	stick,		yo			and	it's	not	his	fault		you	kick		slow	
5		shou-	lda let	your		trick	ho		chick	hold		ya	sick		glow	
6		plus	no-	bod-	y	coul- dn't	do	nut-	tin	once	he	let	the	brick		go
7				and	you	know	I	know	that's	a	bunch	of	snow			
8	the	beat	is	so		but-	ter			peep	the	slow		cut- ter	as	he
9	ut-	ter	the	calm		flow			don't	talk a-	bout	my	moms,		yo	
10		some-		times		he rhyme		quick	some-	times		he	rhyme		slow	or
11	vice		ver-	sa			whip up	a	slice	of	nice		verse		pie	
12	hit	it	on	the	first		try		vil-	lain		the	worst		guy	
13	spot		hot		tracks		like		spot	a	pair	(of)	fat	ass-		es
14	shots		of	the	scotch	from	out	the	square		shot		glass-		es	
15			and	he	won't		stop		'til	he	got	the	mass-		es	
16		'n	show 'em	what	they	know	not	thru	flows	of	hot	mol-	ass-		es	
17	do	it	like	the	ro-	bot		to	head-	spin		to	boo-	ga-	loo	
18		took	a few		min-	utes	to	con-	vince	the	ave-	rage	bug-	a-	boo	
19			it's	ug-	ly		like	"look	at	you	it's	a	damn		shame"	
20		just	re-	mem-		ber	all		caps		when	you	spell	the	man['s]	
21	name															

	1	x	y	z	2	x	y	z	3	x	y	z	4	x	y	z
1	So		nas-		ty	that	it's	prob'-	ly	some-	what	of a	tra-	ves-	ty	hav-
2	ing	me	dai-	ly	to-	tal	peo-	ple	you	can	call	me	your		ma-	je-
3	sty			keep	your	batt-	er-	y	charged				you	Know	it	won't
4	stick,		yo			and	it's	not	his	fault		you	kick		slow	
5		shou-	lda let	your		trick	ho		chick	hold		ya	sick		glow	
6		plus	no-	bod-	y	coul- dn't	do	nut-	tin	once	he	let	the	brick		go
7				and	you	know	I	know	that's	a	bunch	of	snow			
8	the	beat	is	so		but-	ter			peep	the	slow		cut- ter	as	he
9	ut-	ter	the	calm		flow			don't	talk a-	bout	my	moms,		Yo	
10		some-		times		he rhyme		quick	some-	times		he	rhyme		slow	or
11	vice		ver-	sa			whip up	a	slice	of	nice		verse		pie	
12	hit	it	on	the	first		try		vil-	lain		the	worst		guy	
13	spot		hot		tracks		like		spot	a	pair	(of)	fat	ass-		es
14	shots		of	the	scotch	from	out	the	square		shot		glass-		es	
15			and	he	won't		stop		'til	he	got	the	mass-		es	
16		'n	show 'em	what	they	know	not	thru	flows	of	hot	mol-	ass-		es	
17	do	it	like	the	ro-	bot		to	head-	spin		to	boo-	ga-	loo	
18		took	a few		min-	utes	to	con-	vince	the	ave-	rage	bug-	a-	boo	
19			it's	ug-	ly		like	"look	at	you	it's	a	damn		shame"	
20		just	re-	mem-		ber	all		caps		when	you	spell	the	man['s]	
21	name															

Colorplate 4 (Figure 11.3)
Above, Kyle Adams, "On the Metrical Techniques of Flow in Rap Music," *Music Theory Online* 15, no. 5 (2009): ex. 5b, http://www.mtosmt.org/issues/mto.09.15.5/mto.09.15.5.adams.html; *below*, redrawing.

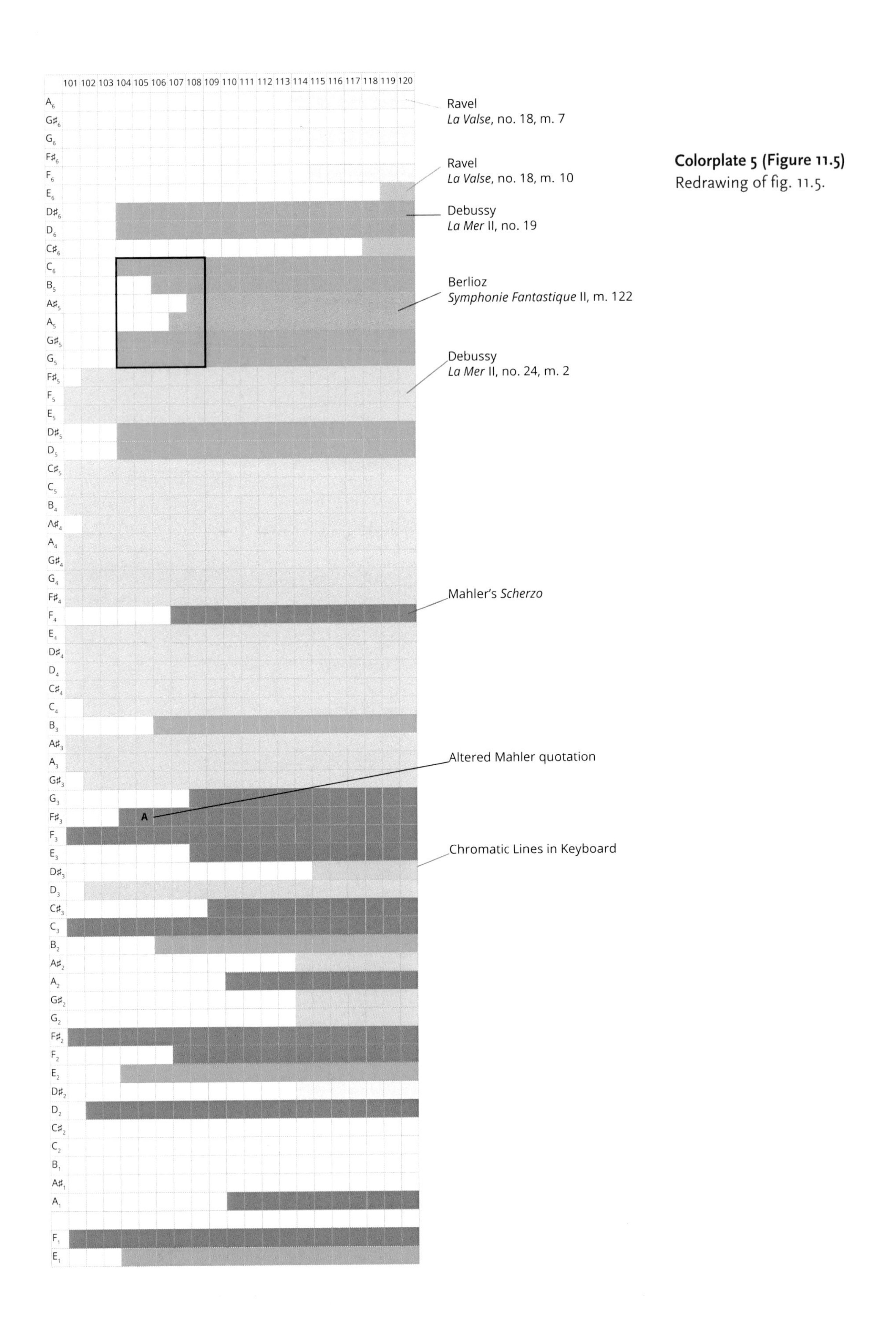

Colorplate 5 (Figure 11.5)
Redrawing of fig. 11.5.

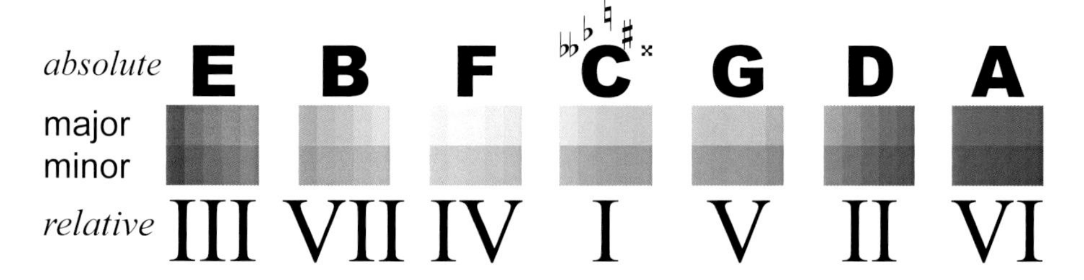

Colorplate 6 (Figure 11.7)
Above, Craig Stuart Sapp, personal communication, December 19, 2020; *below*, Craig Stuart Sapp, "Computational Methods for the Analysis of Musical Structure" (PhD diss., Stanford University, 2011), 74.

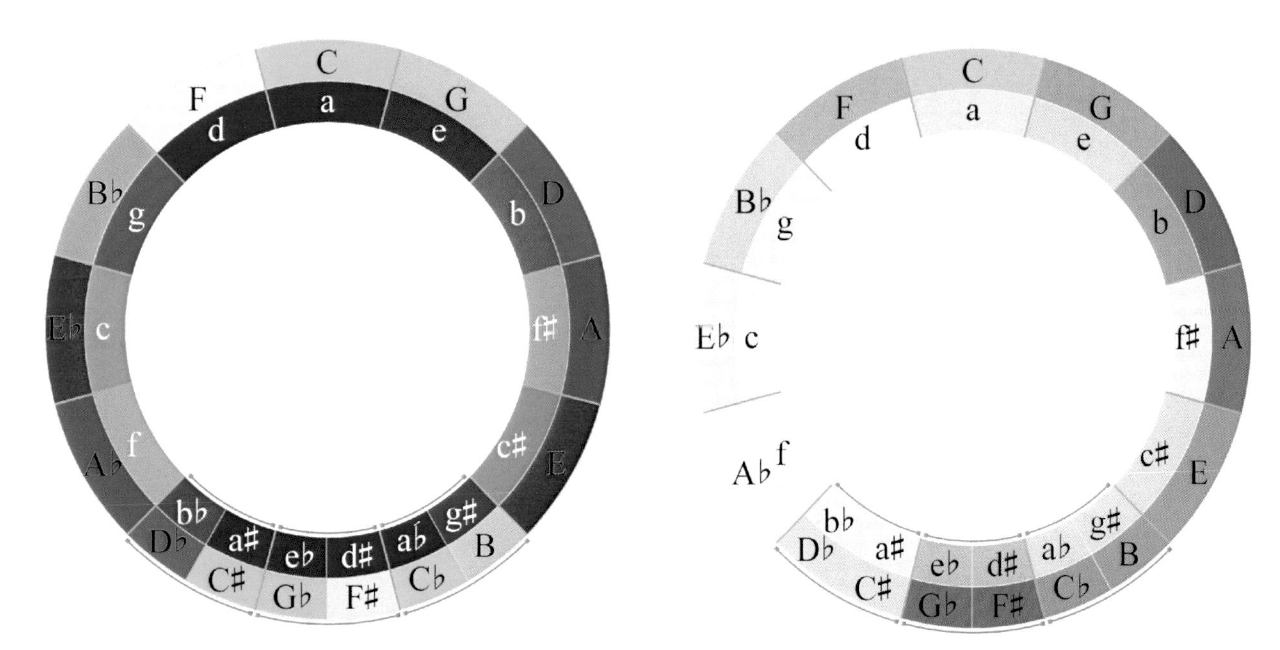

Colorplate 7 (Figure 11.8)
Left, pitch class / color palette from fig. 11.7 (Colorplate 6) displayed in a circle of fifths; *right*, redrawing by Loida Raquel Garza, 2018, unpublished.

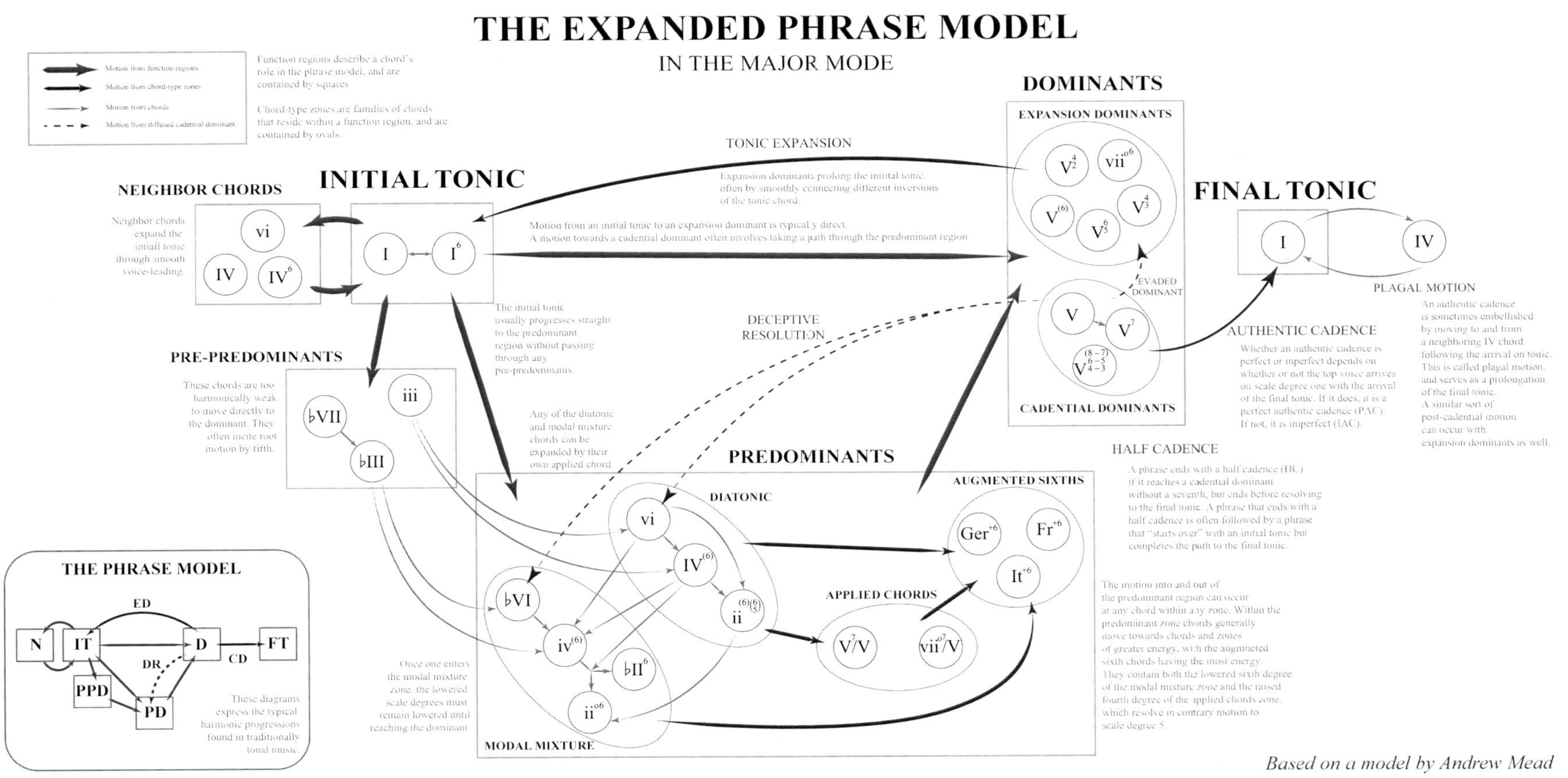
THE EXPANDED PHRASE MODEL
IN THE MAJOR MODE
Motion from function regions
Motion from chord-type zones
Motion from chords
Motion from diffused cadential dominant
Function regions describe a chord's role in the phrase model, and are contained by squares
Chord-type zones are families of chords that reside within a function region, and are contained by ovals.
NEIGHBOR CHORDS
Neighbor chords expand the intiail tonic through smooth voice-leading.
vi
IV
IV6
INITIAL TONIC
I
I6
TONIC EXPANSION
Expansion dominants prolong the initial tonic, often by smoothly connecting different inversions of the tonic chord.
Motion from an initial tonic to an expansion dominant is typically direct.
A motion towards a cadential dominant often involves taking a path through the predominant region
DOMINANTS
EXPANSION DOMINANTS
V4/2
vii°6
V(6)
V6/5
V4/3
EVADED DOMINANT
V
V7
CADENTIAL DOMINANTS
FINAL TONIC
I
IV
PLAGAL MOTION
An authentic cadence is sometimes embellished by moving to and from a neighboring IV chord following the arrival on tonic. This is called plagal motion, and serves as a prolongation of the final tonic. A similar sort of post-cadential motion can occur with expansion dominants as well.
AUTHENTIC CADENCE
Whether an authentic cadence is perfect or imperfect depends on whether or not the top voice arrives on scale degree one with the arrival of the final tonic. If it does, it is a perfect authentic cadence (PAC). If not, it is imperfect (IAC).
HALF CADENCE
A phrase ends with a half cadence (HC) if it reaches a cadential dominant without a seventh, but ends before resolving to the final tonic. A phrase that ends with a half cadence is often followed by a phrase that "starts over" with an initial tonic but completes the path to the final tonic.
The initial tonic usually progresses straight to the predominant region without passing through any pre-predominants.
DECEPTIVE RESOLUTION
PRE-PREDOMINANTS
These chords are too harmonically weak to move directly to the dominant. They often incite root motion by fifth.
♭VII
iii
♭III
Any of the diatonic and modal mixture chords can be expanded by their own applied chord.
PREDOMINANTS
DIATONIC
vi
IV(6)
ii
AUGMENTED SIXTHS
Ger+6
Fr+6
It+6
APPLIED CHORDS
V7/V
vii°7/V
MODAL MIXTURE
♭VI
iv(6)
♭II6
ii°6
Once one enters the modal mixture zone, the lowered scale degrees must remain lowered until reaching the dominant.
The motion into and out of the predominant region can occur at any chord within any zone. Within the predominant zone chords generally move towards chords and zones of greater energy, with the augmented sixth chords having the most energy. They contain both the lowered sixth degree of the modal mixture zone and the raised fourth degree of the applied chords zone, which resolve in contrary motion to scale degree 5.
THE PHRASE MODEL
N
IT
ED
D
FT
CD
DR
PPD
PD
These diagrams express the typical harmonic progressions found in traditionally tonal music.
Based on a model by Andrew Mead

Colorplate 8 (Figure 20.11) Andrew Mead, "Expanded Phrase Model," graphical design by John Heilig, 2016, unpublished.

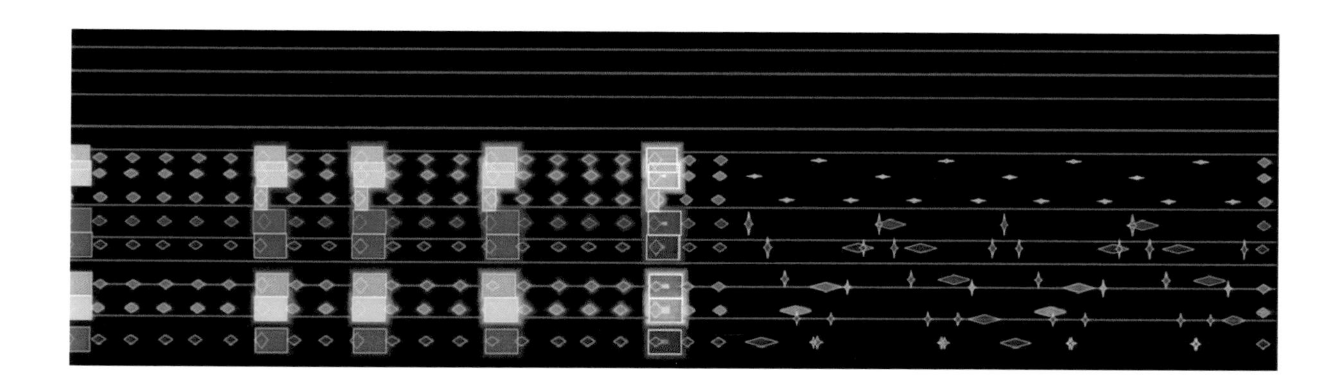

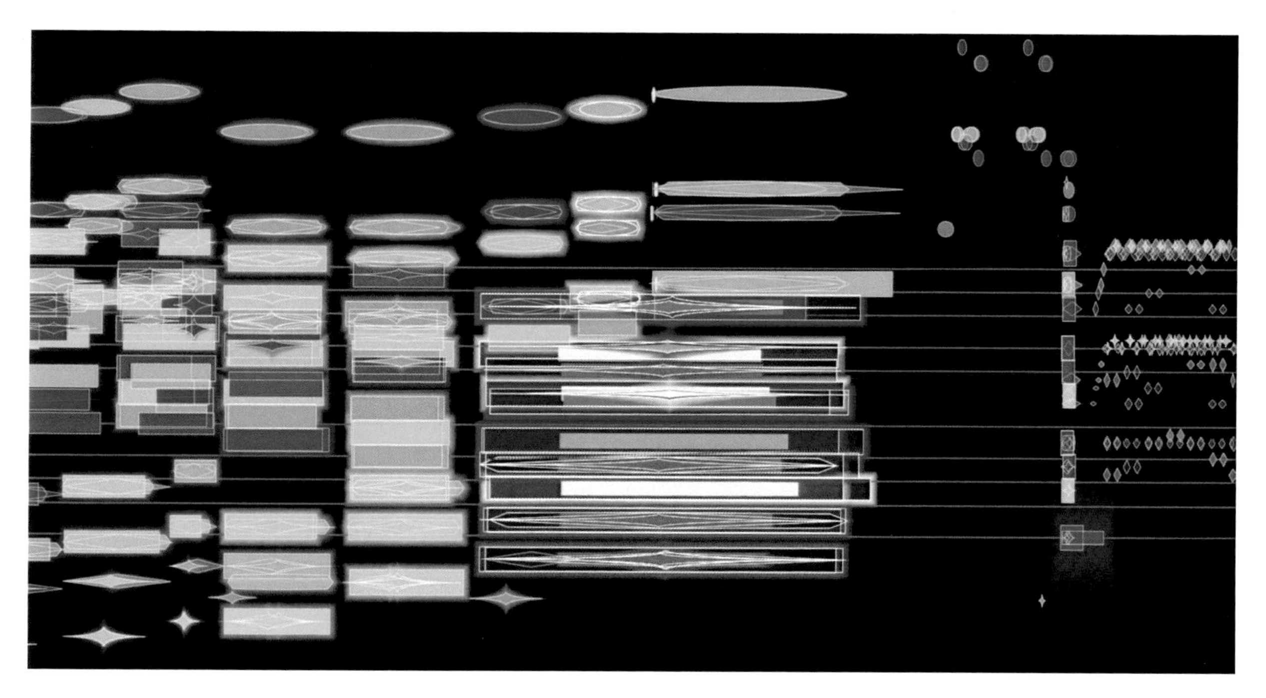

Colorplate 9 (Figure 28.5)
Screenshot of Stephen Malinowski, "Animated Graphical Score of Stravinsky, *Rite of Spring*," YouTube, 2013, accessed February 10, 2015, 3:32 (*above*), 10:33 (*below*), https://www.youtube.com/watch?v=5IXMpUhuBMs.

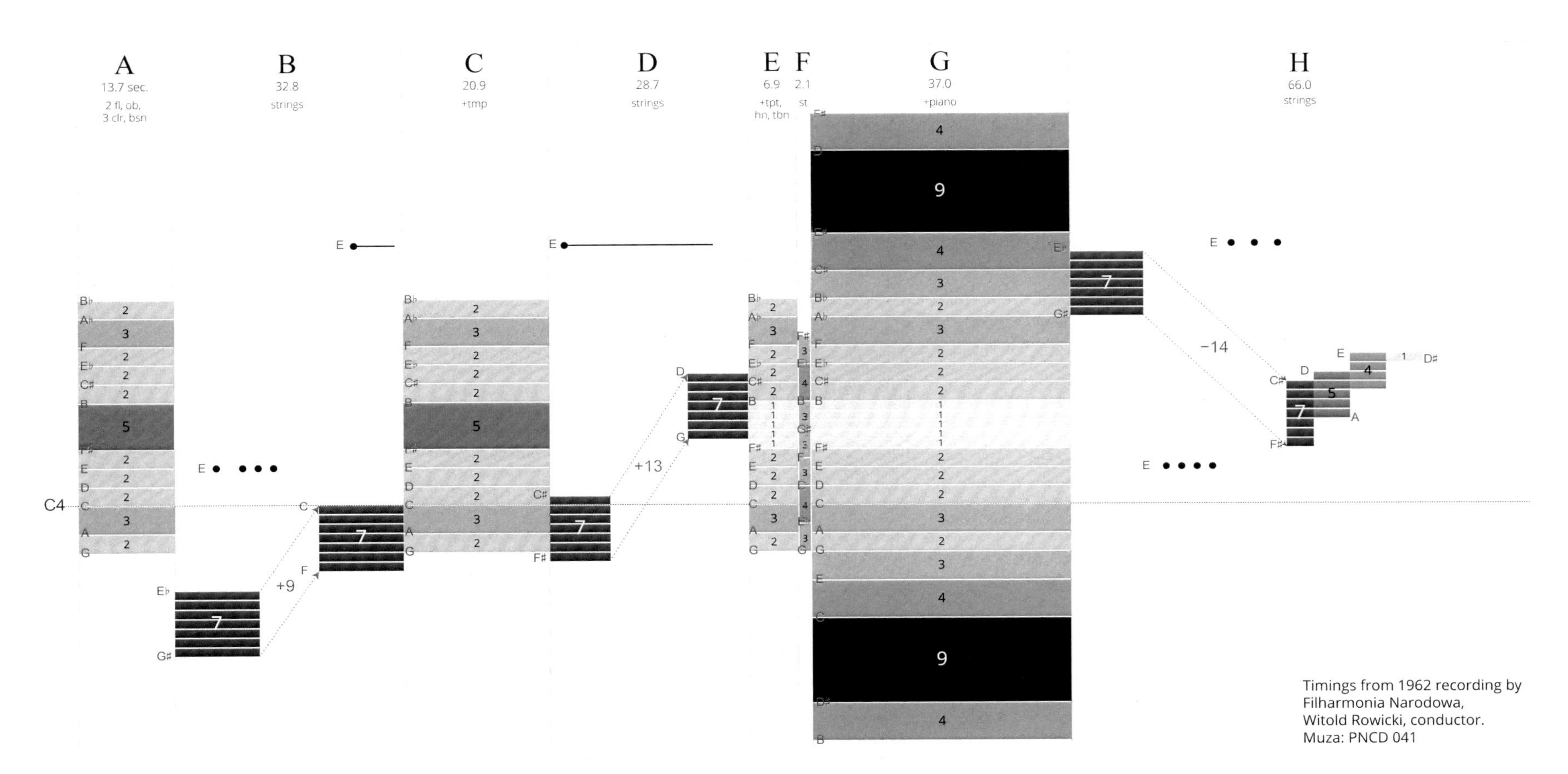

Colorplate 10 (Figure 40.4) Redrawing of fig. 40.1.

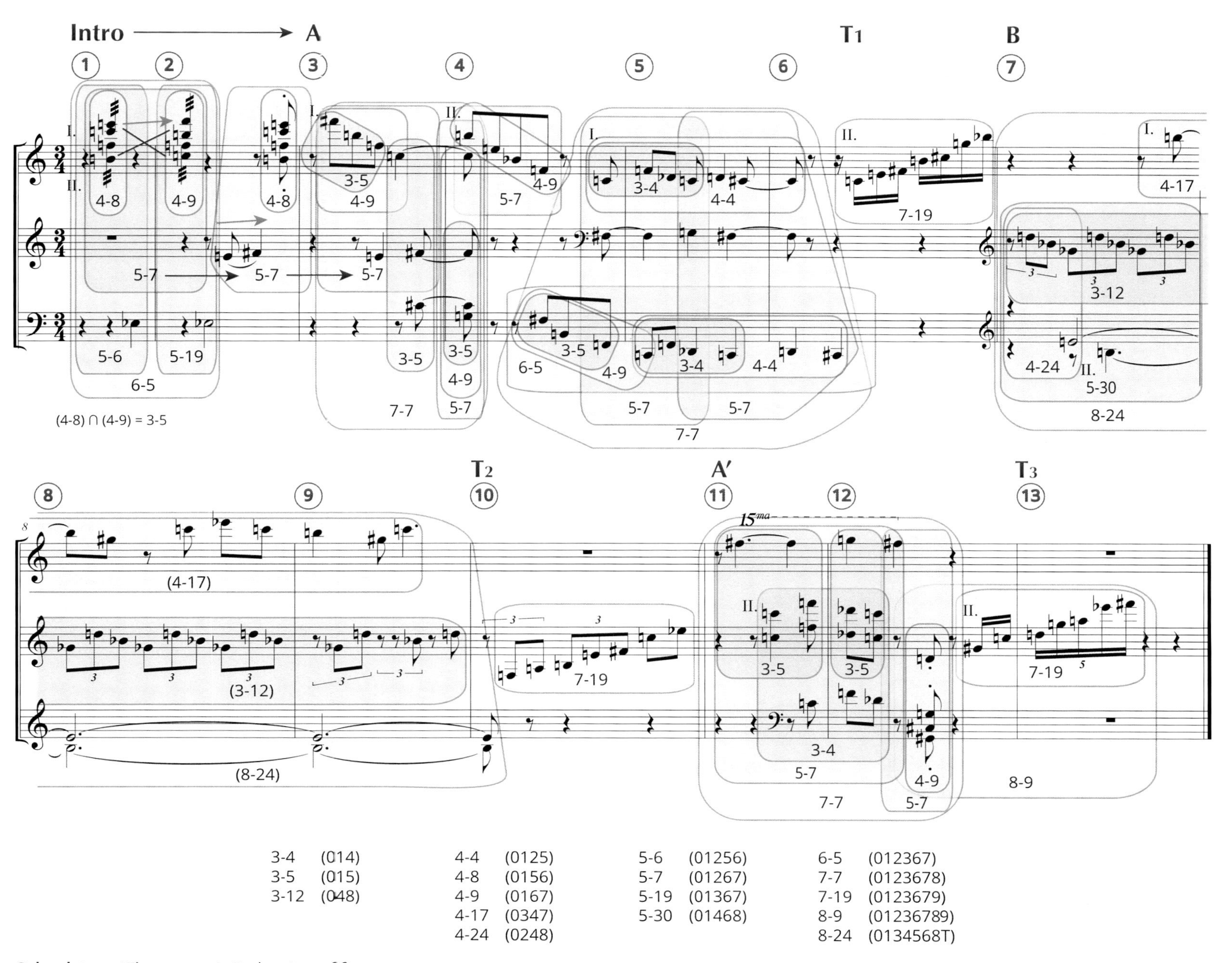

Colorplate 11 (Figure 41.13) Redrawing of fig. 41.1.

Colorplate 12 (Figure 47.16) Hannah Chan-Hartley, "Listening Guide to Antonín Dvořák, Symphony no. 9, mvt. 1," 2018, unpublished manuscript.

LISTENING GUIDE

Antonín Dvořák

Symphony No. 9 in E Minor, Op. 95 "From the New World"

FIRST MOVEMENT

Introduction (Adagio – "slowly")

Exposition (Allegro molto – "very fast"): initial presentation of thematic material

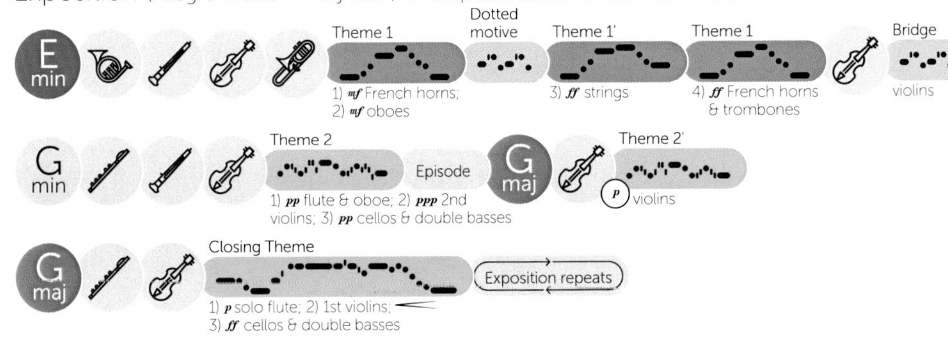

Development: thematic material developed and extended

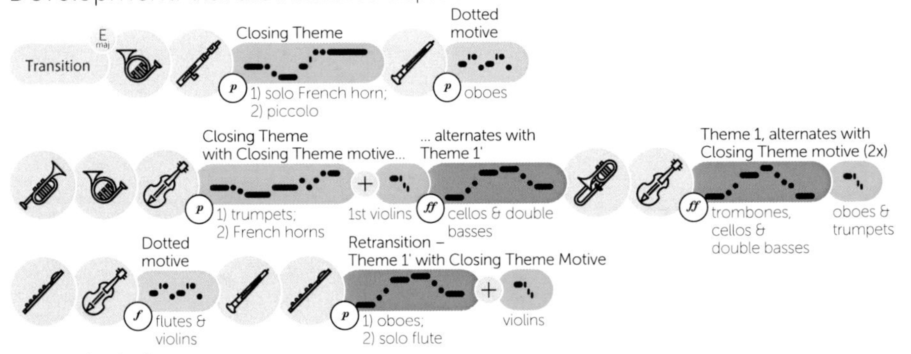

Recapitulation: return of thematic material from the exposition

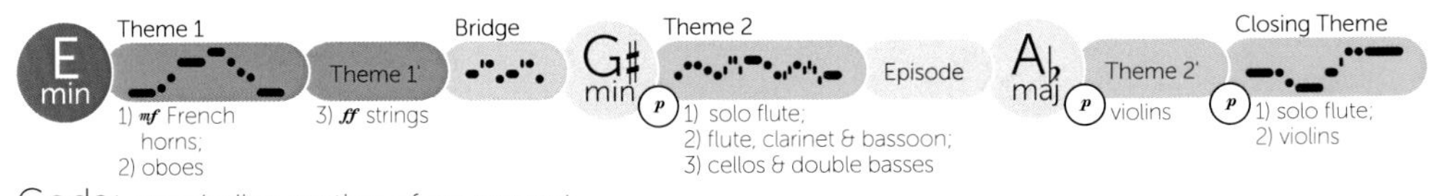

Coda: concluding section of movement

Chapter 12 Additional General Principles

In this chapter we look at examples of other techniques that can be used effectively or that are better avoided. Figure 12.1 illustrates that it is best to align related materials when possible. The image's effort to keep the words next to the music is laudable but is counterproductive for the performer. Although musicians are accustomed to notes appearing in different vertical positions, the vertical axis

12.1 Henri-Louis Choquel, *La Musique Rendue Sensible Par La Mécanique*, nouv. éd. (1762; repr., Geneva: Minkoff Reprints, 1972), 159.

12.4 Erkki Huovinen, *Pitch-Class Constellations: Studies in the Perception of Tonal Centricity* (Turku: Finnish Musicological Society, 2002), 134.

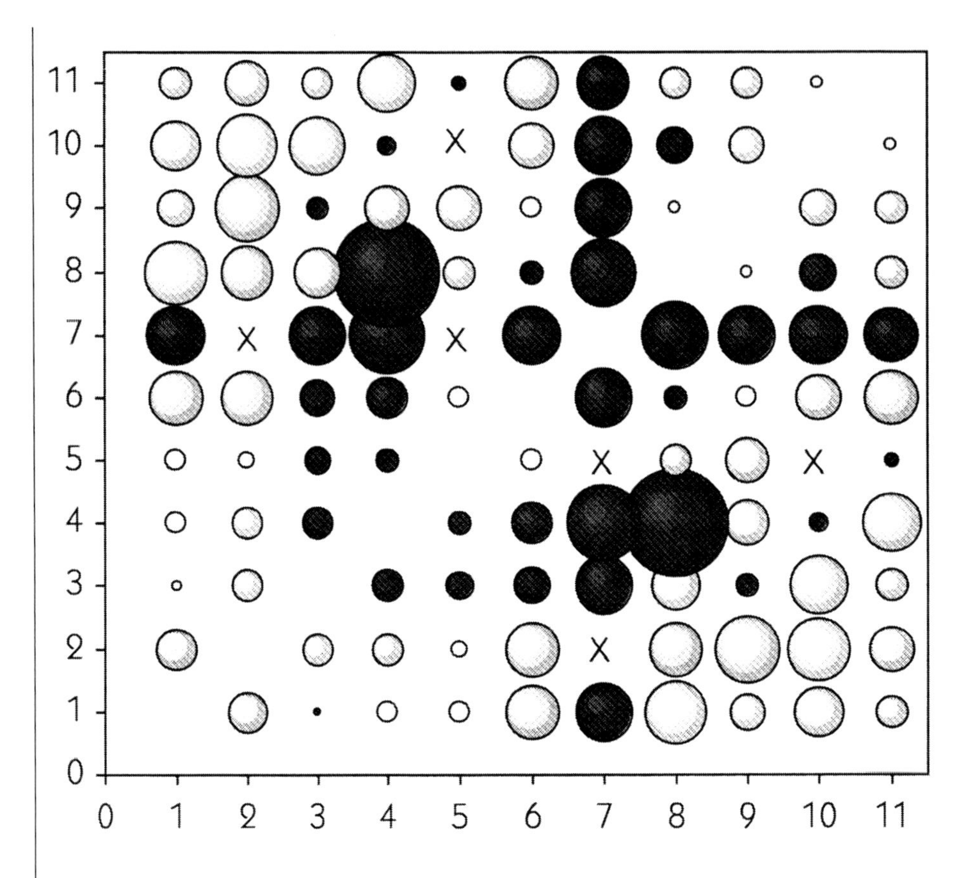

provides information in that case. No meaning is added when text jumps, and keeping it aligned allows for smoother reading. See also online figures 12.2 and 12.3.

The image in figure 12.4 has multiple issues. While the underlying details are more complicated to explain than necessary for this discussion, the image shows deviations from a mean value for pairs of pitch classes, using size and color. The size of the circle indicates how far a value deviates from the average, while color indicates whether the deviation is in the positive (black) or negative (white) direction. Among the problems with the image is that the values represented are not themselves shown. While the data appears elsewhere in the source, as does the formula that generates the values graphed here, the values themselves are absent. Viewers thus get no sense of what the sizes mean and no opportunity for independent verification. In the absence of values, there is also no sense of scale. Analysis of the data reveals that the two largest circles have the value 40.7 percent, while the values of the sequence of black circles in row/column 7, many of which are virtually indistinguishable visually, range from 17.3 percent to 10.3 percent. The image would be stronger if it conveyed this.

In the source, the image's caption says that the values correspond to the *area* of the figures. Yet the use of shading (more apparent in the white bubbles than in the black) implies that these are three-dimensional images—in fact, the caption refers to the figures as bubbles. If we see the figures as spheres and not circles, we will be comparing them in terms of implied volume rather than area. This distorts the data since volume increases by the cube of the radius rather than the square of the radius as is the case for a circle. Since the data are one-dimensional, a three-dimensional representation is even less appropriate than a two-dimensional one. Finally, because the display is symmetrical around the upward diagonal, half of the data could be eliminated without loss of information. The redrawings in online figure 12.5 propose two alternative approaches. The image in online figure 12.6 has similar issues.

Figure 12.7 tackles a nearly impossible task. Fourteen line styles connect twenty-eight musical segments to sixteen "motive forms" in Beethoven's Symphony no. 5, movement 1. Some of the line styles are hard to distinguish from one another. Some even look different from themselves, depending on the angle they are drawn at (for example, see the lines leaving from mm. 10–11 or from mm. 12–13). The image tries to do too much in too little space. While the image would benefit

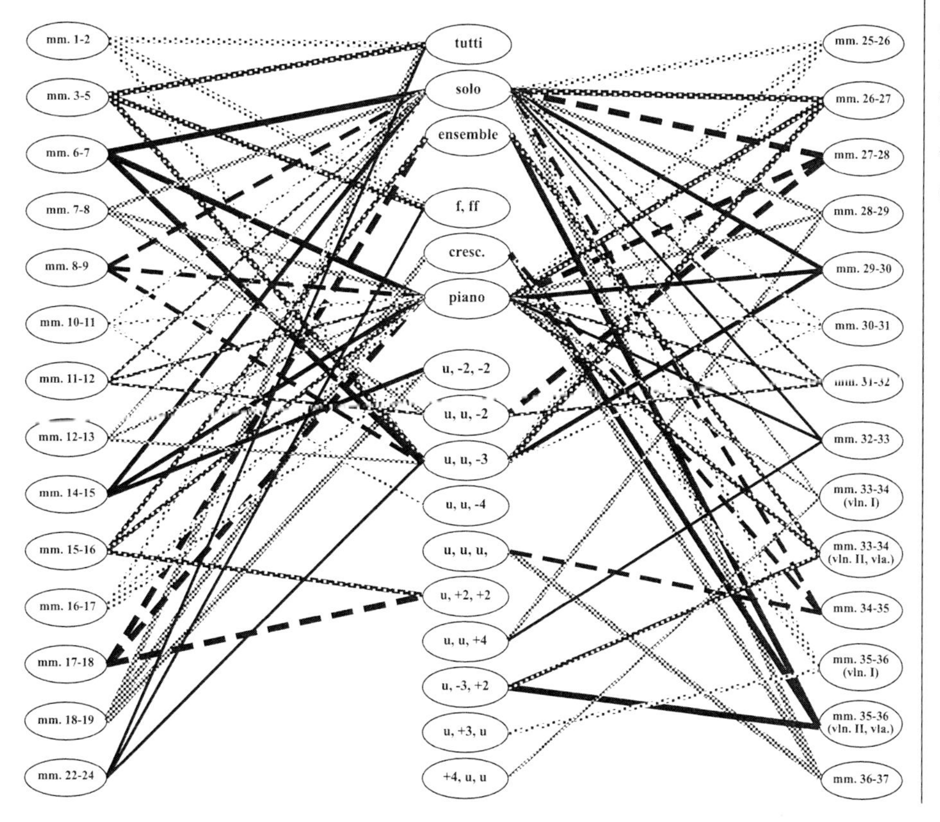

12.7 Lawrence Zbikowski, *Conceptualizing Music: Cognitive Structure, Theory, and Analysis* (New York: Oxford University Press, 2002), 45.

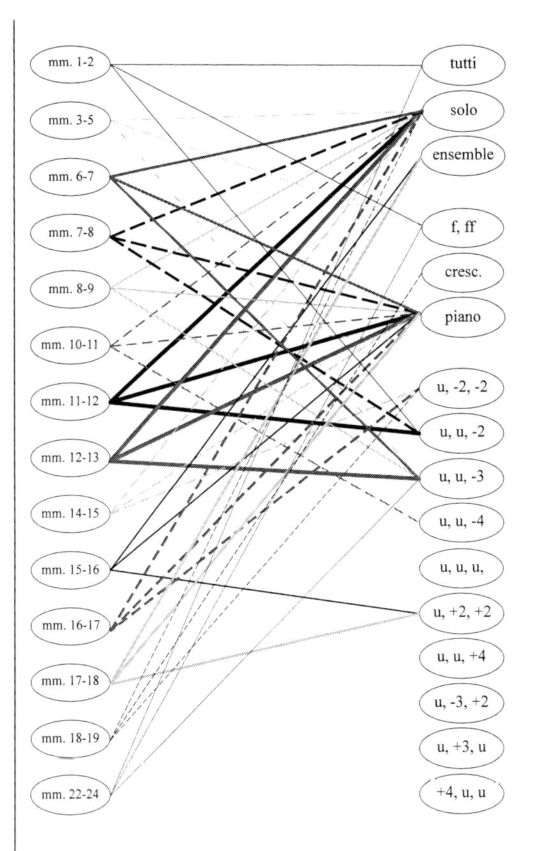

12.8 Left side of fig. 12.7, redrawn.

from a wholesale redesign, some adjustments to the line styles would improve the present design. A redrawing of the left side of the image (fig. 12.8) uses four line widths, three shades of gray (black, gray, light gray), and two styles (solid, dashed). Those lines that need to travel farther (those coming from the upper and lower ends) always lie above those traveling less far (those coming from the center). It doesn't alleviate the graphical density problems altogether, but one can now trace the paths of all the lines. (Online figure 12.9 poses a similar problem with drawn lines.)

William Cleveland (1993) has shown that line-graph images such as figure 12.10 are best designed so the average point-to-point slope is 1.0, meaning the average line is angled forty-five degrees. In figure 12.10, which has eighty-two data points horizontally and twelve vertically, the average point-to-point slope is nearly 10, meaning the image should in fact be close to ten times wider than it is tall. Instead, it is only about 1.25 times wider. The redrawing in figure 12.11 adjusts the scales of the axes to match Cleveland's recommendation, making the shape of the data much clearer. The image's increased width has required the addition of horizontal gridlines, which are rendered here in a very light gray. Online figure 12.12 provides an additional example of a higher-than-necessary scale.

Sometimes the data we want to display in an *x*-*y* chart reside entirely within a quadrant remote from the (0,0) origin. Figure 12.13 provides such an example. It shows percentages of correct rhythmic and pitch responses in "amusics," a term sometimes used to describe people who lack certain basic musical aptitudes, and in a control group of people with normal musical ability. All the data points are greater than 49 percent in the melodic dimension and 50 percent in the rhythmic. All data points would therefore reside in the upper right quadrant of a graph that runs from 0 percent to 100 percent in both dimensions. The image uses two slashes (//) near the origin to indicate the break in the graph between 0 percent and 50

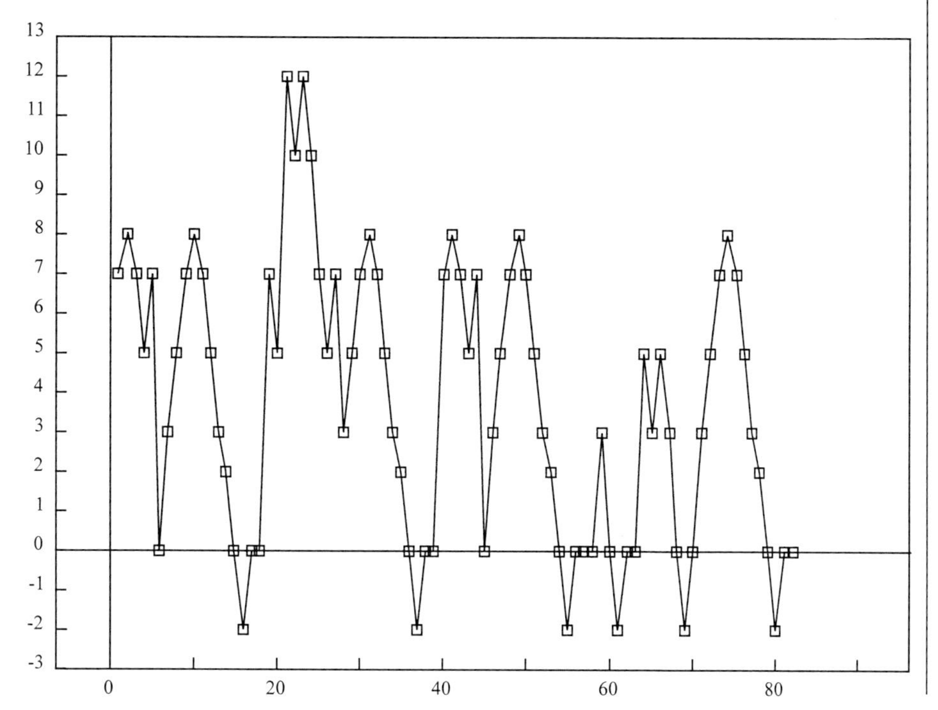

12.10 Charles Madden, *Fractals in Music: Introductory Mathematics for Musical Analysis* (Salt Lake City: High Art, 1999), 183.

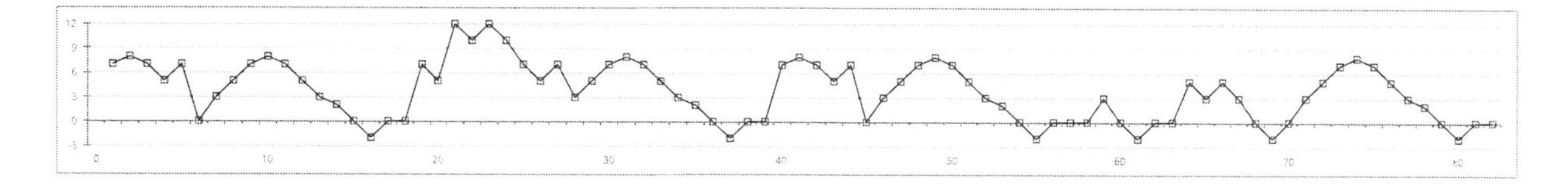

***Below,* 12.11** Redrawing of fig. 12.10.

percent. Visually, however, this is too subtle, and viewers can too easily get the mistaken impression that amusic subjects scored as low as almost zero on the tasks, not about 50 percent. Plotting the data using a full 0–100 chart would result in three of the four quadrants being empty, however. Figures 12.14 and 12.15 provide alternatives. In figure 12.14, the grid begins with the maxima at the upper right rather than the traditional minima from the lower left, so only the top right quadrant is shown. Figure 12.15 tweaks this design by restricting the axes so they show the maximum and minimum values for both tasks, for both populations. In this way, the axes themselves contextualize the data.

Two final examples make clear the relative merits of tabular versus graphical presentation of numerical data. The table in figure 12.16 shows beat durations in five performances of Robert Schumann's *Träumerei*, mm. 30–32. Each performer ("subject") has a row in the table, with columns representing each of the eight beats in the passage. Numbers in each cell represent a third dimension: the lengths of the beats as performed, in seconds. Inspection of the data reveals significant

12.13 Isabelle Peretz, "Brain Specialization for Music: New Evidence from Congenital Amusia," in *The Cognitive Neuroscience of Music*, ed. Isabelle Peretz and Robert J. Zatorre, 192–203 (New York: Oxford University Press, 2003), 197.

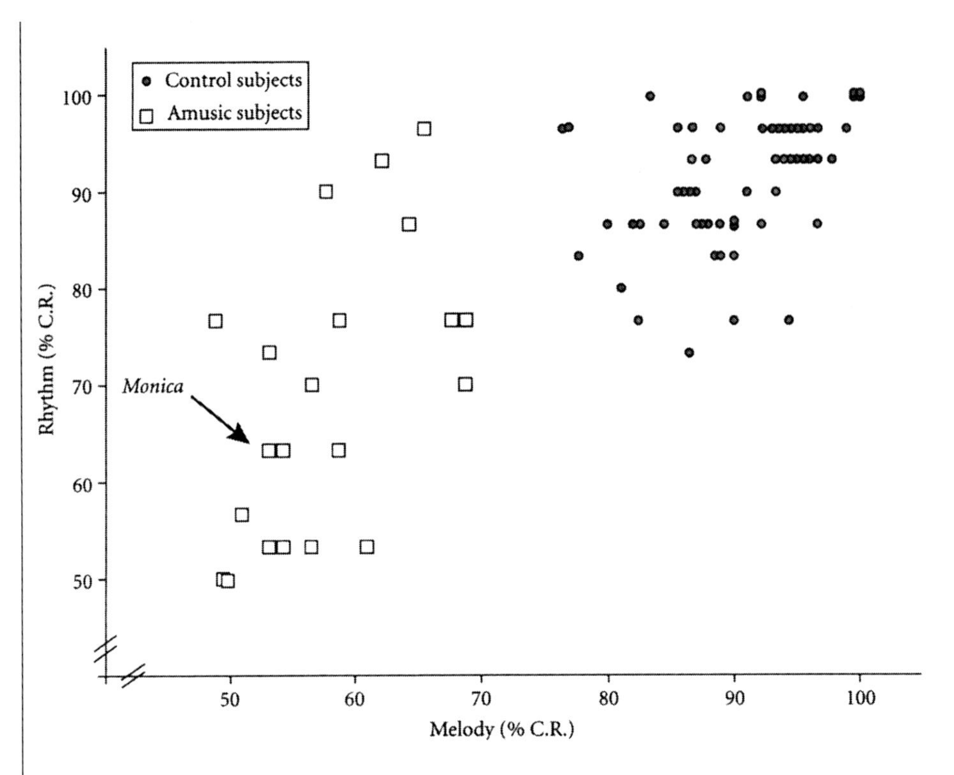

12.14 Redrawing of fig. 12.13.

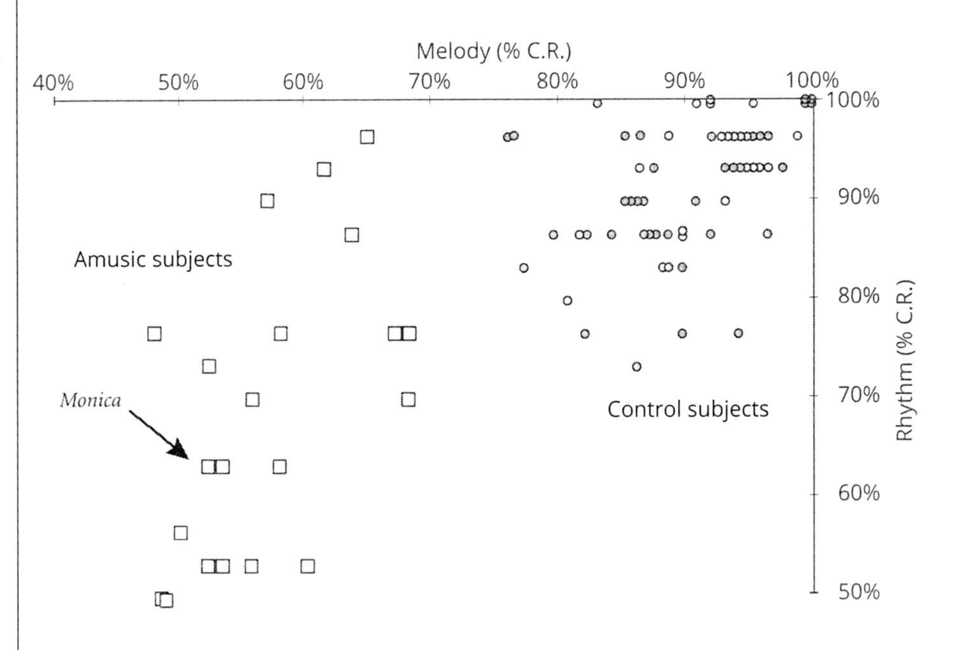

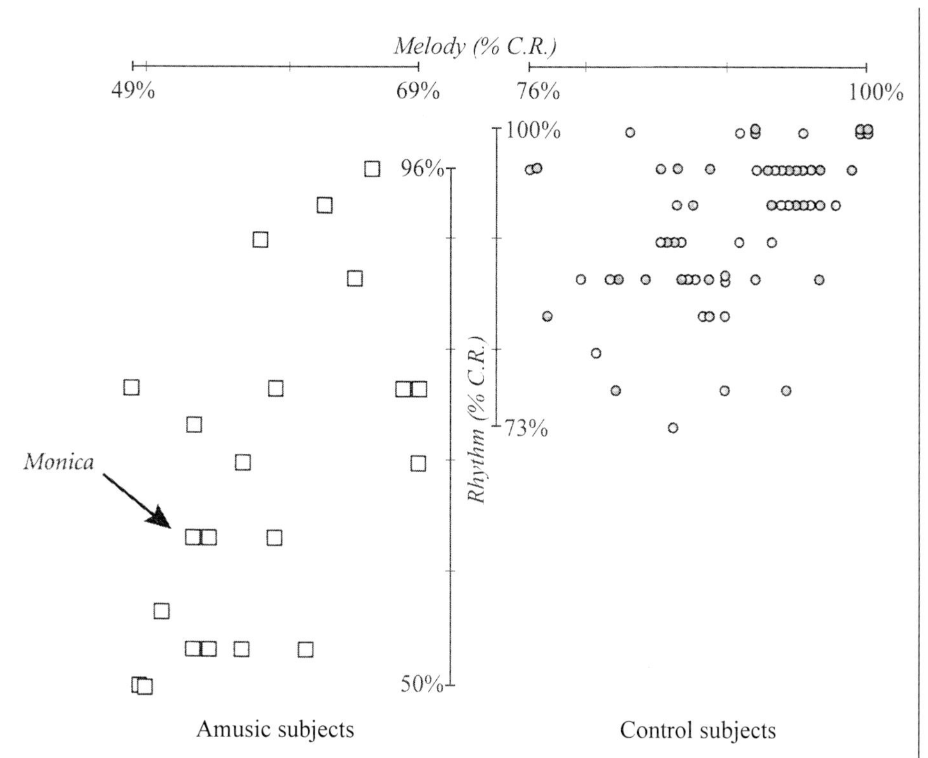

12.15 Redrawing of fig. 12.13.

differences within each performer (scanned horizontally) and between performers (read vertically), but the significance of those differences is not clear. After all, it is the relationship between beat number and beat length that is of interest. We can understand relative magnitude much more readily in graphical form than in numerical form.

	Beat number							
Subject	**1**	**2**	**3**	**4**	**5**	**6**	**7**	**8**
1	1.151	1.068	1.224	1.641	1.896	2.828	3.505	6.485
2	.473	.600	463	.673	.577	.783	1.327	3.094
3	1.054	1.122	1.030	1.214	1.222	1.578	1.879	4.814
4	.405	.410	.414	.462	.487	.543	.722	2.200
5	.982	.958	.842	.918	.778	.998	1.186	1.550

12.16 David Epstein, *Shaping Time: Music, the Brain, and Performance* (New York: Schirmer Books, 1995), 442.

Indeed, the data comes alive when graphed, as in figure 12.17. Now the "duration" dimension appears on the vertical axis, and the "subject" dimension is realized through the small-multiple format: there is one graph for each subject. The design facilitates easy comparison of the interpretations. The only problem is that the scales of the vertical axes differ dramatically for each performer; employing the same scale for each would make the comparison even more vivid. See online figure 12.18 for a similar comparison of tabular and graphical presentations of data.

12.17 David Epstein, *Shaping Time: Music, the Brain, and Performance* (New York: Schirmer Books, 1995), 443.

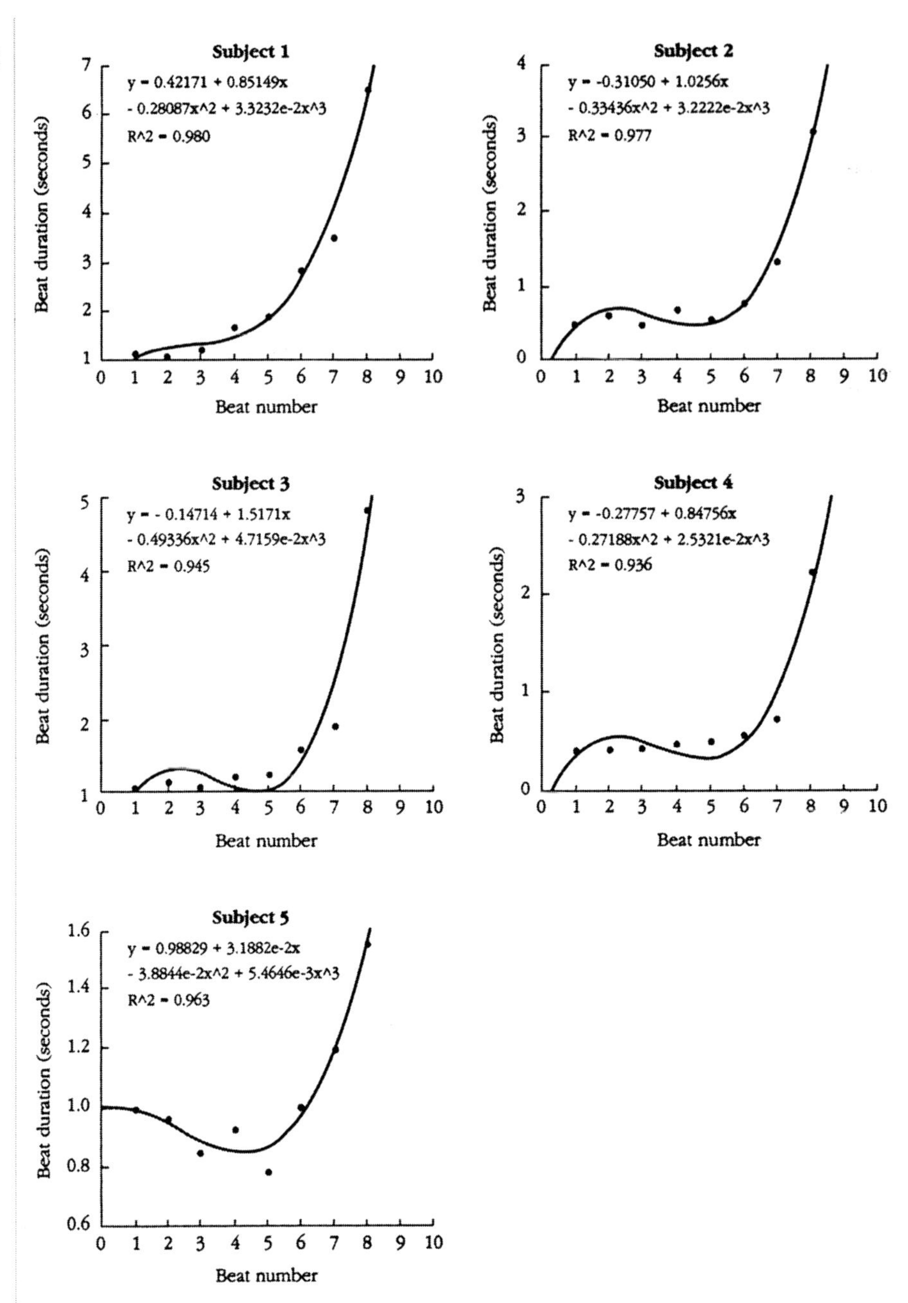

In this chapter, we have seen the merits of keeping related information aligned; of representing numerical data with enough dimensions; of integrating graphics, numbers, and text; of using contrasting line styles; and of representing numerical data graphically. As always, images should take advantage of our natural visual processing system to communicate meaning immediately and unambiguously.

CHAPTER 13

Case Study

WESTERN NOTATION

Many of the principles outlined in part 1 of this book are at work in the notational system for common Western music. The system features an elegant and efficient set of conventions intended to provide direction for the performance of a musical work (though as Morton Feldman [2000, 144] has observed, "The degree to which a music's notation is responsible for much of the composition itself, is one of History's best kept secrets"). Over many centuries, numerous factors have prompted changes in notational practice. Changes in musical style or blendings of concurrent styles are of course responsible for the most significant developments. Other factors include the desire for more, or sometimes less, specific direction from composer to performer about performance practices; matters of convenience to instrumentalists (such as those playing transposing or fretted instruments); the desire to better facilitate coordination among performers; and composers', calligraphers', and typesetters' innovations in communicating the composers' intentions visually. Sometimes these interests are at odds with one another, and the system in many ways balances competing factors. On the whole, however, notational practice has evolved in the direction of an efficient system for the communication of complex, multivariate information. Distinctive treatment of shape and size, the choice of graphic versus text, font characteristics, and consistency in layout allow for the display of an immense amount of varied information. We will examine in detail some key elements of the system as a case study in effective visual communication about music.

13.1 The staff.

First, our notational system uses a convention familiar to those who can read. It is notated in the same way as the prose of European languages (left to right, top to bottom) and thus requires no special scanning strategies. In fact, music notation evolved from annotations of text.

The modern five-line musical staff (fig. 13.1) is a perceptually optimized basis for the visual representation of musical pitch. Its origin lies in notation found in the mid-ninth century (online fig. 13.2 is an example from the mid-tenth century), in which shapes called neumes are thought to have represented gestural information associated with the recitation of chant. The neumes thus served as a memory aid. By the early eleventh century, neumes were drawn at different distances from the text. These "heightened neumes" demonstrate a clear mapping between pitch height and what we now think of as melodic height, even though they represent melodic shape only roughly.

Not long after, a line was added, with a letter clef to serve as a reference point. A four-line staff followed quickly (online fig. 13.3). This allowed pitches to be represented unambiguously. The four-line staff was common well into the Renaissance. However, during the Renaissance, the number of lines was not fixed, and it is possible to find music with four-, five-, or six-line staves. In the end, the five-line staff prevailed. Why? We can only speculate—and I shall.

The staff serves as a grid marking off diatonic steps on a vertical axis. Its design represents a compromise between two aims: to represent as many notes as possible and to facilitate reading by performers. By using both the lines and the spaces between them, the staff can accommodate eleven diatonic pitches (bracketed in fig. 13.1). The addition of three ledger lines, which extend the staff on an as-needed basis, increases this range to twenty-three. That is just over three octaves, nearly half the fifty-two-step range of the piano. Putting bass and treble staves together in a grand staff extends the representational range (again assuming three ledger lines) to thirty-five notes, nearly five octaves (from A1 to F6), or two-thirds of the piano's gamut.

The staff also helps performers interpret music quickly. We can perceive notes on a five-line staff holistically, as a kind of gestalt, without having to count lines to identify where a note lies. The five-line staff allows a trained musician to scan a musical score with virtually the same fluency as an experienced reader can scan printed text. This would not be the case with, say, a ten-line staff (fig. 13.4), and I propose that the five-line staff became conventional because it lies in the sweet spot between perceptual efficiency and representational completeness.

13.4 Richard Strauss, *Till Eulenspiegel* theme on five- and ten-line staves.

The fact that the staff aligns with diatonic rather than chromatic pitch space is an artifact of the modal musical traditions from which it emerged. In addition, because the earliest instruments developed while the modes formed the active musical language, they and their descendants have the same diatonic physical layouts. The diatonic notational system maps directly onto the physical layout of the piano and other keyboards, where the nondiatonic keys are differentiated by color, length, and height; those of woodwind instruments, whose "default" fingering layouts also follow the diatonic scale; and the harp. Efforts to devise notation systems that treat all twelve pitch classes as equal have failed to garner traction.

To help us explore other cognitive optimizations of the Western notation system, two versions of the same musical passage are provided: one that violates many notational conventions (fig. 13.5) and one that adheres to those conventions (fig. 13.6). The former is intended as a straw figure, meant to illustrate in a negative way the significance of the many design details of standard notation. Comparing the images while reading the following discussion should make the points clear.

Staff lines serve as a grid. It is better when they are thin, as on graph paper. Making them gray rather than black allows the musical information laid over them to pop out even more effectively. In figure 13.5, background and foreground are visually indistinguishable, while in figure 13.6, the staff recedes to a less prominent place, ironically making it more effective.

The symbols that make up the noteheads stand out clearly against the staff lines. Because noteheads are slightly elongated and most of them are oriented on a diagonal, they pop out from the horizontal staff lines visually even more than they already would if they were round. I believe it is not an accident that the quarter note has emerged as the most common beat unit. It is the point in the durational spectrum between longer values that feature open noteheads, which do not pop out as readily as solid ones, and shorter durations, whose flags and beams make them more complex. I suspect this provides just enough of a cognitive processing advantage over other beat values that, overall, composers have tended to favor it.

Accidental shapes stand out from noteheads and staff lines because the flat (♭) is rounded and the horizontal components of both the natural (♮) and sharp (♯) are

13.5 Frédéric Chopin, *Mazurka*, op. 17, no. 1, mm. 1–10. Poorly notated.

13.6 Frédéric Chopin, *Mazurka*, op. 17, no. 1, mm. 1–10. Traditionally notated.

slanted slightly and are thicker than the staff lines. The consistent placement of accidentals relative to noteheads also helps prime us to their potential presence. The natural and sharp symbols are not as distinct from each other as one might ideally want, however (both derive from the "hard B" of medieval solmization practice), and this may add some cognitive processing time when they are intermixed.

Stems serve three useful purposes: First, as we scan left to right, they provide visual stopping points, helping to alert us to the presence of noteheads. By default, they lead from the noteheads toward the center of the staff, where the eye is more likely to look. The primary exception is when two parts appear on the same staff; then stems are drawn in opposite directions to keep the two voices visually distinct. Second, stems connect the components of chords (except for those involving whole notes) and reinforce the cues we obtain from the alignment of a chord's noteheads that they form a group. Finally, they connect noteheads to flags or beams, thereby linking the information about a note's pitch and its duration. Because they are perpendicular to the staff, stems are easy to distinguish from the staff lines. Because stems generally carry no musical meaning themselves, it is appropriate that they be thin, so they remain unobtrusive. Since they are an integral part of the representation of a note, however, they should have the same color as the noteheads, not be rendered in gray the way staff lines should be.

Beams and flags provide information about duration. Each additional flag or beam halves a note's duration. I will focus on beams here. The design and use of beams support rapid cognitive processing. Because it is rare to see more than three beams at a time, the performer can quickly ascertain the number present. In fact, it generally suffices to know that a note has one more or one fewer beam than the previous note; musicians learn through experience to adjust the duration by a multiple or division of two accordingly.

Beams gather notes into groups. Because our visual processing system can quickly recognize patterns of three to five objects, the use of beams facilitates more efficient cognitive processing, not only of durations but also of rhythmic patterns. That the groupings usually align with the underlying metrical grid provides an added bonus for cognitive processing. Because beams carry so much useful information, they should stand out visually. This is likely why they are thicker than all other lines and are generally slanted, to ensure that they contrast with the orientation of staff lines.

The fact that rhythmic information is conveyed through shapes (open and closed noteheads, stems, flags, and beams) and not though horizontal spacing reflects our stronger aptitude for learning the hierarchical relationships among these durational symbols than for measuring differences in spatial distance. Although notes of longer duration conventionally receive more horizontal space, and such relative duration does in fact help in the perceptual processing of the notation, that spacing is only approximate and is highly contextual.

Bar lines indicate the ends of measures. Like stems, they mark off sections of a horizontal grid that is part of the background (metric) structure of the music. Because they have a consistent orientation and are not attached to noteheads, they are easy to distinguish visually from other shapes. It is thus appropriate that they be thin. Bar lines with special meaning (repeat signs and final bars) have an additional thick line, which ensures they stand out from other bar lines in particular and from other lines in general.

Curved lines provide two connecting functions. Ties join notes together to form longer durations. The visual representation is metaphorical: they resemble threads that join two things into a single entity. Despite their horizontal orientation, we can easily distinguish them from other lines that may be horizontal: they differ from staff lines by shape (tapered and curved vs. flatand straight), thickness, and typical length (staff lines either extend the width of the page or, in the case of ledger lines, are very short). Musicians easily learn to ignore orientation and to treat upward and downward ties as equivalent. Slurs use the same symbol as ties and operate with a similar metaphor involving the joining of notes. It is the manner of joining that differs. This difference in meaning between slurs and ties must be learned, but once we do so, we have no difficulty distinguishing the meaning of the symbols because their orientation is generally different (endpoints of slurs are rarely on the same horizontal plane) and a slur can connect more than two notes, whereas a tie cannot, and of course slurs are often longer. We easily see the distinction in meaning because the symbol's functions are clearly distinct to our cognitive apparatus. The one exceptional situation involves portato, in which a slur connects two or more notes that also have an articulation mark: ♩ ♩.

Some categories of performance direction can be indicated through either graphical or textual means. Although articulations are sometimes indicated using words, particularly when they are assigned to a stretch of music (*stacc.*, *ten.*), they are more often specified through pictograms that evoke duration •. -, intensity > Λ, or both, which appear immediately above or below the notehead or chord. Other symbols used above or below a note are visually distinct from the articulation marks. They can include bowing symbols for string players V ⊓, fingerings for keyboard players (*1*, *2*, *3*), and the ° symbol for harmonics.

Fixed dynamics appear in a bold italic font ***pp*** that is distinct from other notational elements, using abbreviations based on Italian terms. Hairpins provide graphical direction to gradually grow louder or softer over a specific period of time. Changes of dynamic are sometimes indicated with words set in a nonbold italic font (*cresc.*, *dim.*, *morendo*, etc.). Ornaments are neume-like shapes that resemble the desired melodic figuration. Each category of information is thus visually unique.

Key signatures involve a compromise between providing information at the point of need (a standard mantra among designers) and avoiding the visual overload that can result from using accidentals on every note (as those who read music of the Second Viennese School find). Even the convention of allowing an accidental to apply to each instance of a note until it is canceled by a bar line seems to be a cognitive optimization aimed at reducing the amount of information that must be processed visually. A key signature prompts performers to play in a key they are already familiar with and therefore serves only as a reminder. The signature is repeated in each system, suggesting that performers need that information readily available. Indeed, when music features frequent accidentals, it can be helpful to be reminded which accidentals are "normal" in the passage. Why is the same not true of meter signatures? This is likely because the notation itself provides a regular reminder of the meter and, once entrained into a metrical pattern, musicians tend to think in that meter even against prevailing forces.

Of course, much more could be said about the notational system, but the foregoing should make clear that the system seems to have evolved to take maximum advantage of how our visual system works, at both the lower perceptual level and the higher cognitive level.

As noted above, our notational system involves trade-offs and compromises. It fits some instruments better than others. Brass and stringed instruments (both fretted and otherwise) are not naturally diatonic in the same way as keyboards, woodwinds, or harps are. In the case of lute and guitar, composers sometimes use tablature rather than pitch notation, a topic examined in chapter 29. A diatonic notational space would also seem a less obvious choice for fully chromatic music, though in the more than one hundred years since the introduction of free atonality, no proposal for a chromatic notation system has threatened the primacy of common Western notation.

It is worth noting that the system is optimized for quickly communicating *performance* information, not musical information of the sort that we glean through analysis. In a sense, this is very much like spoken language, for which the printed "instruction" also uses symbols of varying visual, physical, and phonological similarity, little of which has any bearing on the meaning of the words into which the symbols are grouped.

Those interested in just how far the notational system can be stretched will enjoy Don Byrd's website, "Extremes of Conventional Notation" (http://homes.soic.indiana.edu/donbyrd/CMNExtremes.htm). Those interested in learning more about current standards in notational practice should see the exceptional Gould (2011). For a marvelous history of early notation, see Kelly (2015).

* * *

Part 1 has introduced principles that will serve as the foundation for the rest of the book. We started by describing the human cognitive system. We first examined the human visual processing system and ways to leverage that system to create effective visual designs. We then discussed the role of metaphor in designing visualizations and introduced several principles and techniques associated with effective information design, drawn from the writings of Edward Tufte. These include conceptual principles such as the value of presenting multivariate images, of constructing narratives in images, and of integrating images and text. They also include design principles such as the representation of information in tabular form, the use of small multiples, the effective use of color, and the importance of increasing the data-ink ratio. Part 1 ended with a case study showing how the system of Western music notation illustrates many of these principles. In parts 2 through 5, we will turn our attention to the application of these principles in a wide variety of musical contexts, beginning with musical spaces.

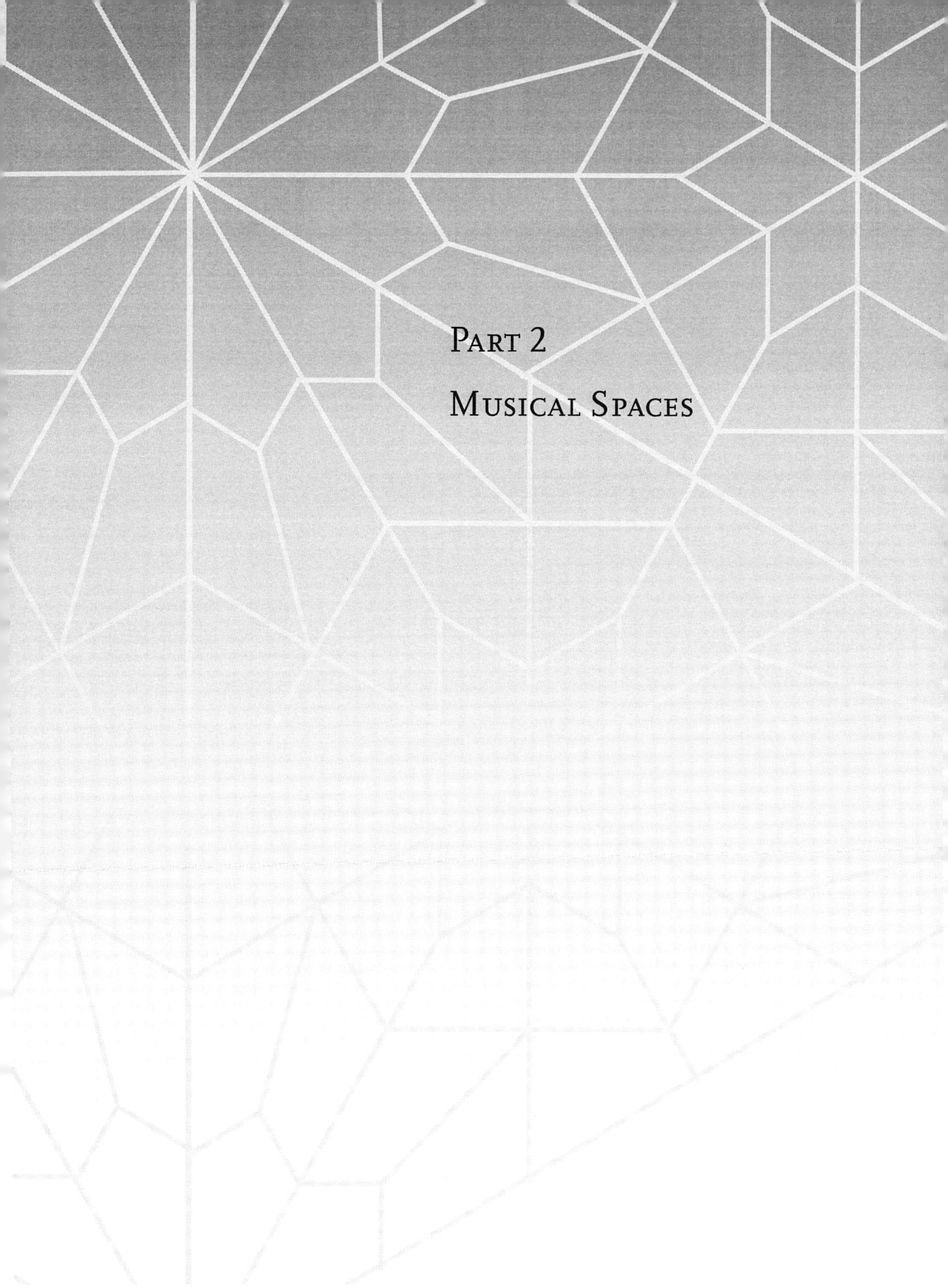

Part 2

Musical Spaces

Part 1 discussed general principles for the design of information visualization, illustrated with musical examples. The rest of the book will focus on specific musical topics. Each chapter will explore the issues that affect decisions about how to express a particular topic visually and will look at examples that compare approaches. The basic principles enumerated in part 1 will be reinforced throughout.

In part 2, the focus is on musical spaces. Musical spaces arrange musical "objects" of some kind into micro-realms within which music can be thought to operate. Musical spaces are theoretical abstractions, purely conceptual in nature. Sometimes the nature of a musical space makes it possible to define relationships among the objects or collections of objects within that space. Even within such mini-worlds, visualization requires careful thought and intention. The first chapter will summarize some of these concerns. The remainder of part 2 will look at some familiar musical spaces, including scales and modes, the circle of fifths, the *Tonnetz*, and atonal pitch spaces. It will also consider the system of Western tonality, tendency tones, harmonic progression, and the overtone series. Among the issues that will come up are the representation of continuous versus modular spaces and the representation of spaces that serve as subsets of larger spaces (such as the various familiar scales vis-à-vis chromatic space). Because symmetry appears regularly in music theoretical writing, we will explore the visualization of symmetrical structures as well. As we will see, the depiction of musical spaces often rewards clear, and sometimes clever, design.

CHAPTER 14 Pitch Spaces

The prose in figure 14.1 describes two familiar musical spaces, which are sometimes referred to as pitch space and pitch-class space, and defines a mathematical function for each to describe intervals between objects in those spaces. Those who have read the book the image comes from know that these representations are entirely appropriate in that context and would never propose changing them. Nevertheless, they provide a useful foil for this question: Would a visual representation of these spaces make them easier to understand? And if so, how might such representations be designed? The first of these spaces is visualized in figure 14.2. Since the description in figure 14.1 explicitly refers to twelve-tone equal temperament, objects in the space are spaced evenly in figure 14.2. It renders the example intervals in gray, since they reside in a different conceptual layer than the space

> 2.1.2 EXAMPLE: The musical space is a gamut of chromatic pitches under twelve-tone equal temperament. Given pitches s and t, int(s, t) is the number of semitones one must move in an upwards-oriented sense to get from s to t, not counting s itself. Thus int(C4, D4) = 2, int(C4, G4) = 7, int(C4, C5) = 12, int(C4, F3) = –7, and int(C4, F2) = –19.
>
> 2.1.3 EXAMPLE: The musical space comprises the twelve pitch-classes under equal temperament. If we arrange the pitch classes around the face of a clock following the order of a chromatic scale, then int(s, t) is the number of hours that we traverse in proceeding clockwise from s to t. For instance, if s is at 8 o'clock and t is at 1 o'clock, int(s, t) = 5. Note that the number int(s, t) does not depend on which pitch class is positioned at 12 o'clock. In any case, int(E, E) = 0, int(E, F) = 1, and int(F, E) = 11.

14.1 David Lewin, *Generalized Musical Intervals and Transformations* (New Haven, CT: Yale University Press, 1987), 17.

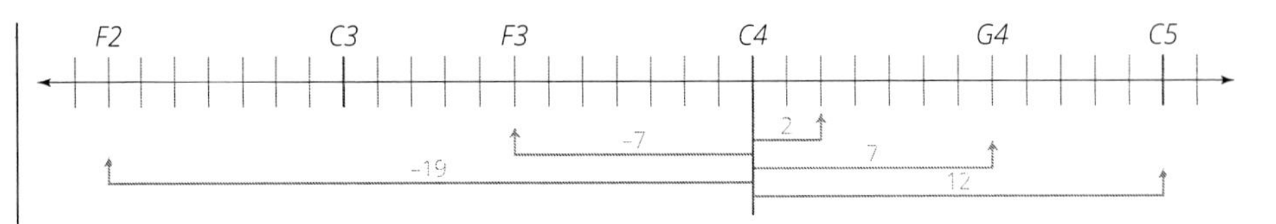

14.3 Pitch space, oriented horizontally.

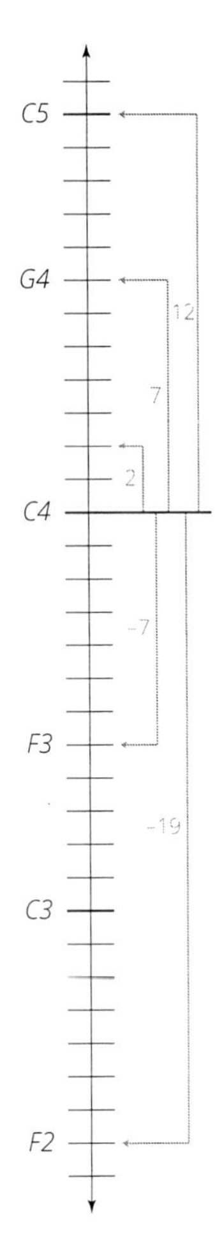

14.2 Pitch space, oriented vertically.

itself. Only a few of the pitches in the space have been labeled here, mainly those referenced by name in figure 14.1. The exception is C3, which fills out the octave Cs and helps anchor F2. The line has been rendered vertically here because the description refers to "upward" motion. This conforms to a metaphor of height that we often associate with pitches.

Despite that metaphor, the image survives a rotation to a horizontal orientation (fig. 14.3) with no loss of comprehensibility. Why? The "right = higher" metaphor is something many of us learned on number lines in school. It corresponds to the left-to-right reading orientation, paired with our tendency to count from low to high, rather than the opposite. Counting down from one hundred is considered counting backward, after all, and the *x-y* grids in math always place higher values above or to the right, depending on the axis. "Higher to the right" is also familiar from the layout of keyboard instruments and the numbers on the computer keyboard. To decide whether this explanation is plausible to you, see if you think the reversed orientation of online figure 14.4 works as well as figure 14.3.

Since pitch names repeat in each octave, it makes sense to wrap the line of pitches around a central axis, so octave-related pitches are on the same spoke of a circle, as in figure 14.5. Here, proximity to the center of the spiral represents what we think of as pitch height. To represent pitch height as spatial height requires pulling the center of that spiral upward like a Slinky as in figure 14.6. This image represents both pitch height (as in fig. 14.2) and the octave-repeating pitch names. Each traversal of the coil contains all twelve pitch classes. (A spiral representation similar to this, but arranged by circle of fifths, is used in fascinating ways, including in an interactive projection during a live performance, in Chew and Francois 2005).

Viewed from above, and ignoring the effects of perspective, octave duplicates go away and we are left with the circle of pitch classes in figure 14.7. This represents the second space described in figure 14.1. The parallel between the twelve-pitch-class musical space and the twelve-hour clockface makes the clock a useful metaphor for understanding at least some aspects of a space consisting of pitch classes and makes the clock a handy pedagogical tool for teaching pitch-class set theory. (We will see examples elsewhere in the book.) The analog clock is a powerful analogue because we know how to tell time using one. We are accustomed to traversing hours, so traversing the isographic pitch-class space is comparatively easy. A

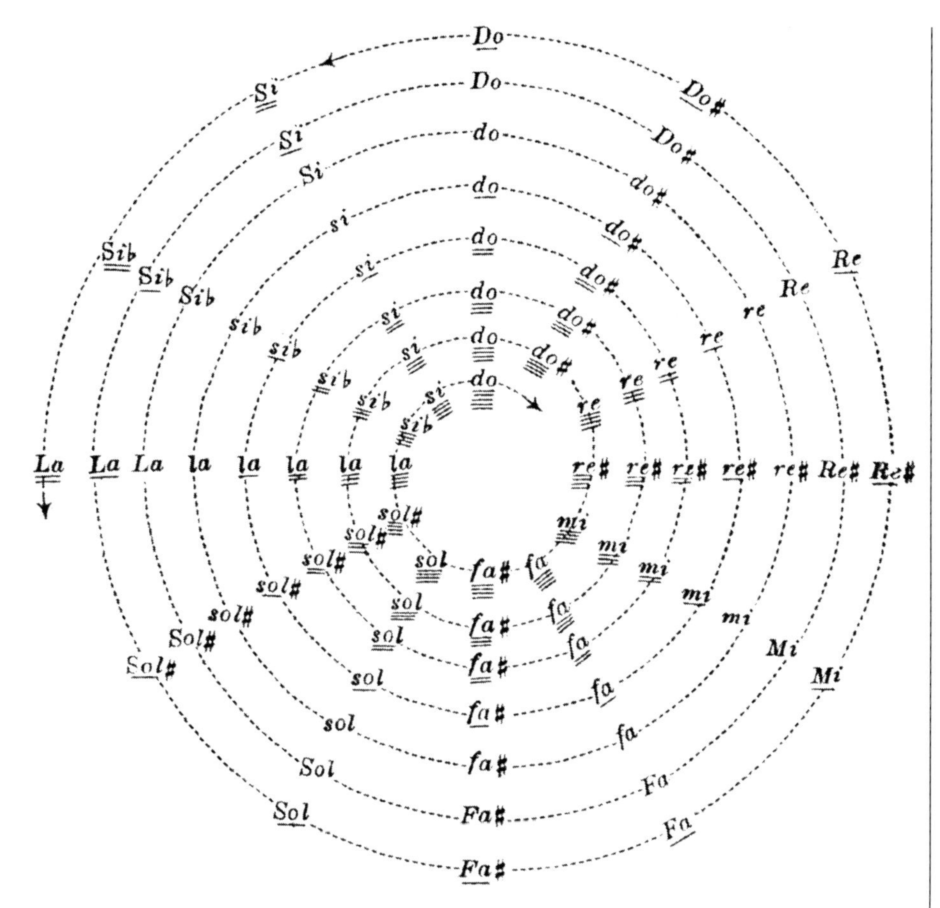

Figure 14.5 Sergei Protopopov, *Элементы Строения Музыкальной речи* [Elements of the structure of musical speech] (Moscow: Izdatel'stvo Muzykal'nyĭ Sektor, 1930), 1:19.

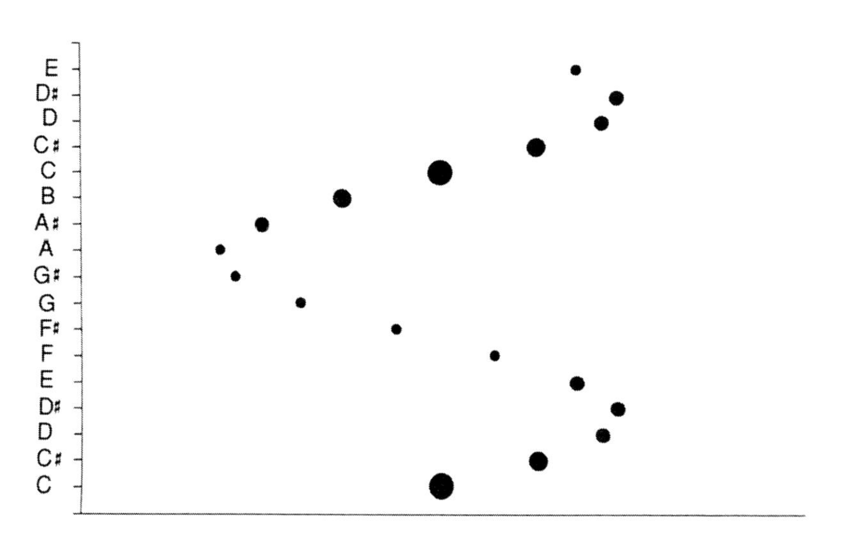

14.6 W. Jay Dowling, "Pitch Structure," in *Representing Musical Structure*, ed. Peter Howell, Robert West, and Ian Cross (London: Academic 1991), 37.

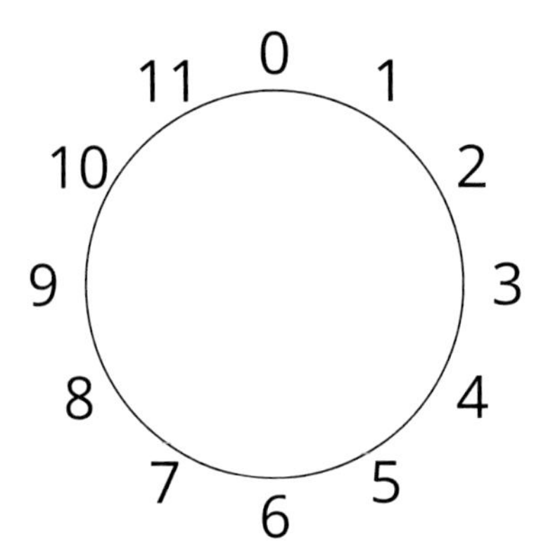

14.7 Modular pitch-class space.

sixty-pitch-class universe would be easy to learn because of our comfort with the minute hand on the clock. We are also familiar with another twelve-unit cycle: the months of the year. I am unaware of the use of the calendar as a metaphor for musical time, however. Whether this is because the year passes on too long a scale or because our visual representations of the calendar are less congenial to musical applications is unclear.

Pitch-class spaces can be made up of nonchromatic collections, including the whole-tone, hexatonic, diatonic, and octatonic (fig. 14.8). Conceptually, objects within each of these spaces are equally distant from one another in their respective spaces. Our notation system reflects this with regard to the diatonic pitch space; each line and space on the staff signifies a different member of this space. The modular spaces shown in figure 14.8 preserve this equidistant arrangement within each collection. When working solely within these collections, viewing them in this way can be useful. However, sometimes we want to consider these spaces as subsets of the chromatic space of figure 14.7, in which case representations such as those in figure 14.9 may be more appropriate. This decision should be made deliberately.

The representation of pitch and pitch-class spaces underlies representation of more complex spaces and of other musical phenomena. Decisions made at this fundamental level can have wide-ranging effects. Hook (2002) provides an excellent and more extensive discussion of various visual pitch spaces.

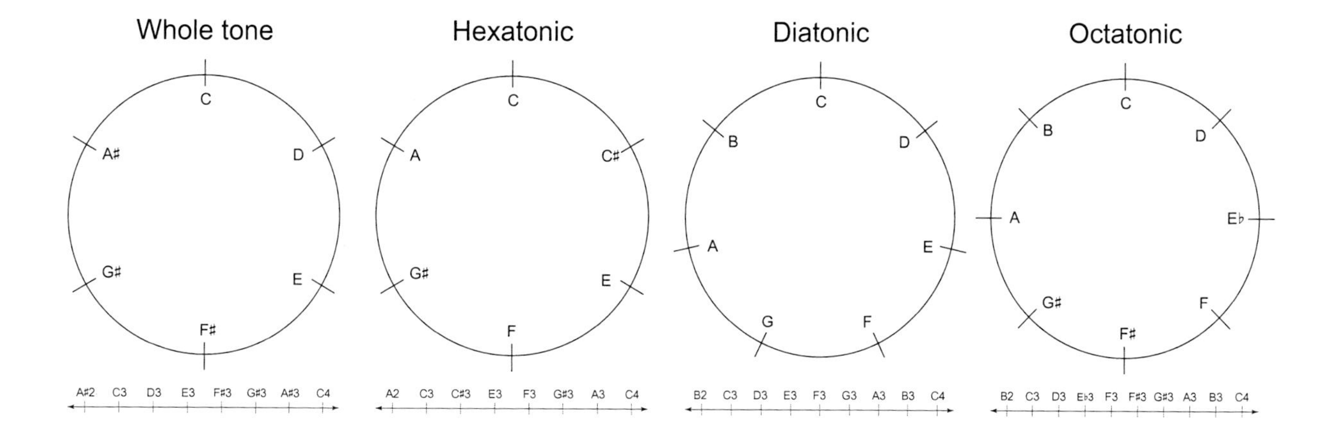

14.8 Various linear and modular pitch spaces.

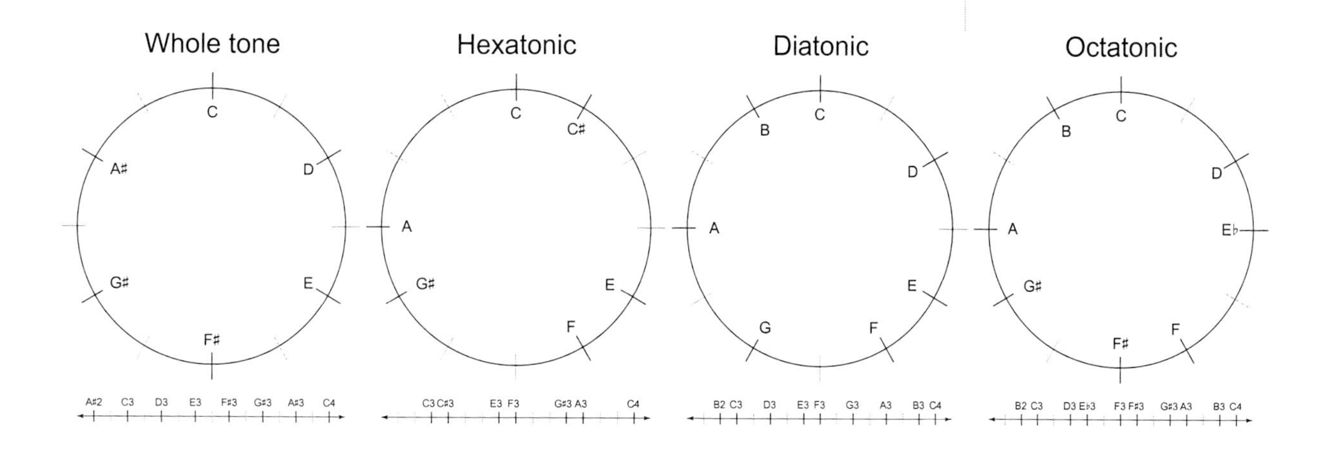

14.9 Various linear and modular pitch spaces, mapped onto an underlying chromatic space.

Chapter 15 Collections, Scales, and Modes

By *collection*, I mean a set of pitch classes from which scales are derived. We might speak generally about the "diatonic collection" or the "whole-tone collection," or more specifically about the "three-sharp collection" or the "odd-numbered whole-tone collection." *Scales* are collections in which one member has been given priority, such as the A major scale (vs. the three-sharp collection). Although the term *mode* is sometimes used almost synonymously with *scale*, it often implies additional qualities, such as a characteristic range and special functions.

Even the visual depiction of items as seemingly simple as collections, scales, and modes involves several decisions. For instance, one must decide whether to emphasize the ladder metaphor implied by the word *scale* (*scala* is Latin for ladder) by showing the pitches as ascending, descending, or both (say, on a staff) or to emphasize their repeating (that is, modular) nature. One must also decide whether to show the intervals between elements in the collection and, if so, how. Another decision is whether to represent the components of the collections by letter name, noteheads, or some other means. Finally, one must decide what context to show the collection in—for example, whether to show all the church modes together. Doing so leverages the principle that meaning arises through the study of comparisons and differences. The images in this chapter illustrate many combinations among these choices.

We start with the church modes. Depictions almost always show the modes in linear, not modular, format. Depictions of the modes as used in the medieval and Renaissance periods usually emphasize the variety of arrangements within

MODI PRINCIPALI ET AVTENTICI.																									
CHORDE	C		D		E		F		G		a		♮		c		d		e		f		g		aa
Primo.	C		D		E		F		G		a		♮		c										
		Tuono.		Tuono.		Semituono maggiore.		Tuono.		Tuono.		Tuono.		Semituono maggiore.		Tuono.		Tuono.		Semituono magg ore.		Tuono.		Tuono.	
Terzo.			D		E		F		G		a		♮		c		d								
Quito.					E		F		G		a		♮		c		d		e						
Setti-mo.							F		G		a		♮		c		d		e		f				
Nono.											a		♮		c		d		e		f		g		
Vndecimo.											a		♮		c		d		e		f		g		aa

MODI NONPRINCIPALI, O PLAGALI.																									
CHORDE	Γ		A		♮		C		D		E		F		G		a		♮		c		d		e
Secõdo.	Γ		A		♮		C		D		E		F		G										
		Tuono.		Tuono.		Semituono maggiore.		Tuono.		Tuono.		Semituono maggiore.		Tuono.		Tuono.		Tuono.		Semituono magg ore.		Tuono.		Tuono.	
Quarto.			A		♮		C		D		E		F		G		a								
Sesto.					♮		C		D		E		F		G		a		♮						
Ottauo							C		D		E		F		G		a		♮		c				
Decimo									D		E		F		G		a		♮		c		d		
Duodecimo.											E		F		G		a		♮		c		d		e

15.1 Gioseffo Zarlino, *Dimostrationi Harmoniche* (Venice, 1571), 301, 303.

a single underlying diatonic collection, such as the white-note collection. Figure 15.1 depicts Zarlino's twelve-mode system. Its two parts segregate authentic and plagal modes (they are two pages apart in the original). It displays the total gamut of pitches at the top of each and then copies downward the pitches that belong to a particular mode. (The initial note G in mode 9, "nono.," is missing.) The representation reveals the rotational relationship among the modes and makes the considerable pitch overlap between adjacent modes clear. As a bonus, the image shows the intervals between the pitches, though the ninety-degree rotation of the text is less than ideal. Considerably simplified depictions of mode appear in online figures 15.2 and 15.3.

Figure 15.4 provides a compact representation of the modal scales that uses staff notation instead of note names. It eschews the authentic/plagal distinction and uses the now familiar mode names rather than mode numbers. It features a two-octave white-note diatonic collection, with brackets showing where the scales occur within the pitch sequence. The display is more efficient than the one in figure 15.1. The figure lists notes just once rather than multiple times, though this efficiency comes at the cost of clarity. The characteristic interval patterns are not shown here, which is also a loss. Clarity is further compromised because the later brackets move so far from the notation that it becomes more difficult to see exactly where the start and end points are, particularly for the Locrian and Ionian modes.

When the seven modes are combined with the twelve diatonic collections, not just the white-note collection, the resulting combinatorial explosion makes visualizing the possibilities more challenging. Figure 15.5 is one such attempt. It lays out the modal scales that can be built from each diatonic collection. The image is unwieldy. Since there are three variables (collection, starting pitch, and mode name), it would be better to create a fully two-dimensional table, with two values facilitating the lookup of the third. Figure 15.6 allows one to select a collection and a starting pitch to see which mode (if any) can be built on the combination. Figure 15.7 allows one to select a collection and mode and find the starting pitch. The latter is more compact and works better to answer questions like, "What is the

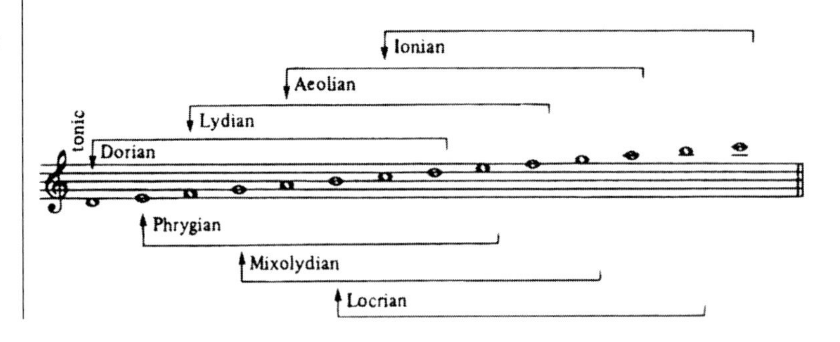

15.4 Bryan R. Simms, *Music of the Twentieth Century: Style and Structure*, 2nd ed. (New York: Schirmer Books, 1996), 54.

first note of the five-sharp Phrygian scale?" The former better reveals the "gapped" nature of the diatonic collection and also makes more apparent the fact that a given pitch can belong to only seven collections that are adjacent on the circle of fifths.

Collection name	*Possible orderings (scales)*
0-sharp (or 0-flat)	C-Ionian, D-Dorian, E-Phrygian, F-Lydian, G-Mixolydian, A-Aeolian, B-Locrian
1-sharp	C-Lydian, D-Mixolydian, E-Aeolian, F♯-Locrian, G-Ionian, A-Dorian, B-Phrygian
2-sharp	C♯-Locrian, D-Ionian, E-Dorian, F♯-Phrygian, G-Lydian, A-Mixolydian, B-Aeolian
3-sharp	C♯-Phrygian, D-Lydian, E-Mixolydian, F♯-Aeolian, G♯-Locrian, A-Ionian, B-Dorian
4-sharp	C♯-Aeolian, D♯-Locrian, E-Ionian, F♯-Dorian, G♯-Phrygian, A-Lydian, B-Mixolydian
5-sharp	C♯-Dorian, D♯-Phrygian, E-Lydian, F♯-Mixolydian, G♯-Aeolian, A♯-Locrian, B-Ionian
6-sharp (or 6-flat)	C♯-Mixolydian, D♯-Aeolian, E♯-Locrian, F♯-Ionian, G♯-Dorian, A♯-Phrygian, B-Lydian
5-flat	C-Locrian, D♭-Ionian, E♭-Dorian, F-Phrygian, G♭-Lydian, A♭-Mixolydian, B♭-Aeolian
4-flat	C-Phrygian, D♭-Lydian, E♭-Mixolydian, F-Aeolian, G-Locrian, A♭-Ionian, B♭-Dorian
3-flat	C-Aeolian, D-Locrian, E♭-Ionian, F-Dorian, G-Phrygian, A♭-Lydian, B♭-Mixolydian
2-flat	C-Dorian, D-Phrygian, E♭-Lydian, F-Mixolydian, G-Aeolian, A-Locrian, B♭-Ionian
1-flat	C-Mixolydian, D-Aeolian, E-Locrian, F-Ionian, G-Dorian, A-Phrygian, B♭-Lydian

15.5 Joseph Straus, *Introduction to Post-tonal Theory*, 3rd ed. (Upper Saddle River, NJ: Prentice Hall, 2005), 142.

We turn our attention to other collections now. Figure 15.8 displays the Javanese Pelog and Slendro scales in the context of the twelve-note equal-tempered collection. The close vertical alignment facilitates easy comparison with the Western scale, though it would have been helpful to highlight the diatonic and pentatonic collections, with which the Javanese scales are more naturally compared than the full chromatic. Online figure 15.9 does this.

The next images show scales in modular spaces, depicted on circles. Such representations make it clear that the collections wrap around instead of ending and also that these scales tend to exist conceptually within a larger pitch-class space. Figure 15.10 puts various scales on a chromatic pitch wheel, with the presumed centric pitch in the twelve o'clock position. (Messiaen's second mode is incorrectly rendered; it shows a chromatic rather than an octatonic collection, and Messiaen's

	C	C♯	D	D♯	E	F	F♯	G	G♯	A	A♯	B
0♯ / 0♭	Ionian		Dorian		Phrygian	Lydian		Mixolydian		Aeolian		Locrian
1♯	Lydian		Mixolydian		Aeolian		Locrian	Ionian		Dorian		Phrygian
2♯		Locrian	Ionian		Dorian		Phrygian	Lydian		Mixolydian		Aeolian
3♯		Phrygian	Lydian		Mixolydian		Aeolian		Locrian	Ionian		Dorian
4♯		Aeolian		Locrian	Ionian		Dorian		Phrygian	Lydian		Mixolydian
5♯		Dorian		Phrygian	Lydian		Mixolydian		Aeolian		Locrian	Ionian
6♯		Mixolydian		Aeolian		Locrian	Ionian		Dorian		Phrygian	Lydian
	C	**D♭**	**D**	**E♭**	**E**	**F**	**G♭**	**G**	**A♭**	**A**	**B♭**	**B**
6♭		Mixolydian		Aeolian		Locrian	Ionian		Dorian		Phrygian	Lydian
5♭	Locrian	Ionian		Dorian		Phrygian	Lydian		Mixolydian		Aeolian	
4♭	Phrygian	Lydian		Mixolydian		Aeolian		Locrian	Ionian		Dorian	
3♭	Aeolian		Locrian	Ionian		Dorian		Phrygian	Lydian		Mixolydian	
2♭	Dorian		Phrygian	Lydian		Mixolydian		Aeolian		Locrian	Ionian	
1♭	Mixolydian		Aeolian		Locrian	Ionian		Dorian		Phrygian	Lydian	

***Above,* 15.6** Mode table, providing mode name, given a collection and starting pitch.

***Right,* 15.7** Mode table, providing starting pitch, given the collection and mode name.

Collection	Ionian	Dorian	Phrygian	Lydian	Mixolydian	Aeolian	Locrian
0♯/0♭	C	D	E	F	G	A	B
1♯	G	A	B	C	D	E	F♯
2♯	D	E	F♯	G	A	B	C♯
3♯	A	B	C♯	D	E	F♯	G♯
4♯	E	F♯	G♯	A	B	C♯	D♯
5♯	B	C♯	D♯	E	F♯	G♯	A♯
6♯/6♭	F♯	G♯	A♯	B	C♯	D♯	E♯
5♭	D♭	E♭	F	G♭	A♭	B♭	C
4♭	A♭	B♭	C	D♭	E♭	F	G
3♭	E♭	F	G	A♭	B♭	C	D
2♭	B♭	C	D	E♭	F	G	A
1♭	F	G	A	B♭	C	D	E

third mode is rendered by Messiaen himself rotated two semitones in either direction from what is shown here.)

The characteristic interval patterns would pop out more if the open circles were replaced with a shape that contrasted more with the solid circles, such as the short hatch markers in figure 15.11. The latter image shows various collections arranged according to the circle of fifths rather than on a chromatic pitch circle.

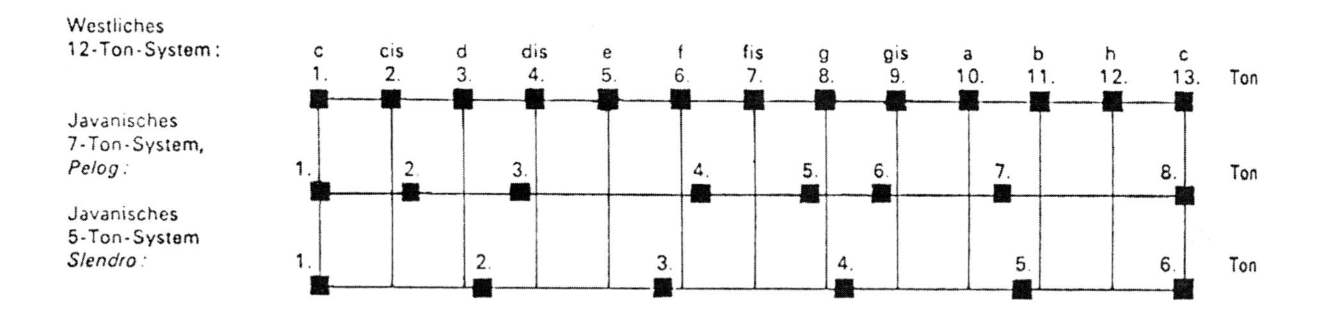

15.8 Werner Breckoff, *Musik Aktuell: Informationen, Dokumente, Aufgaben* (Kassel: Bärenreiter, 1971), 252.

The layout makes the bilateral symmetry of these collections apparent. It would be nice if the image consistently showed the axes and labeled the scale members. In pentatonic and acoustic, all notes are labeled; only three notes of the major scale are labeled; and in the "leading notes" (that is, octatonic) collection, only the missing notes are labeled.

Online figure 15.12 provides another, problematic, approach.

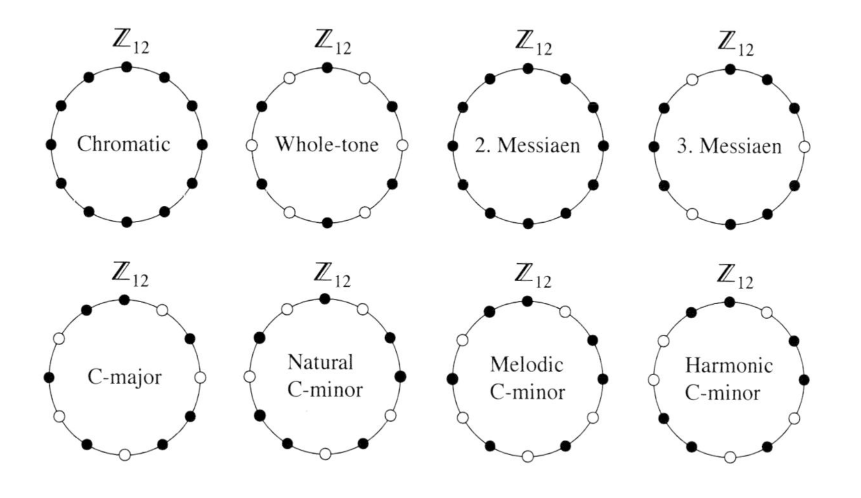

15.10 Guerino Mazzola, Stefan Göller, and Stefan Müller, *The Topos of Music: Geometric Logic of Concepts, Theory, and Performance* (Basel: Birkhauser, 2002), 114.

The novel perspective of figure 15.13 illuminates properties of scales not seen in the prior examples. The image depicts scales along a row of perfect fifths. With an assumed tonal center C in white, it reveals how the various collections lean toward the sharp side (leftward) or flat side (rightward). The image also uses arrows (🠅) to show the "center of gravity" for each collection. The display is representative rather than comprehensive; not all the modes are listed.

The system of Indian ragas is far more complex than the modal and other Western scales. The South Indian *melakarta* system defines seventy-two seven-note parent scales, on which the much more elaborate system of ragas is built. Figure

15.11 Ernő Lendvai, Mikløs Szabø, and Mikløs Mohay, *Symmetries of Music: An Introduction to Semantics of Music* (Kecskemét: Kodály Institute, 1993), 110.

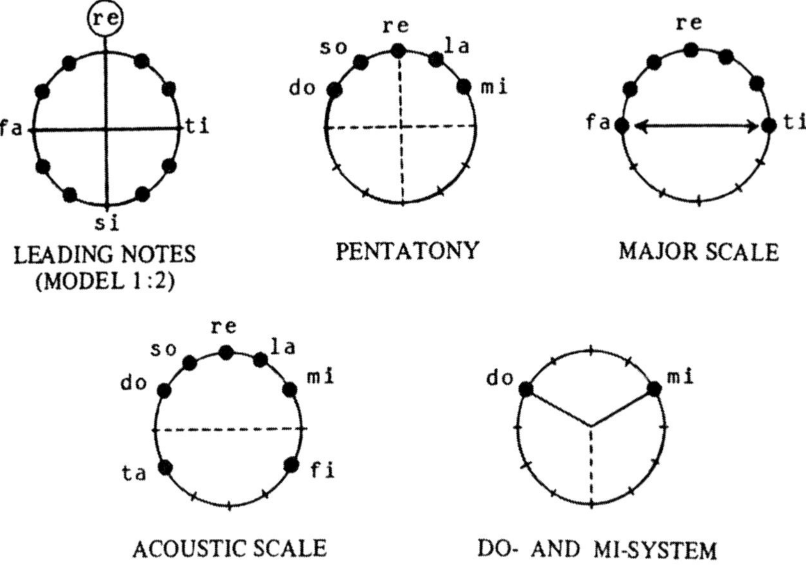

15.13 David Temperley, *The Cognition of Basic Musical Structures* (Cambridge, MA: MIT Press, 2001), 342.

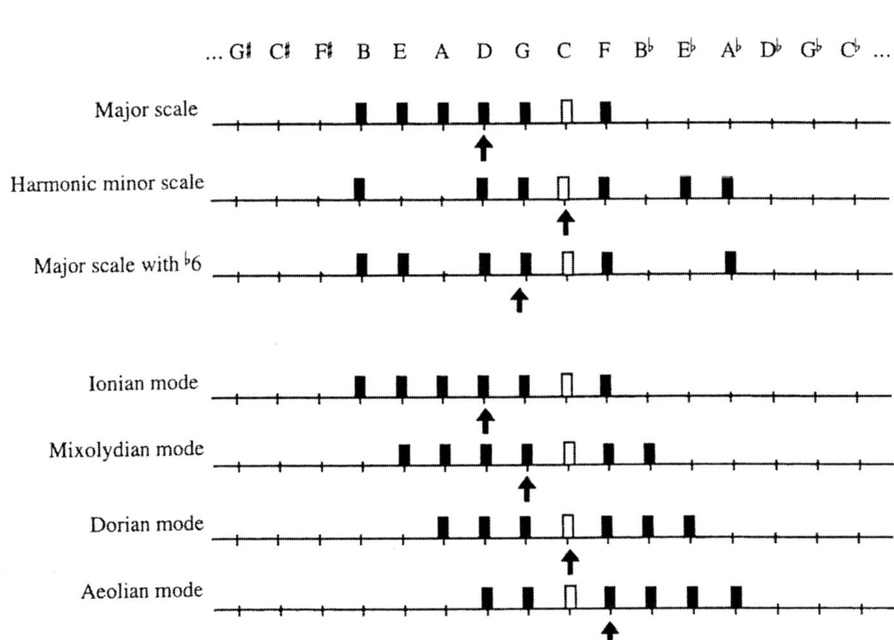

15.14 collates all of these scales in a single image. Although it does not name the scales, it permits their discovery and invites exploration of the varied pitch structures of a rich and venerable musical style. The duplication of several Western pitch names for scale steps *ga* and *ni* prevents the need for crisscrossing lines that would hamper the image's clarity.

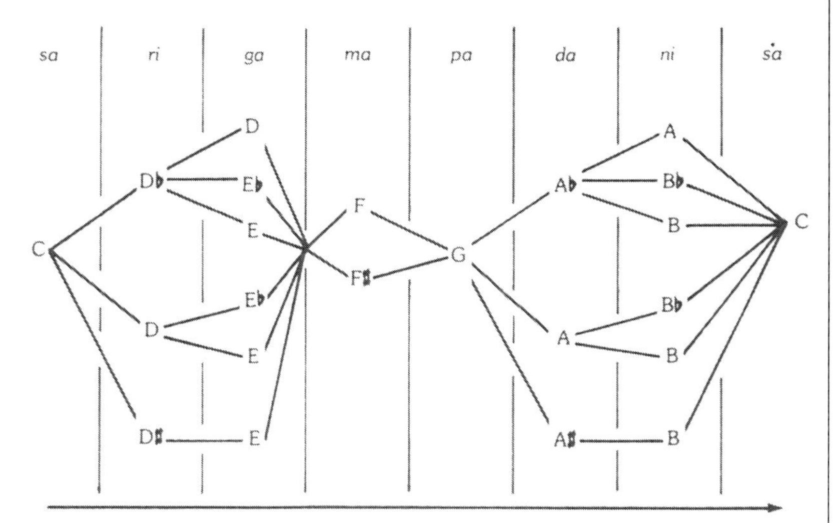

15.14 David B. Reck, "India/South India," in *Worlds of Music*, ed. Jeff Todd Titon, 252–315 (New York: Schirmer, 1996), 305.

Whereas images of Western modes tend to focus on rotations of a single (usually the all-natural) collection, images of major and minor scales, and of tonality in general, often focus more on relationships between collections. The ambitious image in figure 15.15 draws together multiple facets of the tonal system, focusing on the major scale and related keys. It shows how the lower tetrachord of one major scale is also the upper tetrachord of the scale a fifth lower (F), which contains one flatted note, and how its upper tetrachord is also the lower tetrachord of the scale a fifth higher (G), which contains one sharped note. This process repeats in both directions like a fan, adding flats (left branch) or sharps (right branch) one at a time until all the pitches are altered. The image lists keys and the number of accidentals next to the scales and gathers the accidentals into key signatures at the center. Numbers indicate the order in which the accidentals appear both in the key signatures and in the scale. The basic layout works well. One might replace the empty staff segments with a single connecting line (the empty staff segments involve a lot of wasted ink) and move the accidentals within the key signatures closer together, as they would appear in an actual key signature. See online figure 15.16 for a different approach to the same topic.

Musical collections, scales, and modes are not especially complicated, but images involving them invite many decisions, such as what to compare them to and how, and whether to emphasize their linear nature, which is easier to process visually, or their modular nature, which is more accurate conceptually.

15.15 Renée Longy-Miquelle, *Principles of Musical Theory* (Boston: E. C. Schirmer Music, 1925), 36.

Chapter 16 The Circle of Fifths

Depictions of the succession of fifths are similar in many ways to those showing successions of semitones with which we associate pitch or pitch-class space. When dealing with equal-tempered pitches, both involve twelve pitch classes and both cycle within a twelve-note modular space. Therefore, strong similarities exist between images of the two.

When pitches are ordered by perfect fifth, they can be oriented as in the upper image in figure 16.1 (where the fifths are descending). The image does not indicate explicitly which octave the pitches belong to, but it clearly treats enharmonic pitches as distinct entities. Were we to treat enharmonic pitches as equivalent, such that the F𝄪 at the left end and E𝄫 at the right end were equivalent to the G and D near the center, then of course the pitch classes would repeat every twelve items, a fact this representation would hide. Meanwhile, because it omits octave information, in either absolute or relative terms, the image does not convey that seven octaves separate those potentially enharmonic pitches.

The second image in figure 16.1 shows the familiar circle of (equal-tempered) fifths. Here, enharmonic spellings are provided for pitches represented by black keys on the keyboard. As in the pitch-class space images seen earlier, C appears at twelve o'clock, a position privileged in analog timekeeping as the initiator of morning and afternoon and here, metaphorically, as a "default" pitch or key.

The simplicity of the two representations mirrors the simplicity of the information they convey. The line and the circle invoke the desired metaphors: an unbounded succession above, closed-loop circularity below. Spacing of the markers,

16.1 David Temperley, *The Cognition of Basic Musical Structures* (Cambridge, MA: MIT Press, 2001), 118.

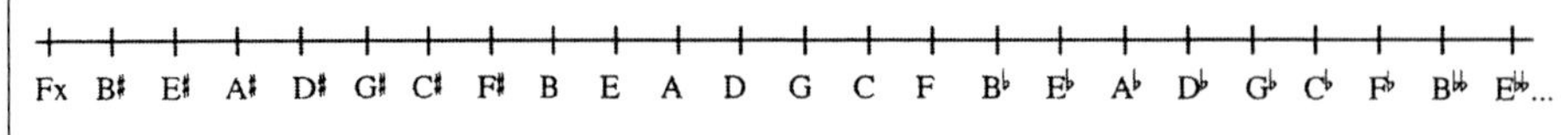

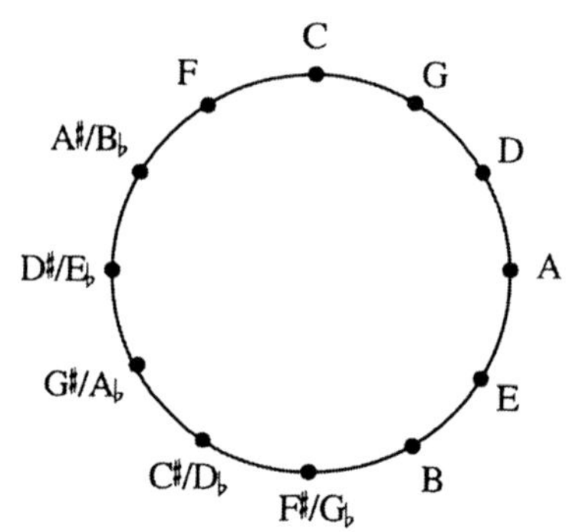

with vertical lines above and dots below, suggests equal spacing of objects within the conceptual space. Online figure 16.2 is similar.

Figure 16.3 ingeniously combines the line of fifths and the circle of fifths. With the line of fifths wrapped around a mod-7 space and as a spiral rather than as a circle, pitches that share a letter name (sometimes said to have the same *chroma*) align along seven spokes. Online figure 16.4 further enhances the image through the use of color.

16.3 Julian Hook, unpublished.

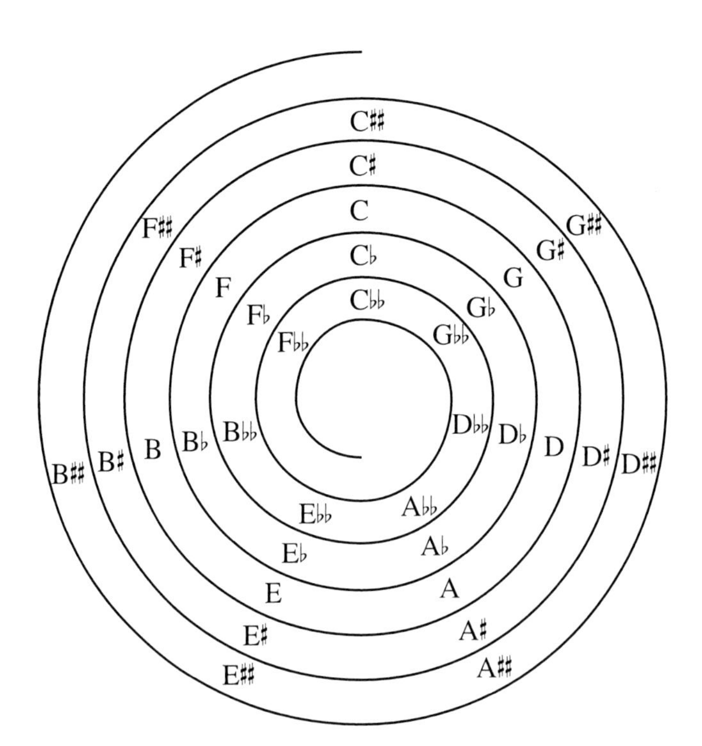

Additional information often accompanies circle-of-fifths images. A look at several examples provides an opportunity to see where additions cross the line from appropriate to inappropriate. A common use of the circle of fifths is to show the pattern of key signatures in major and minor keys, as in figure 16.5. Here, C stands in its conventional twelve o'clock position. The twelve pitch classes are arranged clockwise by ascending perfect fifth around the outside of the diagram. Because these pitch names represent tonics of keys and not pitch-class names, the image provides only labels associated with the major keys through seven sharps and flats and so requires only three enharmonic spellings. To this it adds descriptions of the key signatures for each major key (for example, 4♯) drawn in circles inscribed on the larger circle, a design that presupposes a knowledge of key signatures, including the order in which the accidentals appear. The visually prominent circles overwhelm the information inside them and also draw attention away from the information outside them. Finally, the tonics of the corresponding minor keys appear inside the larger circle. They are visually inferior, implying a perhaps unintended judgment on their relative significance. This is exacerbated by the rendering of the minor key names in lowercase letters, a quasi-conventional approach that the

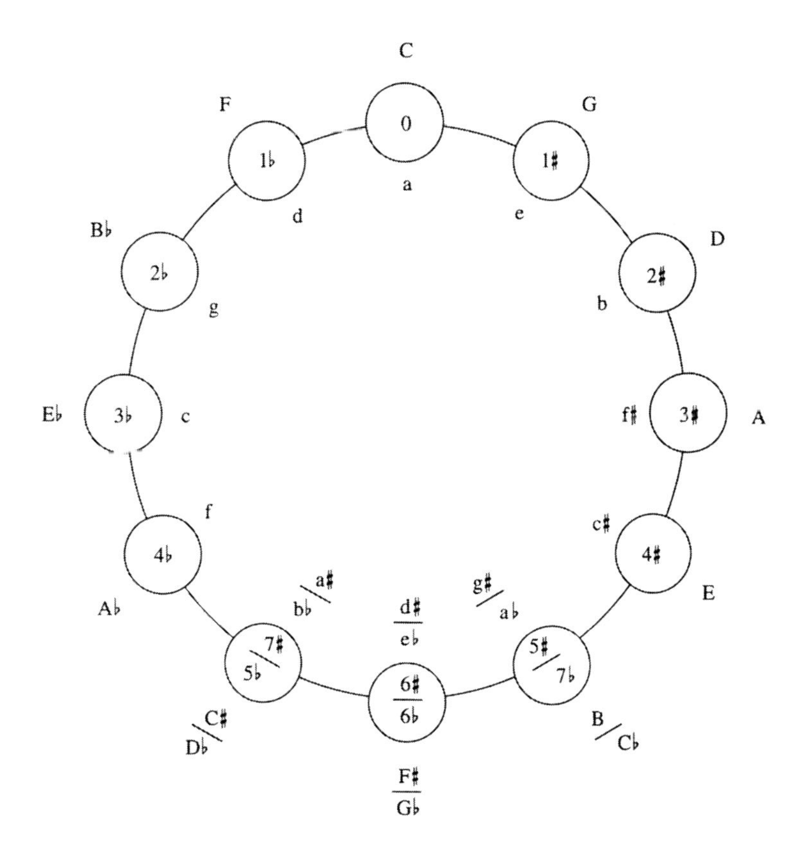

16.5 Stefan M. Kostka and Dorothy Payne, *Tonal Harmony: With an Introduction to Twentieth-Century Music*, 6th ed. (New York: McGraw-Hill, 2009), 14.

uninitiated would need to have explained. The obvious resemblance between the circle of fifths and an analog clockface is made explicit in online figure 16.6.

The circle-of-fifths diagram in figure 16.7 renders key signatures on a treble staff as they will be seen in music, unlike in figure 16.5. This will be helpful for those learning key signatures. Tonics of major keys appear in open noteheads and those of minor keys in closed noteheads. This fact is indicated both in the upper half of the diagram (with text and seven dotted arrows) and in a redundant footnote. The numbers of accidentals appear in tiny numbers around the perimeter of the circle, while the diagram seems to yell *sharps* and *flats*, the words accompanied by visually overpowering arrows. Online figure 16.8 provides another historic look at the circle of fifths, while the circle-of-fifths image from Wikimedia (online fig. 16.9) has some attractive features.

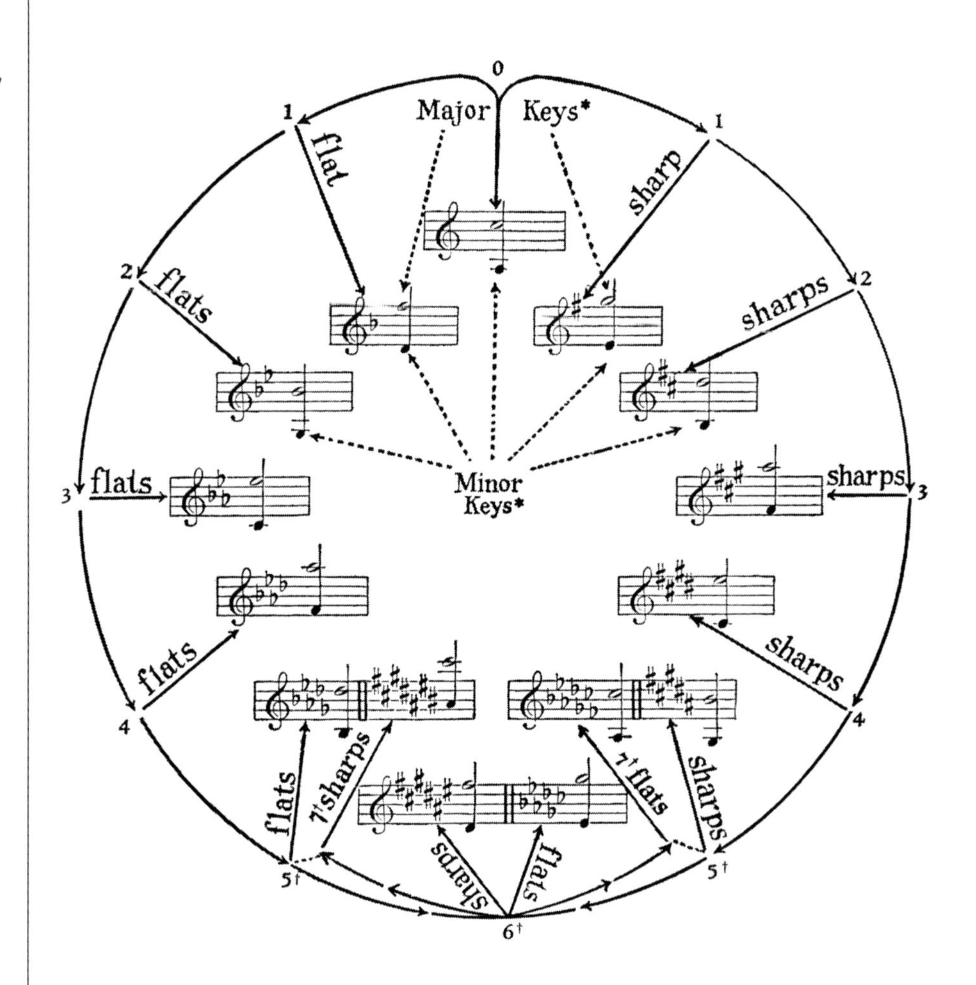

16.7 Renée Longy-Miquelle, *Principles of Musical Theory* (Boston: E. C. Schirmer Music, 1925), 51.

Figure 16.10 effectively extends the circle of fifths to a jazz context. The image posits a Two-Five (TF) transformation, which changes a minor seventh into a dominant seventh and a dominant seventh into a major seventh, as shown on each of the spokes of the image. Changing the quality of the third of a V^7 (from major to minor) or the seventh of a major seventh chord (from major to minor) moves the chord one "hour" to the right and one level inward. Four distinct line styles make the various operations clear. As "$ii^7 \rightarrow V^7$" promenades around the circle, the progression turns upside-down as it nears the "A o'clock" position. While this makes conceptual sense, it would be more user-friendly to keep the roman numbers upright throughout, like the pitch names. Here, listing just one enharmonic spelling is appropriate. The image tells a story of a circle of fifths; we need to flip to the flat-side pitches only once.

In these examples, we have seen that the addition of *information* enhances the images, while the addition of non-information-bearing ink often has the opposite effect. This truth pervades the design of visualizations of all kinds.

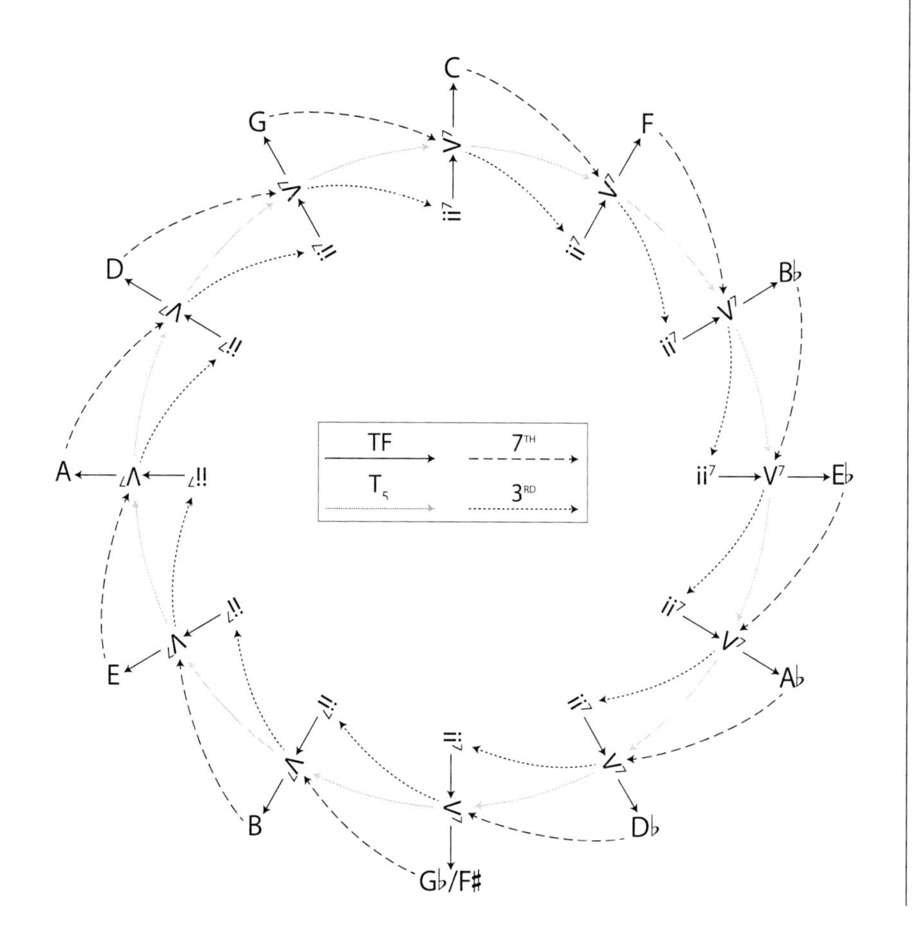

16.10 Michael McClimon, "Transformations in Tonal Jazz: ii–V Space," *Music Theory Online* 23, no. 1 (2017): fig. 6, http://mtosmt.org/issues/mto.17.23.1/mto.17.23.1.mcclimon.html.

Chapter 17 The *Tonnetz*

By *Tonnetz*, a German word meaning "tone net," I refer to a general category of representation that extends the concept of the line of pitches to two dimensions (for an extensive discussion of the Tonnetz, see Tymoczko 2012). The Tonnetz is generally attributed to Euler (fig. 17.1), who relates each pitch in his 1739 image to those a perfect fifth and major third above and below them (in contemporary parlance). The layout creates a chain of fifths in one direction (here roughly between the one o'clock and seven o'clock positions) and a chain of major thirds (roughly ten o'clock to four o'clock). The combination of these intervals produces (unlabeled) chains of minor thirds, found on the horizontal (nine o'clock to three o'clock) line.

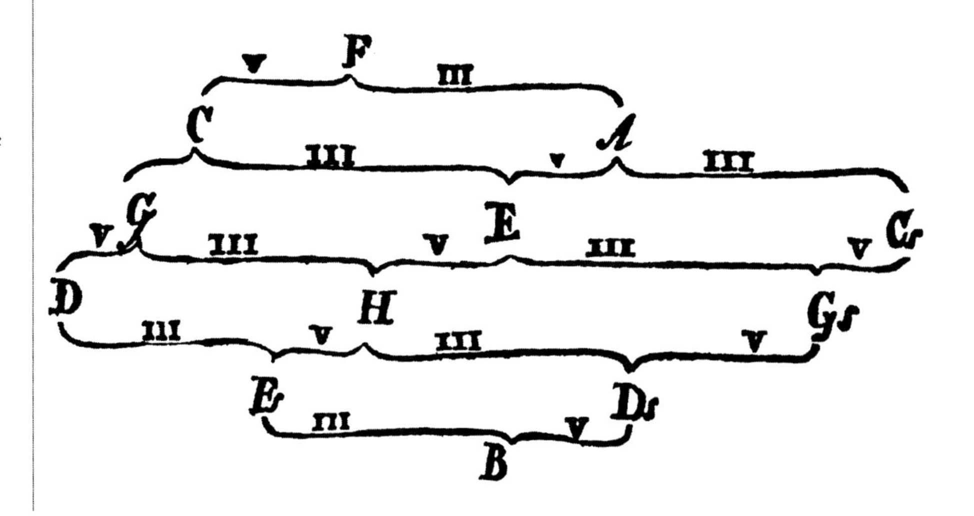

17.1 Leonhard Euler, *Tentamen novae theoriae musicae: Ex certissismis harmoniae principiis dilucide expositae* (Petropoli: Ex typographia Academiae scientiarum, 1739), 147.

The Tonnetz has been represented and extended in a number of ways, which we will sample in this chapter. When representing a Tonnetz, one must decide how to represent the three primary intervals they embed—minor third, major third, and perfect fifth—and whether to represent pure or equal-tempered intervals. Yet more decisions are required when grouping clusters of pitches into triads and moving the *Tonnetz* into the realm of neo-Riemannian theory. Whatever the use, things to watch for include how to provide general structural clarity, whether and how to show the relationships among the nodes, whether to represent the inherently tabular data in a visual grid, and, if so, how to handle that.

Figure 17.2 locks notes in a visual lattice that is certainly suggestive of the name Tonnetz. Whereas figure 17.1 suggests that minor thirds arise through a combination of a major third and perfect fifth, figure 17.2 makes the thirds primary: major thirds traverse a diagonal moving southwest to northeast, while minor thirds are arranged in almost-vertical columns. The resulting perfect fifths are arrayed on the horizontal but are not explicitly labeled. Bars above and below the pitch names indicate third-generated pitches that differ from the fifth-generated pitches by a syntonic comma. One must infer that the interval chains extend beyond the bounds of the image as the design does not imply it. Online figure 17.3 partially implies this; it also arranges the intervals differently and draws attention to other interval chains that the Tonnetz embeds.

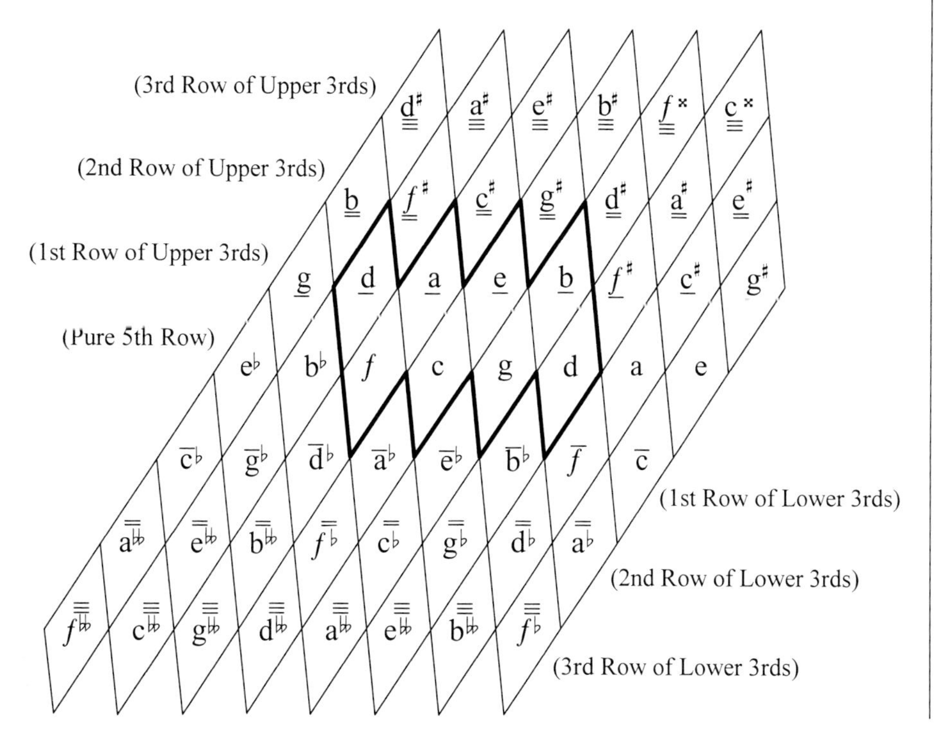

17.2 Hugo Riemann, "Ideen zu einer 'Lehre van den Tonvorstellungen,'" *Jahrbuch der Musikbibliothek Peters*, vol. 21–22 (1914–15), trans. Robert W. Wason and Elizabeth W. Marvin, *Journal of Music Theory* 36, no. 1 (Spring 1992): 102.

Infinite continuations are made explicit in figure 17.4, whose hexagonal nodes grant the minor third, major third, and perfect fifth equal status. To get from one pitch to another, one "steps over" the shared side of the adjoining hexagon. The design renders the other diagonal and knight's-move intervals of online figure 17.3 all but invisible. This image unequivocally treats enharmonic pitches as equivalent. This is the first of the images we will see that gather pitches related by third and fifth into triads. The triangles around the central C form little maps of the triads involving the three pitches they touch. Open triangles indicate major triads; closed ones, minor. These triangles are reduced to open and closed dots at the intersections surrounding other pitches. Despite the explicit infinite continuation, because chains of thirds and sixths repeat after three or four notes (depending on whether they are major or minor), the image also implicitly wraps on itself in those two directions, implying that this is an unfolded version of a three-dimensional representation. Online figures 17.5 and 17.6 make a third dimension explicit.

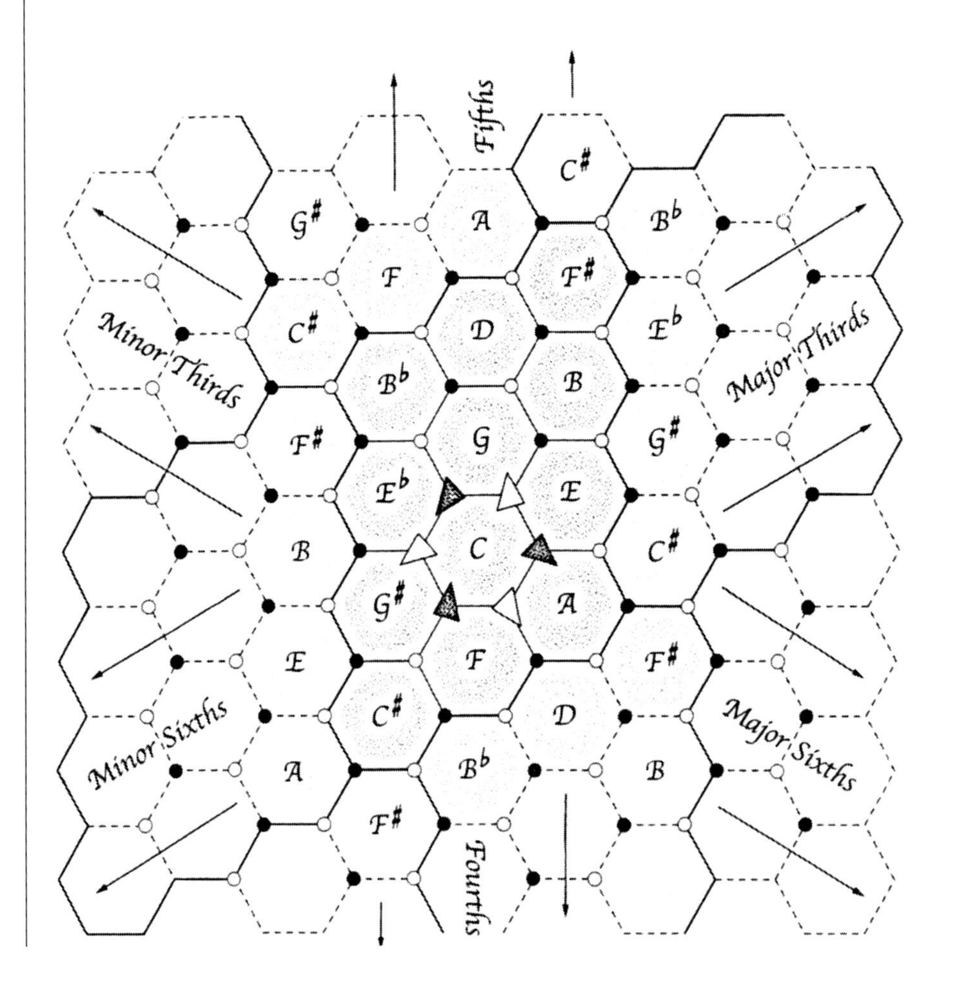

17.4 Anthony Ashton, *Harmonograph: A Visual Guide to the Mathematics of Music* (New York: Walker, 2003), 43.

Figure 17.7 replaces the triads formed by triangles in the various Tonnetz images with triad names. Although we've seen a honeycomb structure before (fig. 17.4), in this one the structure has meaning; each of the three lines leading to or from a note represents one of the neo-Riemannian transformations available from each triad: *P* (parallel) horizontally, *R* (relative) diagonally downward, and *L* (leading-tone exchange) diagonally upward. This is a modular space. Bracketed pitch names along the top edge provide reentry points for the pitches along the bottom edge. Online figure 17.8 presents the same information using the nomenclature of group theory. Online figure 17.9 combines the pitch representation of the traditional Tonnetz and the chordal representation of figure 17.7 in magnificent fashion.

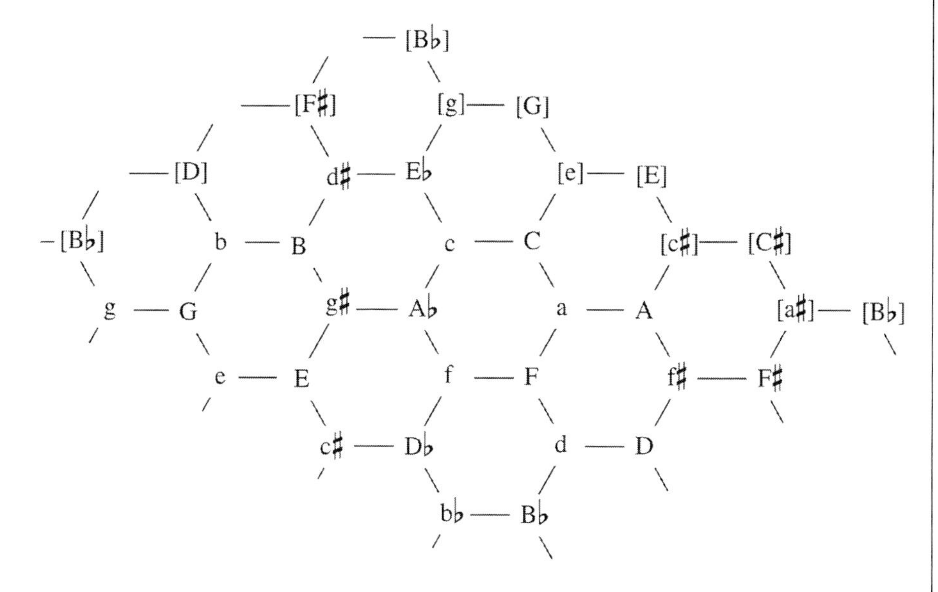

17.7 Dmitri Tymoczko, "The Generalized Tonnetz," *Journal of Music Theory* 56, no. 1 (Spring 2012): 2.

Within the Tonnetz's two-dimensional, minor third plus major third grid lurks a wealth of other relationships that music theorists have explored in a rich literature. The system's modular structure and the way it opens a path to the exploration of relations among triads provide the opportunity for richness and innovation in visual representation.

CHAPTER 18 Atonal Spaces

The analysis of pre-serial atonality has engendered a substantial literature that regularly employs visual techniques. Images relating to pitch-class set theory and analysis appear in many places elsewhere in the book, including chapters 14, 17, 19, and particularly 41. Here we focus on the representation of various post-tonal pitch spaces.

The interval-class cycle generalizes the chromatic pitch wheel and circle of fifths. Figure 18.1 shows the interval-class cycles in a small-multiple format. In addition to providing a comprehensive listing of the cycles, the layout allows us to scan vertically to see the different number of instances of each interval-class cycle and makes particularly striking the unusual qualities of the interval-5 cycle. Scanning horizontally reveals the underlying principles of each cycle. The image would be strengthened by reducing the weight of the circles and adding white space on each side of the numbers, to better separate them from the arcs.

The left image of figure 18.2 builds on this. It nests interval cycles of size 2, 3, and 4, repeating the cycles so they each make twelve stops around the circle. Light-gray spokes gently skewer stacks of pitch classes, which are capped just outside the outer ring by the trichord types they form. Below the graphic, the image tallies the number of times each trichord type results, though the layout is cramped. The image on the right shows how trichord types in the upper half recur symmetrically in the lower half, using solid lines for T_3 and dashed lines for T_9. Here it would be helpful to provide more visual separation between layers, for instance by letting the original image recede into a background gray and making the T_3 and T_9 labels more prominent.

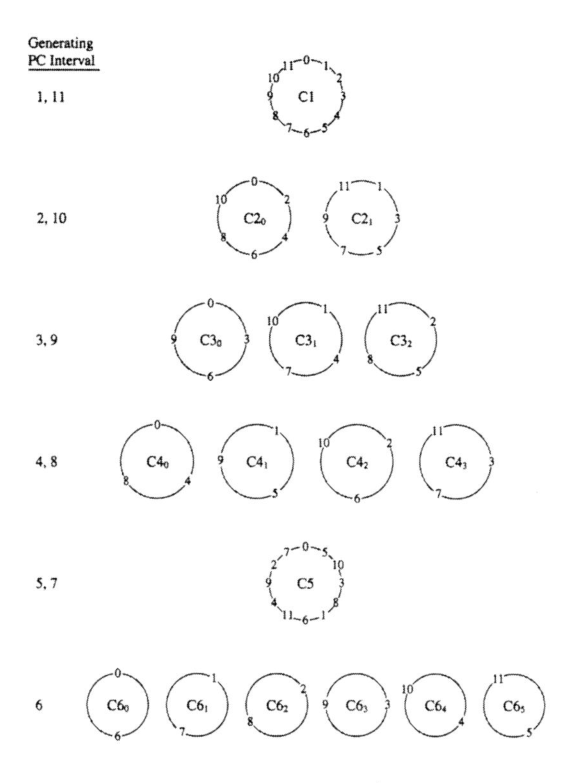

18.1 Joseph N. Straus, *Introduction to Post-tonal Theory*, 3rd ed. (Upper Saddle River, NJ: Prentice Hall, 2005), 154.

Studies of post-tonal theory sometimes involve abstract subset/superset (or inclusion) relationships among pairs of set types. (In this chapter, most uses of the word *set* would more properly be *set class*. I use the simpler term for the sake of brevity.) Figure 18.3 shows an inclusion lattice for the octatonic collection. It uses the set-labeling system devised by Allen Forte (fully worked out in *The Structure of Atonal Music* [1973]), in which the first number indicates the size of the collection and the second its position on Forte's table of sets. The image shows that all seven-note subsets of 8–28 form set 7–31, whose hexachordal subsets come from six different sets. From there, things get messy, as sets begin to share subsets. While the details represented by the lines are not really the image's main point, if they were, the image would demand to be set using highly saturated contrasting colors. Even then, it would look very much like a game of pickup sticks. See also online figures 18.4, 18.5, and 18.6.

As a general rule, prime forms are generally superior to Forte labels. Forte labels are handy for those who know them, and they reveal certain attributes, including their size, the label for their complementary set, and whether they are a member of a Z-related pair, and of course they facilitate finding the collection in a set table to see other attributes. However, the prime form conveys more immediate information about a set than the Forte label. Yet even those who traffic actively in

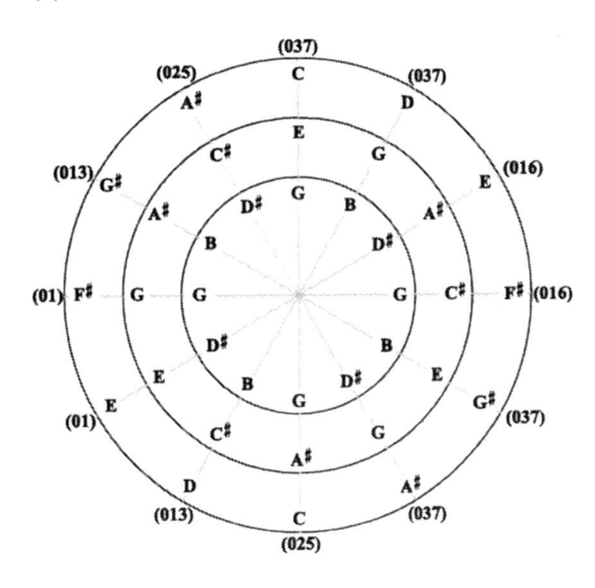

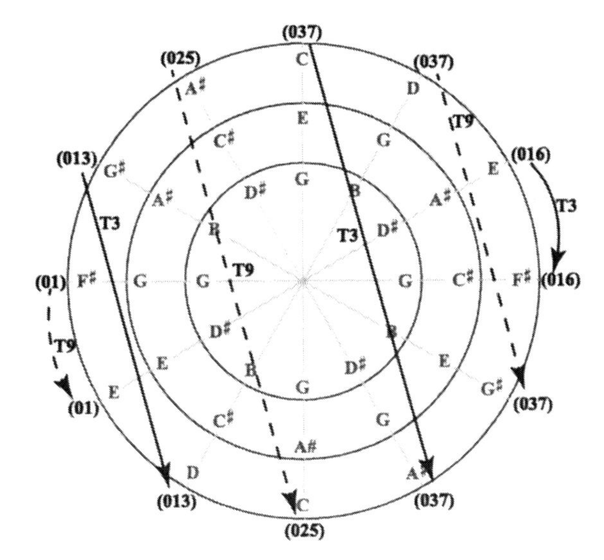

4:(037) 2:(025) 2:(016) 2:(013) 2:(01)

***Above*, 18.2** Philip Stoecker, "Aligned Cycles in Thomas Adés's Piano Quintet," *Music Analysis* 33, no. 1 (2014): 34.

***Right*, 18.3** Robert Morris, *Class Notes for Advanced Atonal Music Theory* (Lebanon, NH: Frog Peak, 2001), ex. 25.1c.

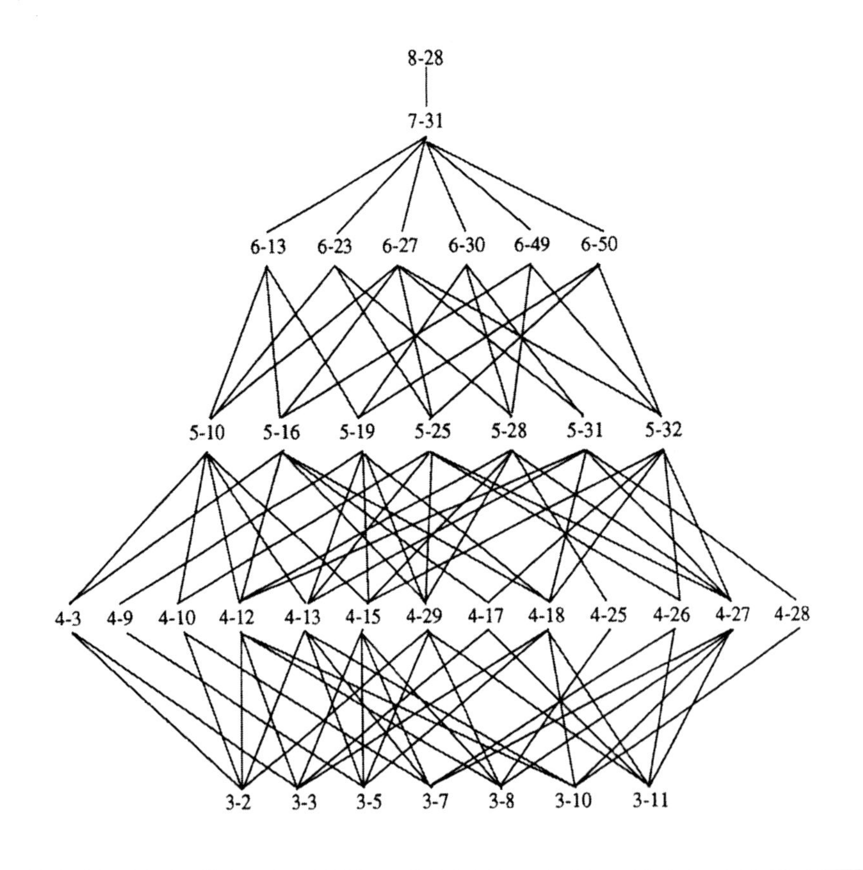

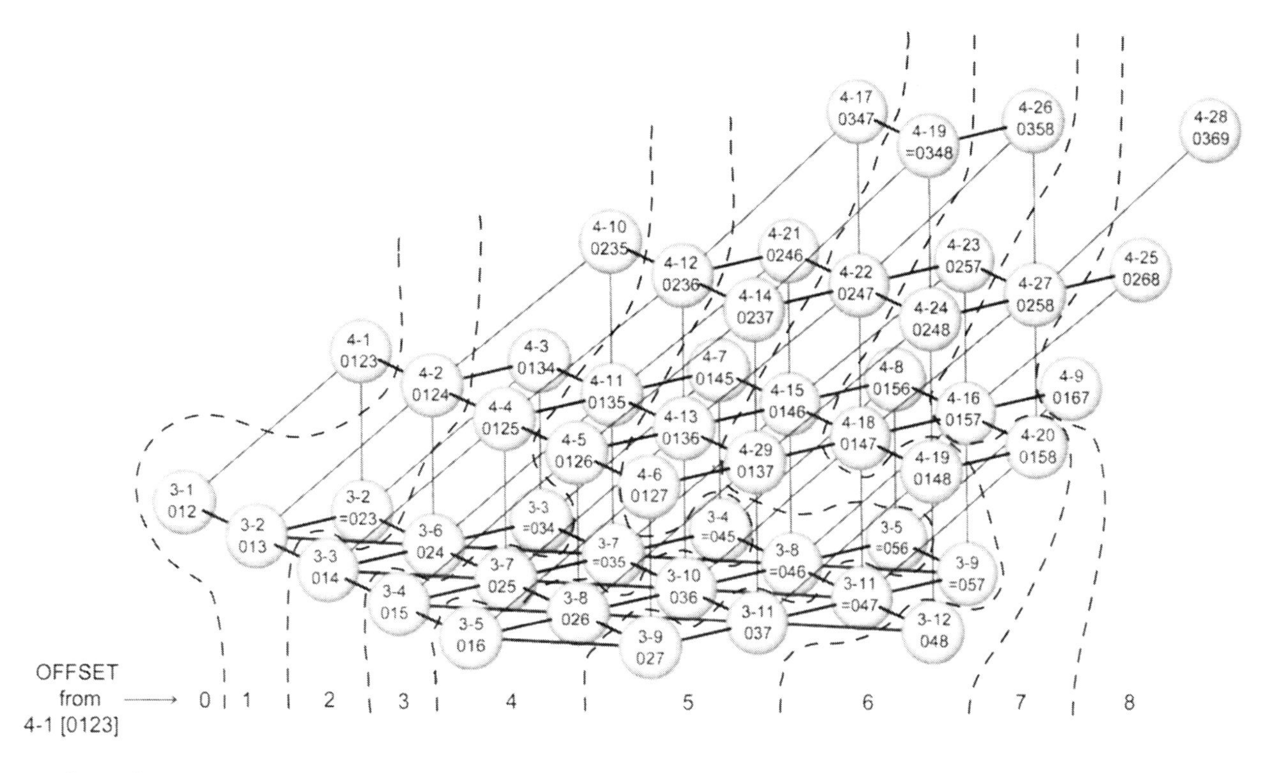

18.7 Joseph N. Straus, "Voice Leading in Set-Class Space," *Journal of Music Theory* 49, no. 1 (2005): 72.

set theory know the prime forms for only a fraction of the 224 Forte labels. In short, unless your target audience is the set of die-hard set theorists, it is better to include prime forms when identifying pitch-class collections.

Voice-leading spaces are configurations of pitch-class sets structured to highlight smooth voice-leading connections between them. These bear a strong resemblance to the Tonnetz (see chap. 17). Figure 18.7 shows the network of all sets that we obtain if we start with four voices arranged as the chromatic tetrachord (0123) and move one voice at a time by a half step in each direction, continuing until all possible combinations have been reached. The result is even messier visually than the previous images. Lines spread in four directions. While the use of bold strokes for lines connecting sets drawn on the same horizontal plane helps, the entanglement is substantial. Adding to the visual clutter, dashed lines group the sets based on the number of total semitonal moves required to reach a set from (0123) (the "offset"). Online figure 18.8 attempts to address some of these point, while online figure 18.9 features a yet more successful approach to the same issue involving only three-note sets.

Another kind of voice-leading space is the visually striking Riemann wreath (fig. 18.10), whose name is an apt metaphor for the structure. Any two connected nodes form a major or minor triad. Placement of the 0 node at the top aligns with the convention used in pitch-class circles and circles of fifths. Each node

18.10 Robert D. Morris, "Voice-Leading Spaces," *Music Theory Spectrum* 20, no. 2 (1998): 194.

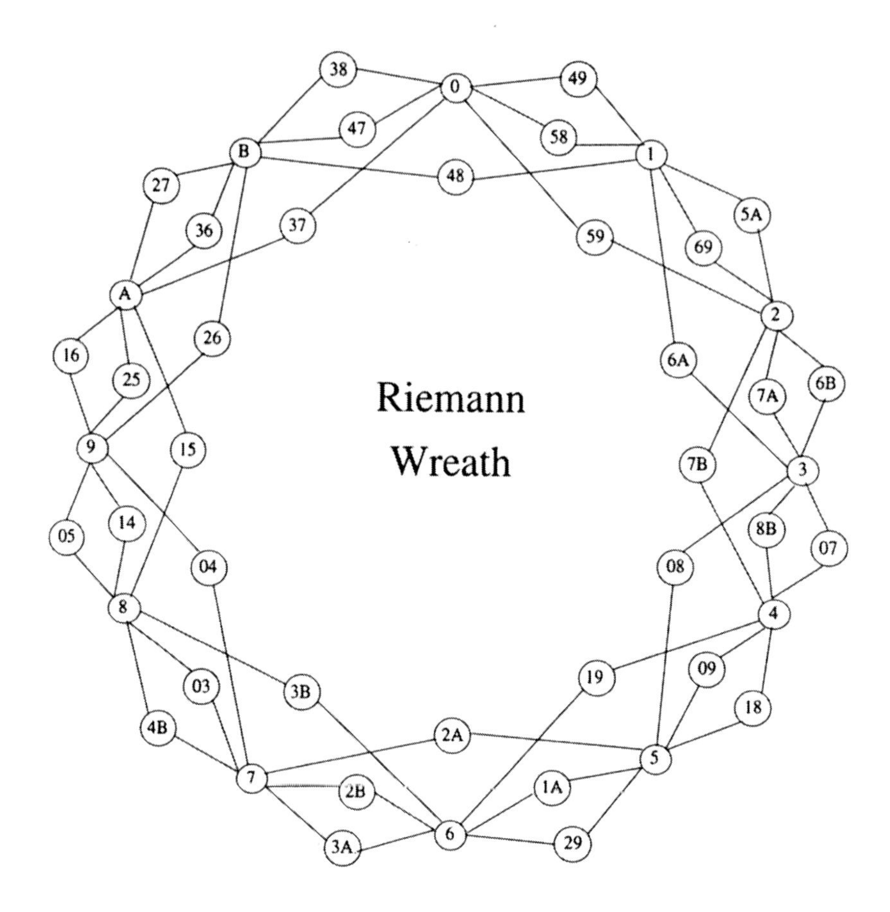

serves as the inversional axis between the two nodes that are connected to it. For instance, on one side of the 0 node at the top center is (38). Inverting these pitch classes over (0) yields (94) = (49). Likewise, flipping that (0) to the other side of (49) yields (1), the next node along that path. The structure is called a Riemann wreath because it captures in one place the neo-Riemannian *P*, *L*, and *R* relations and their inverses in one place. Ordinarily, one might want the image to give greater emphasis to the content of the nodes and less to the organizational apparatus, lines, and circles. In this case, however, in many ways the structure is exactly the point, and the information contained within the nodes simply provides the evidence justifying the visual interpretation.

Visualizing atonal pitch spaces involves many of the same decisions as any other musical space: determining whether the space is linear or modular, choosing an appropriate representation for the elements that make up the space, and then designing an appropriate, and visually effective, connecting structure.

Chapter 19 Symmetrical Pitch Structures

Symmetry involves the projection of an object (real, imagined, or metaphorical) either onto itself or onto a new version, often (but not always) across an axis of some kind. Our daily experience of the physical world, most notably the bilateral symmetry that characterizes our bodies, informs our understanding of symmetry. Many objects we interact with daily are also symmetrical in one or more dimensions—animals, modes of transportation, sports objects (baseballs and footballs, pucks, shuttlecocks), buildings, furniture, and so on. We perceive symmetry better when we experience it through our eyes than with any other sense. Whether a particular symmetry exists in actual space, a conceptual or abstract space, or in time, we can often understand it most effectively through visualization.

Symmetry can be the property of a single object, or a relationship between a pair of objects. There are many kinds of symmetry. We are most familiar with mirror or *reflectional* symmetry, in which an object is a reflection of itself across a central axis. A pair of objects can also be related by reflectional symmetry if one can draw an axis between them such that corresponding points are equidistant from that axis. Many instances of musical symmetry are of this type, including melodic inversion. There are other kinds of symmetry, however, most of which have musical realizations. Transposition in pitch space is an example of *translational* symmetry. Transposition in a pitch-class space depicted on a pitch-wheel involves *rotational* symmetry. Rhythmic augmentation and diminution are types of *scalar* symmetry, and the post-tonal T_nI operation combines *reflectional* (inversion) and *translational* (transposition) symmetry. While there is often no reason to invoke a symmetry-related explanation—such as for musical phenomena for which we already have commonly used terms and for which a symmetrical explanation is secondary—when symmetry is the primary point, it should be represented clearly.

Visualizations of symmetrical structures require foremost that the axis of symmetry be apparent and that the correspondence between points be clear. This does not require that the axis and correspondences actually be shown, but they should be easily apprehended. The images in this chapter show several approaches to the study of symmetrical pitch structures. Chapter 45 examines the study of symmetry in music analysis.

One can display symmetrical collections of pitch classes in a line, as in figure 19.1 and online figure 19.2, or on a circle, as in figures 19.3 and 19.4. Figure 19.1 shows odd- and even-length segments of the circle of fifths (the "cycle-7 complex" referenced by the title of the image's source). Those whose lengths are odd are symmetrical around a single pitch class (hence, the "PC Axis"), and those whose

19.1 Robert Gauldin, "The Cycle-7 Complex: Relations of Diatonic Set Theory to the Evolution of Ancient Tonal Systems," *Music Theory Spectrum* 5 (1983): 41n15.

PC AXIS SYSTEM

Set type	pcs in set	ic of framing pcs from axis
1–1	9	
3–9	5 0 7	5
5–35	10 5 0 7 2	2
7–35	3 10 5 0 7 2 9	3
9–9	8 3 10 5 0 7 2 9 4	4
11–1	1 8 3 10 5 0 7 2 9 4 11	1
(12–1)	6 1 8 3 10 5 0 7 2 9 4 11 6	6

IC AXIS SYSTEM

Set type	pcs in set	ic of framing pcs from axis
2–5	0 7	
4–23	5 0 7 2	5
6–32	10 5 0 7 2 9	2
8–23	3 10 5 0 7 2 9 4	3
10–5	8 3 10 5 0 7 2 9 4 11	4
(12–1)	1 8 3 10 5 0 7 2 9 4 11 6	1

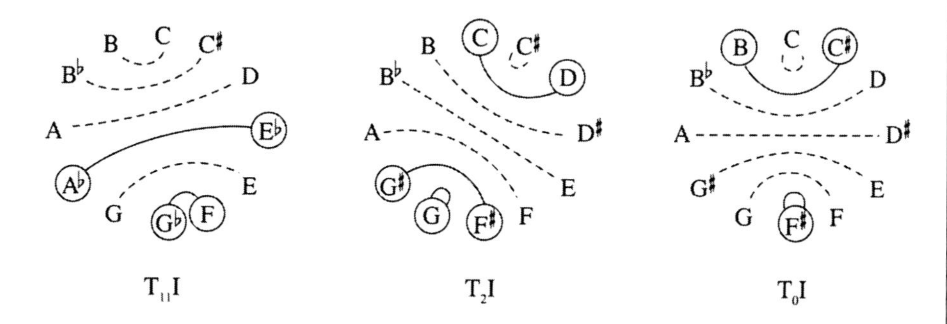

19.3 Joseph Straus, *Introduction to Post-tonal Theory*, 3rd ed. (Upper Saddle River, NJ: Prentice Hall, 2005), 87.

lengths are even are symmetrical around an interval (hence, the "IC Axis"). One can easily see the symmetry of the structures. The brackets on the second and bottom entries of both systems, which seem intended to draw attention to the symmetry, are unnecessary and could be removed. Identifying which elements correspond to each other (for example, 8 and 4 in the lower three lines of the PC Axis system image) requires some visual search, however. The vertical bars marking off the axes of symmetry could safely appear in gray. This image includes a second symmetry: complementary set types are marked with brackets on the left. The brackets show which rows correspond. They also make clear that 6 is the axis of symmetry. (In the first row of the PC Axis system, the listing of pitch class 9 as the axis is clearly a typographical error; it should be 0.)

Figure 19.3 shows the same sets as online figure 19.2 in a different configuration. Pitch classes belonging to a set are circled on a clockface listing all twelve pitch classes. The axis of symmetry is implied, less obviously than in online figure 19.2, by lines that connect corresponding pitches: solid for pitch classes that are present and dashed for corresponding pitch classes that do not appear in the set being depicted. Most readers will likely recognize the symmetry of figure 19.3's third figure more quickly than the others. This is because our vertical orientation provides an implicit axis to everything we see. The other figures require a greater effort to see the central axis.

The gentle redrawing in figure 19.4 offers minor improvements. Pitch classes that are part of the set pop out from the background because the note names appear

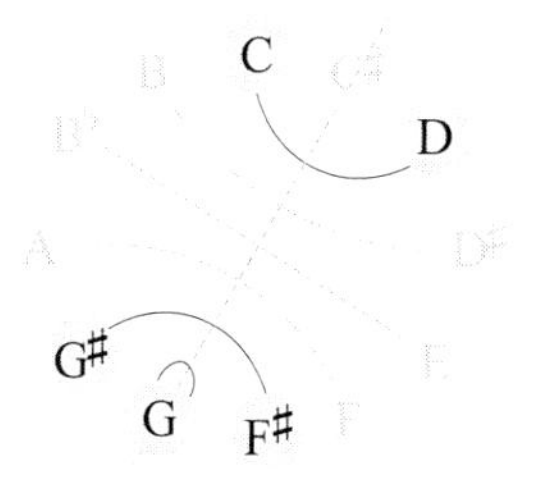

19.4 Redrawing of central portion of fig. 19.3.

in bold and the circles are solid. Names of absent pitch classes are more muted in gray. The two kinds of lines connecting pitch classes opposite the axis of symmetry likewise contrast more. The axis, not sufficiently implicit in figure 19.3, is rendered explicitly in figure 19.4. Even though we are quite adept at recognizing symmetry, subtle design clues often draw attention to the details more quickly.

CHAPTER 20

Tonal Hierarchy, Tendency, Progression

Functional tonality is teleological. Much of its sense of goal-directedness arises from a strong sense of hierarchy. In discussions of tonal music, one often encounters metaphors such as *stability* and *attraction* used to characterize this hierarchy of pitches in the scale. Several images from one book illustrate ways of showing this. The upper part of figure 20.1 depicts degrees of stability in C major: the tonic pitch C is most stable (level *a*), followed by the dominant (*b*) and then mediant scale degrees (*c*). Next come the remaining diatonic pitches (*d*), and finally the chromatic pitches (*e*). The lower portion of figure 20.1 generalizes this hierarchy

a)

level *a* :	C												(C)
level *b* :	C							G					(C)
level *c* :	C				E			G					(C)
level *d* :	C		D		E	F		G		A		B	(C)
level *e* :	C	D♭	D	E♭	E	F	F♯	G	A♭	A	B♭	B	(C)

I/I (= I/C)

b)

level *a:*	0												(12 = 0)
level *b:*	0							7					(12 = 0)
level *c:*	0				4			7					(12 = 0)
level *d:*	0		2		4	5		7		9		11	(12 = 0)
level *e:*	0	1	2	3	4	5	6	7	8	9	10	11	(12 = 0)

I/I

20.1 Fred Lerdahl, *Tonal Pitch Space* (New York: Oxford University Press, 2001), 47.

to any major scale, where 0 is the tonic pitch. Online figure 20.2 depicts the same space, less effectively, on a cone.

Neither image conveys the *dynamic* nature of the tonal system, however. Figure 20.3 more effectively shows the tendency of melodic pitches to move to others, where tendencies, in a sense stepwise within the context of that level, are represented at each level using arrows (the levels are the same as in fig. 20.1). The image conveys how a local point of stability at the lowest level (e.g., F) can simultaneously be an unstable pitch at the next higher level.

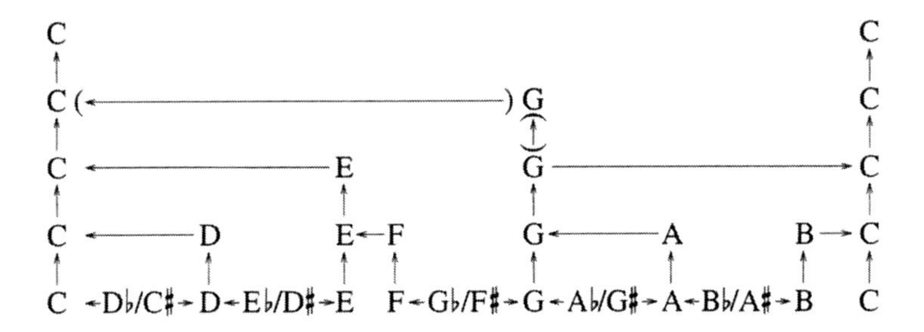

20.3 Fred Lerdahl, *Tonal Pitch Space* (New York: Oxford University Press, 2001), 49.

Figure 20.4 uses notation and arrows to convey similar information. Lines show stepwise resolutions to a more stable pitch. When a pitch can resolve stepwise to two pitches, the resolution that leads to greater stability is a solid line and the other a dashed line. Here the higher the level number, the greater the stability (the note E has stability 6 when it results from the resolution of the dominant seventh, otherwise 5). The direction of resolution is clearer in figure 20.3 than in figure 20.4 because of its use of arrowheads. On the other hand, showing the pitches just once, as in figure 20.4, makes the stability ratings of each pitch clearer. The redrawing of the latter image in online figure 20.5 clarifies things.

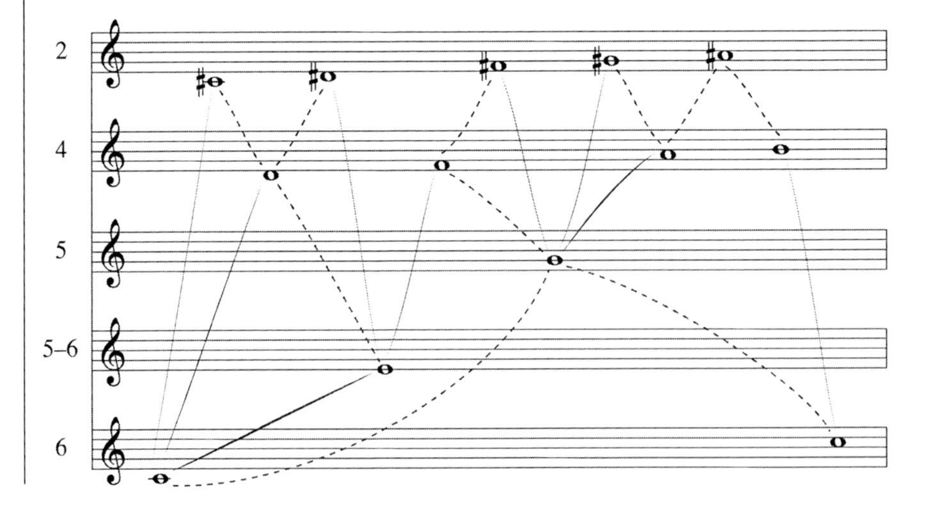

20.4 Steve Larson, *Musical Forces: Motion, Metaphor, and Meaning in Music* (Bloomington: Indiana University Press, 2012), 103. The ratings are derived from Elizabeth Hellmuth Margulis, "Melodic Expectation: A Discussion and Model" (PhD diss., Columbia University, 2003).

Whereas figures 20.3 and 20.4 represent "normative" behavior of melodic scale degrees, figure 20.6 and online figure 20.7 show how these tendencies play out in bodies of actual music, for diatonic and selected chromatic pitches, respectively. Both employ relative line thickness to express the probability that one scale degree will be followed by another. The gross proportions are clear from the images, though attaching a number to each arrow would make more fine-grained comparisons possible.

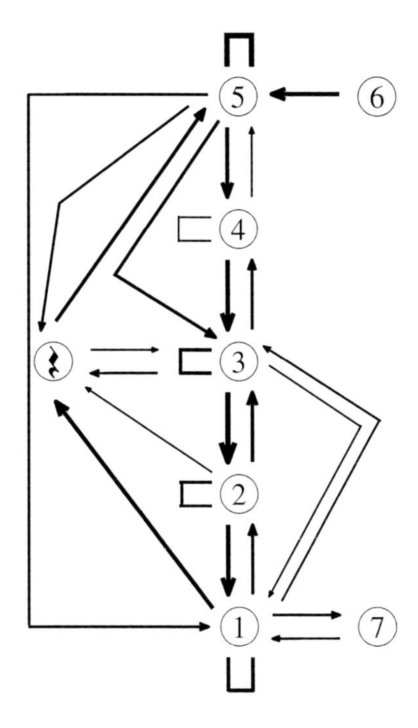

20.6 David Huron, *Sweet Anticipation: Music and the Psychology of Expectation* (Cambridge, MA: MIT Press, 2006), 160.

Melodic stability is closely related to the notion of tonal harmonic progression. Melodic tendencies, which affect voice leading, necessarily also affect how chords tend to progress from one to the next. The next several images show various methods for representing these tendencies. Invoking a metaphor of harmonic gravity, figure 20.8 places diatonic chords in orbit around the tonic triad. The closer a chord is to tonic, the stronger the gravitational pull. While gravity seems like an apt metaphor for this kind of harmonic force, the idea of orbits causes the metaphor to collapse under its own weight. First, objects in orbit have reached a point of equilibrium between their momentum, which would propel them in a straight line, and the pull of gravity. There is no musical force, however, that could counter the gravitational pull toward tonic and produce the harmonic equilibrium that would lock a chord in an orbit. Even if there were, orbits decay continually, not discretely, so the notion of orbital layers is wrong. Furthermore, effecting a soft landing from

20.8 Ralph Turek, *The Elements of Music: Concepts and Applications*, 2nd ed. (New York: McGraw-Hill, 1996), 1:128.

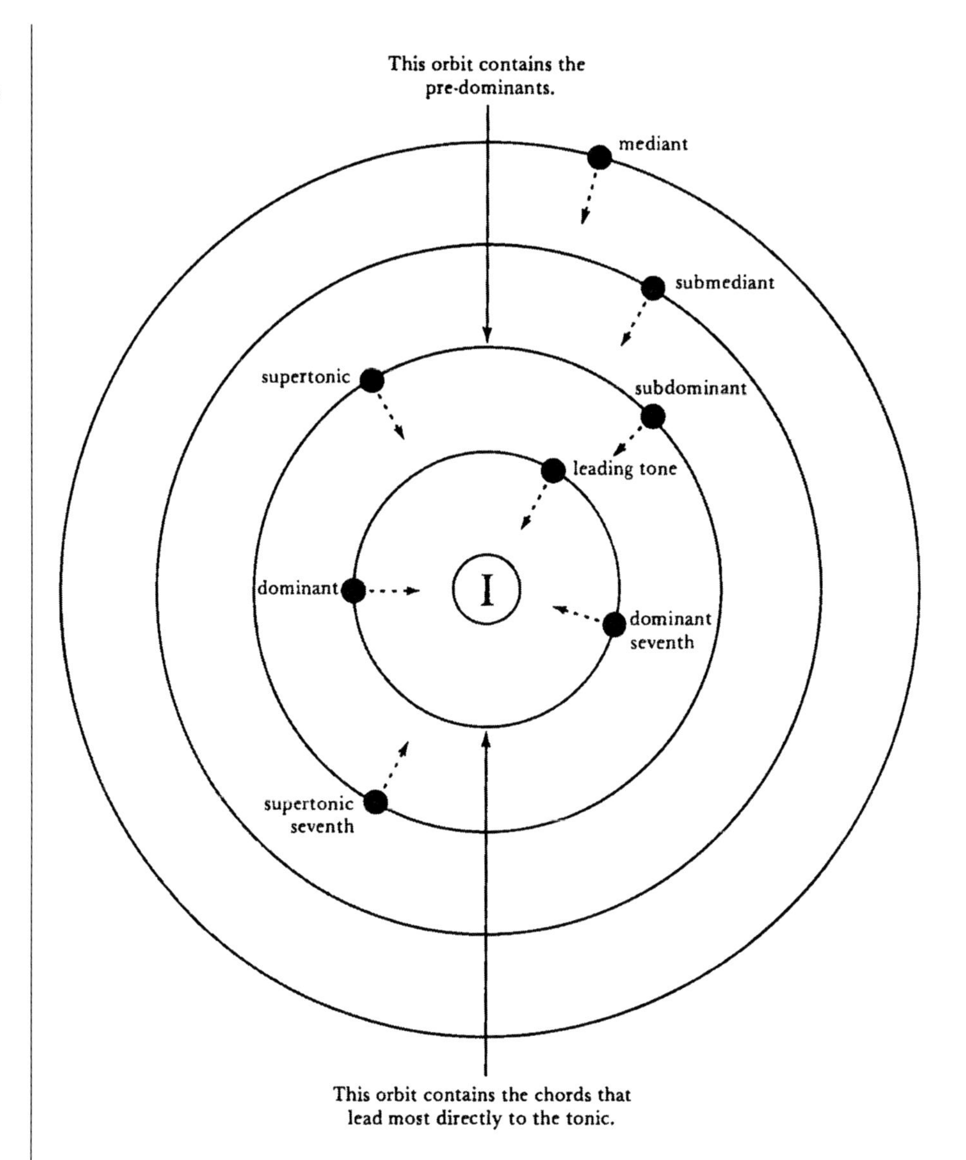

orbit requires the expansion of a great deal of energy. This is very different from the real effect of music, which carries more of a sense of flowing downstream toward the point of maximum stability, such as sea level. (An alternative interpretation of the design is that the orbits are not planets around a star but electrons around a nucleus, which as a metaphor is more apt but less accessible.)

I have not encountered any visual representations of harmonic progression that explicitly invoke a sense of flowing downstream, but online figure 20.9 and figure 20.10 would be consistent with such a metaphor. Like figure 20.8, online figure 20.9 includes V^7 and ii^7 but no other diatonic sevenths. Because they omit sevenths

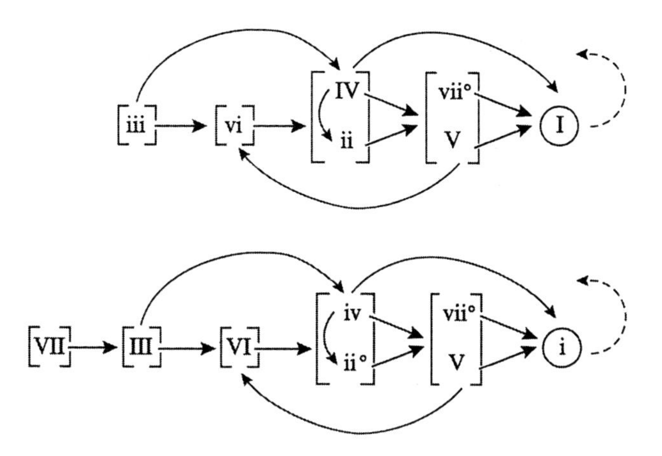

20.10 Stefan M. Kostka and Dorothy Payne, *Tonal Harmony: With an Introduction to Twentieth-Century Music*, 6th ed. (New York: McGraw-Hill, 2009), 113.

altogether, the images in figure 20.10 are in a sense more accurate, as they seem to pertain to chord roots in general. They also paint a more nuanced picture, showing the practice by which (in a major key) IV may be followed by ii, but not the reverse, as well as the equal tendency of iii to progress to either IV or vi, and of course the deceptive resolution. The dashed arrow coming from the final chord allows for the essential departure from tonic, correctly leaving undefined where that initial leap might lead.

Compare these to figure 20.11 (colorplate 8). Beautifully designed and executed, the image could replace an entire chapter in many harmony texts. The image merits closer scrutiny. First, the hierarchy of information is clear. Header text of obviously different sizes and weights tag the relative importance of the various blocks of information. So effectively does the image convey the layers of information that the overview diagram at the lower left, which shows a simplified schematic, is barely necessary. Chords and groups of closely related chords are drawn in circles or ellipses, with major chord families enclosed by a different shape (rectangles). All of the enclosures appear in gray and so do their job without drawing attention to themselves. The image depicts four kinds of harmonic motion using lines that differ in thickness and color. Because these progressions are central to the image's purpose, the use of high-saturation colors that pop out from the rest of the image is appropriate. The arrows turn harmonic progression, antiseptically conveyed in online figure 20.9 and figure 20.10, into a rich walking map of tonal harmonic syntax. Although all routes run left to right, from Initial Tonic to Final Tonic, the map captures a myriad of the twists, turns, and surprises that can occur in between. Along the way, if one sees a place that looks interesting, leaning in brings explanatory text into focus. Because these text blocks occupy a fairly large amount of surface area but represent a different layer of information from the rest of the image, their gray color provides just the right visual separation.

20.12 David Huron, *Sweet Anticipation: Music and the Psychology of Expectation* (Cambridge, MA: MIT Press, 2006), 253.

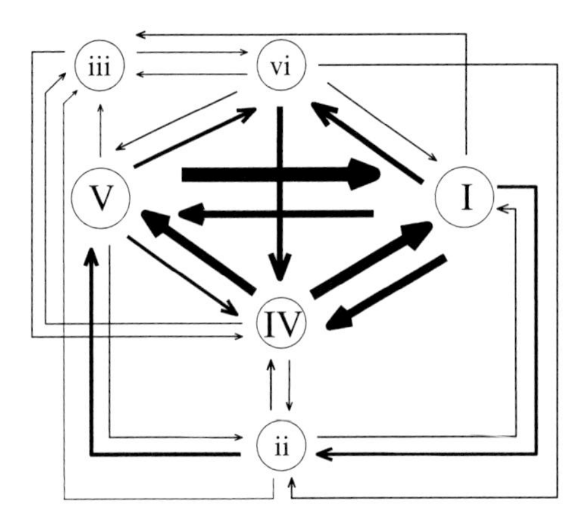

As with figure 20.6 and online figure 20.7, figure 20.12 shows the actual frequency of chord-to-chord interactions, here in a sampling of Western popular music. This enhances the more generic harmonic flow of online figure 20.9 and figure 20.10 by suggesting how often each progression occurs. The layout of the chord symbols does not invite the same left-to-right reading that invokes a flowing-downstream metaphor in the previous images. This is perhaps a weakness in online figure 20.13, which applies the same method to a collection of Baroque music, but for the music represented in figure 20.12, where the idea of paradigmatic progressions is considerably weaker, the more "neutral" layout strengthens the image.

Western functional tonality is intensely goal directed. In one way or another, each of the images in this chapter conveys either the typical flow of harmonic motion or the structural conditions that given rise to that flow, though some succeed more than others. (Tonal hierarchy is of course central to Schenkerian analysis, which is discussed in chap. 48.)

Chapter 21 The Overtone Series

An understanding of the overtone series is essential to brass players and organ builders and is often useful for performers of other instruments, for throat singers, and in the study of acoustics. Although the overtone series consists of whole-number multiples of a fundamental pitch, that spacing does not resonate in our ears arithmetically, so visualizations of the series nearly always treat the series logarithmically by mapping the frequencies as pitches on a staff. The primary decisions one makes when representing the overtone series are (1) how far to carry the series and (2) what to say about the intonation of pitches relative to how they are normally tuned in performance. Regarding the latter, equal temperament of course makes all pitches except the octaves out of tune, but as we will see, most depictions of the overtone series are a bit forgiving on this front.

Figure 21.1 presents a basic representation of the overtone series of the sort found in the early chapters of many undergraduate harmony textbooks. Besides the notation of "out of tune" partials with solid noteheads (though not all sources suggest that the thirteenth partial, A, is out of tune), it conveys no additional information. The image does not specify the direction in which, much less the extent to which, the pitches are out of tune. It also draws no attention to the beginning of new octaves, which would highlight the fact that each new octave contains more overtones than the previous one. Compare this to online figure 21.2.

The 1764 image in figure 21.3 extends the scope of the previous images by two octaves, depicting the first sixty-four partials. By expressing each octave on a separate line, it better conveys that each overtone, once introduced, carries over

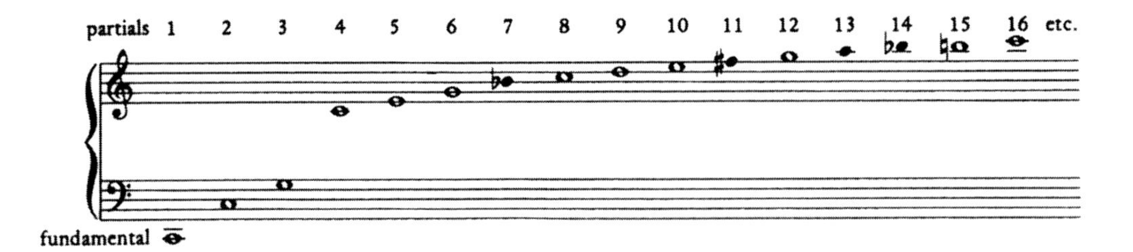

21.1 Edward Aldwell and Carl Schachter, *Harmony and Voice Leading* (Belmont, CA: Wadsworth, 2003), 25.

to subsequent octaves. It also makes clear which notes are new in each octave and where they fit into the previous overtones. The image does not show *fa* to be out of tune, though it gives the out-of-tune flat seventh degree (*sib*) a new name, *za*. In the 16–32 and 32–64 levels, micro-pitches are needed, and labels are provided for those, though details about the internal combinations by which they arise are not.

Figure 21.4, from the turn of the twentieth century, is richer in information than the images above. It interprets the first thirty-two partials, also according to octaves. It marks pitches that are flat relative to where they "should" be ° and those that are sharp ×, though does not show by how much. The image interprets the pitches spanned by the different octaves as resembling (sometimes "analogous" to) another construct: the open fifth in the case of notes 2–3, the major triad (4–6) and dominant seventh (4–7), the major scale ("in 8 tones," 8–15), and the chromatic scale ("in 16 tones," 16–31). In keeping with the then-prevailing dualist view of pitch space, the image posits a corresponding undertone series.

Visualizations of the overtone series are overwhelmingly concerned with representing the series relative to what is familiar: standard notation. The images note that deviations from the familiar exist, but they are largely unconcerned with showing the extent of those deviations. (These issues are discussed more fully in

21.3 Denis Ballière de Laisement, *Thèorie de la musique* (Paris: Chez P. F. Didot le jeune, 1764), cited in Ian Bent, "Momigny's *Type de la Musique* and a Treatise in the Making," in *Music Theory and the Exploration of the Past*, ed. Christopher Hatch and David W. Bernstein, 309–40 (Chicago: University of Chicago Press, 1993), 318.

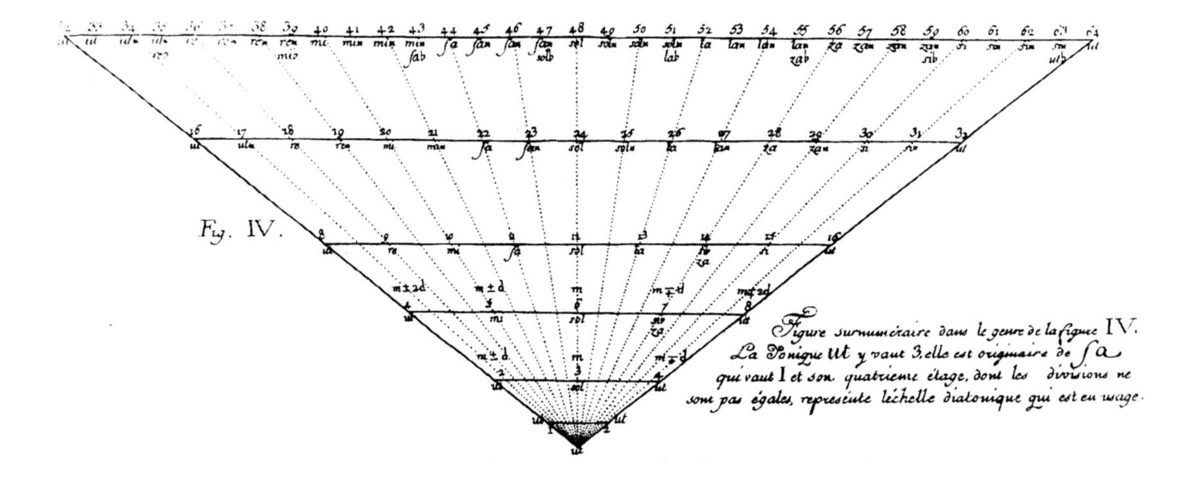

21.4 Hermann Schröder, *Die Symmetrische Umkehrung in der Musik* (Leipzig: Breitkopf und Härtel, 1902), 5.

chap. 30.) And many fail to make explicit what brass players must know and eventually master: the higher one goes, the more "natural notes" get squeezed into each octave. Both omissions are unfortunate.

* * *

In part 2, we have explored a wide variety of musical spaces. Some of these have been basic collections of pitches, from pitch and pitch-class space to collections, scales, and modes, the circle of fifths, and the Tonnetz. We explored the representation of symmetry, before concluding with a look at tonal hierarchy and the overtone series. Visualizations of many of these spaces start as simple conceptions and over time accrete richness as writers find richer structures in the surrounding musical space.

Common to all visualizations of musical spaces should be clarity about which objects belong to that space and a metaphorically sensible way of representing those objects with respect to one another and (potentially) to a larger space of which they are a subset. One must also decide how to represent either the infinite extension of a linear space or the circularity of a modular space. Finally, when designing a space visually, one must remember that it is more important to make the design meaningful to the viewer than to oneself.

Part 3

Musical Time

Musical time has many facets. We are aware of musical time at scales ranging from performance microtiming to proportion across multimovement works. Our metaphors for time likewise vary, from the mechanistic, relating to the clock, for instance, to water-related metaphors like ebb and flow.

Visual representations typically depict musical time moving left to right. This parallels how those in Western cultures read, an activity that also happens in time. It also mirrors time's frequent representation in things like historical timelines. In notation, musical time is conveyed through durational symbols, words (*accelerando* and *ritardando*), and symbols like fermatas, tempo markings, and repeat signs. Images of musical time often use aspects of notation, though there are many other ways to represent time: dots marking time points, lines signifying durations, arrows showing directionality, and so on.

Illustrating musical time requires decisions about how to capture the various ways musical time can unfold. One must consider which aspects of the music to convey (individual events, harmonies, phrases); whether to measure time or simply order it; if it is measured, whether to measure the music as notated or as performed; whether to use analog or digital measurements; if digital, whether to measure it against an external temporal grid, such as minutes, seconds, or milliseconds, or an internal musical grid, such as beats, measures, phrases, sections, or even movements; whether to view musical time hierarchically; how to handle conflicts between music and its underlying grid; and whether to represent time numerically or graphically.

In part 3, we mainly deal with the issues abstracted from musical context. Part 5 covers the visualization of some of these concepts in music analytical contexts. Chapter 22 examines the representation of basic musical durations. Chapters 23 through 25 explore approaches to the measurement of time: chapter 23 looks at representations in which there is no measurement, chapter 24 looks at representations in which musical units such as beats measure musical time, and chapter 25 summarizes externally measured musical time, as with a clock. Finally, chapter 26 looks at the representation of proportion, which plays such an important role in music scholarship.

Chapter 22 Basic Durations

The representation of duration in Western music notation grew out of (or evolved alongside) musical traditions in which a basic pulse came to be divided (and subdivided) into and grouped (and further grouped) by twos or threes. Even when combined with a metronome marking, the notation system relates to clock time loosely at best. It marks states more than it measures. It quantizes generously; the notation freely accommodates performance traditions including the Chopin rubato, the irregular beats of the Viennese waltz, and jazz's swung eighth notes. Those familiar with these traditions have little trouble reproducing the rhythmic notation within that tradition.

In this chapter, we look at representations of the system of basic durations. In showing proportional relationships among the basic musical durations, one must first decide whether to emphasize the accumulation of short durations into longer ones or the division of longer values into shorter ones. Figure 22.1 takes the latter approach, showing the rabbitlike multiplication of notes as each duration is divided by two. It shows how each duration can be notated, employing both upward and downward stems and both flags and beams. The image has some minor design problems: The numbers down the center column tempt the eye to read the topmost symbol as a zero rather than a whole note. In addition, the braces add visual clutter that could be omitted. The image also does not make it explicitly evident that each individual note is divided into two. The similar online figure 22.2 is more successful in some ways, less in others.

Figure 22.3 is from the facing page in the same source as online figure 22.2. It takes the opposite perspective of that image and of figure 22.1. As one scans from

22.1 William Turner, *Sound Anatomiz'd* (London: William Pearson, 1724), 17.

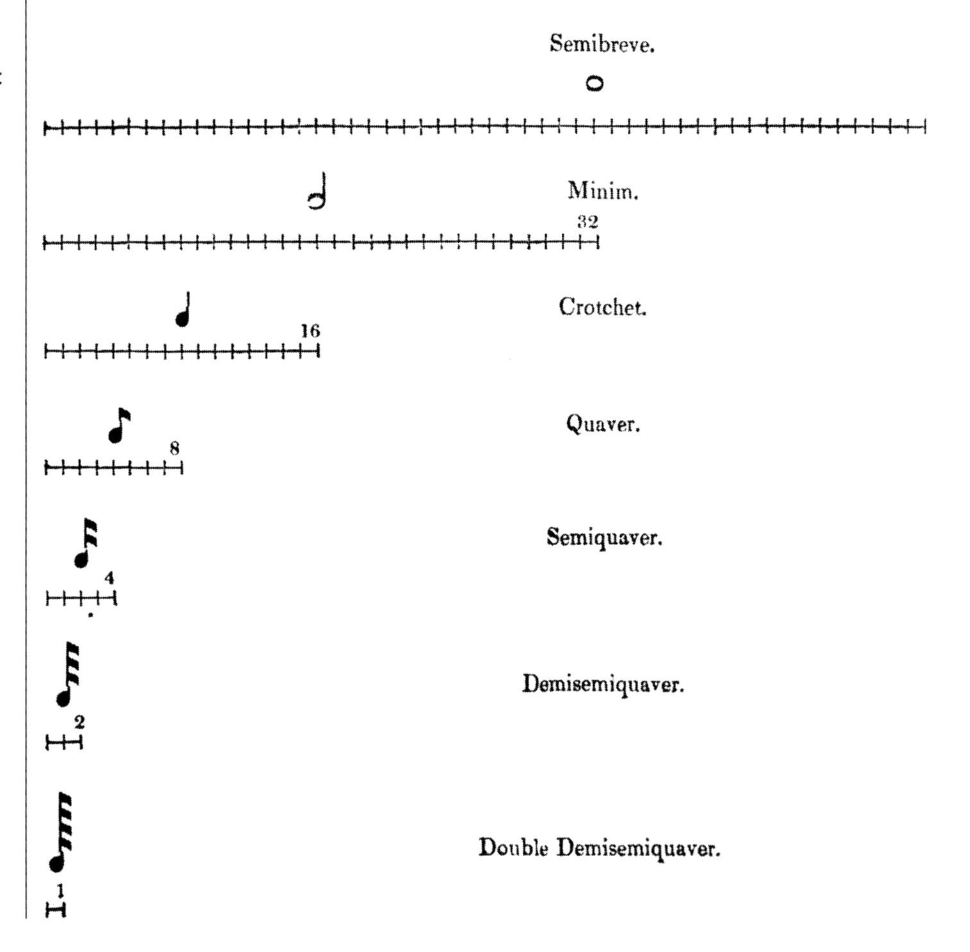

22.3 Thomas A. Busby, *Grammar of Music* (London: J. Walker, 1818), 68.

top to bottom, the length of each duration is halved relative to the one above it, with the duration of each measured by the shortest value given, a sixty-fourth note. This image conveys the differences in duration better than the first two, though with the loss of the visual sense of how many notes of one value are needed to make up a longer value. Both approaches convey useful information: online figure 22.2 and figure 22.1 show the proliferation of durations as they get shorter; figure 22.3 shows how short they get. An ideal image would do both.

Figure 22.4 tries, though its implementation undermines its success. Unlike online figure 22.2 and figure 22.1, it makes the division of each duration into two halves explicit. This, combined with the fact that each line is left-justified, helps create an implicit grid that strongly suggests the relative duration of each note, in the manner of figure 22.3. Therefore, the image succeeds at showing that the whole note is equivalent in duration to that of the long line of sixty-fourth notes, while at the same time showing clearly how many sixty-fourth notes it takes to equal a whole note. Unfortunately, the lines that show the division of each duration

22.4 Renée Longy-Miquelle, *Principles of Musical Theory* (Boston: E. C. Schirmer Music, 1925), 8.

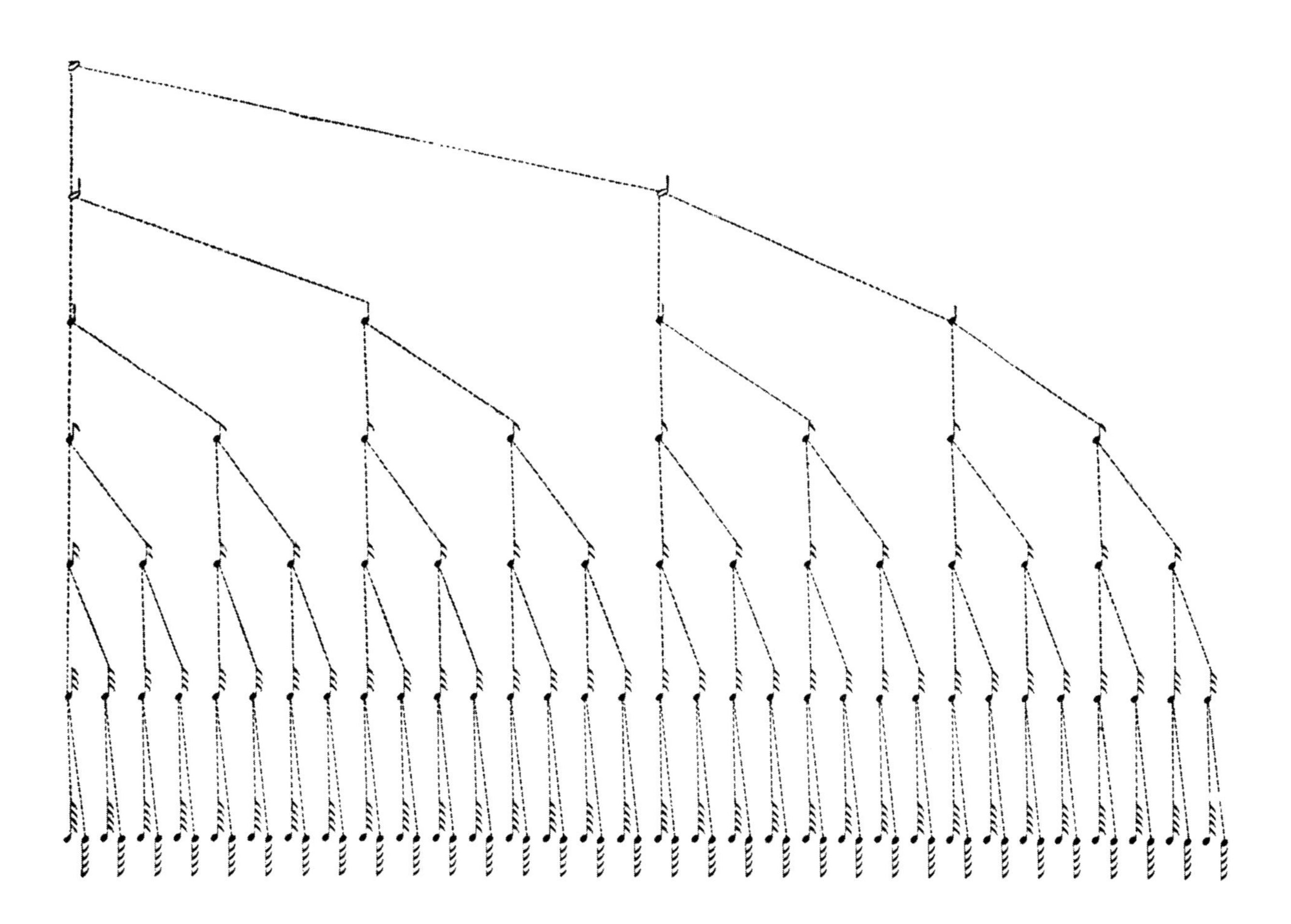

overwhelm the tiny durational symbols. Aside from those lines, this image otherwise conveys in a single image the inverse relationship that online figure 22.2 and figure 22.3 both need.

Contemporary textbooks tend to emphasize the basic relationship between durations rather than the broad sweep seen in the historical images we've looked at so far. As is typical of these, figure 22.5 simply shows the binary division of each value and includes the names of the durations (except for the thirty-second note). It fails to convey visually the sense of exponential growth or division that characterizes the images above, however. The inclusion of rests is a plus. Online figure 22.6 shows an image from another contemporary textbook that employs prose at the expense of clarity.

Value	Note	Rest
Breve		
Whole		
Half		
Quarter		
Eighth		
Sixteenth		

22.5 Stefan M. Kostka and Dorothy Payne, *Tonal Harmony: With an Introduction to Twentieth-Century Music*, 6th ed. (New York: McGraw-Hill, 2009), 27.

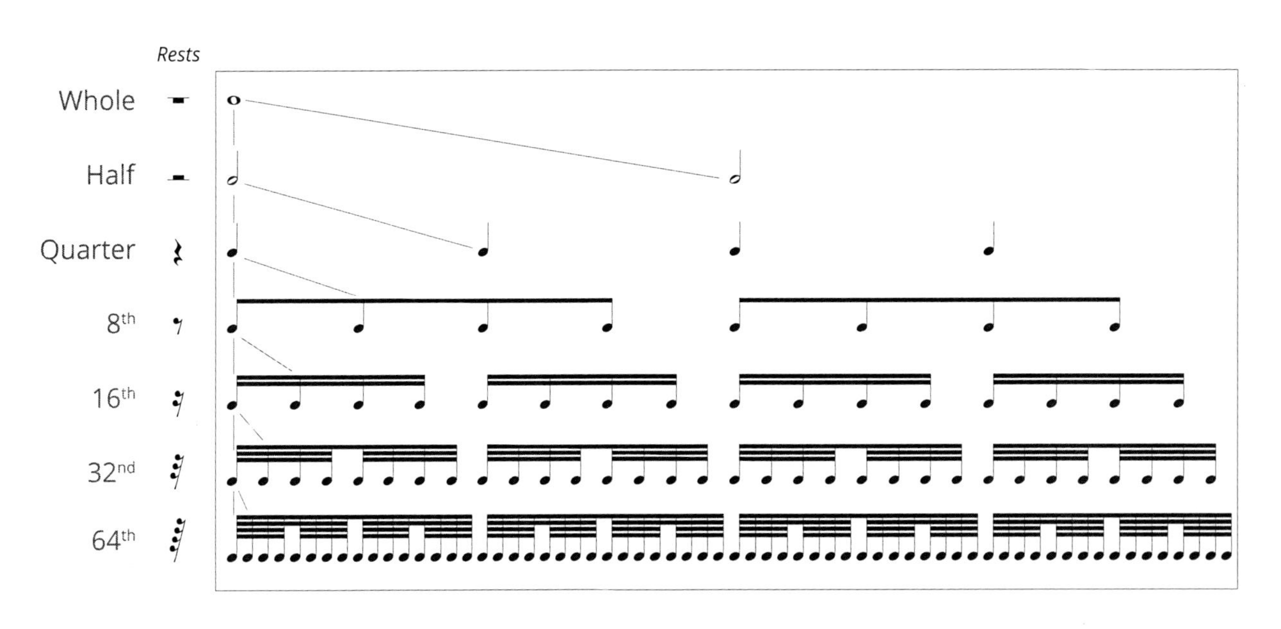

***Below,* 22.7** Adaptation of fig. 22.4.

The drawing in figure 22.7 preserves the essential layout of figure 22.4, addresses its main flaws, and adds the duration names and the symbol for the corresponding rests, inspired by figure 22.5 and online figure 22.6. The box, which does not convey any information, nevertheless reinforces the fact that all of the durations fill the same horizontal temporal space. Only the first division of each duration is shown explicitly. The lines are thin and use a different shade of gray, which bumps them into a background layer and keeps primary attention on the durations. The image presumes that the question of stem direction and the use of flags and beams will be addressed separately, as will the grouping of durations by threes and the use of dotted notes.

Sometimes we are forced to choose between two representations, but as we have seen here, sometimes we need only minimal ingenuity to have it both ways.

Chapter 23 Unmeasured Musical Time

Musical images sometimes convey a sense of ordering, in which one cannot discern anything about the timing of the events beyond which come before and which come after. In such images, even if the timing of events is unknown, the ordering must be as clear as possible.

Until the twentieth century, few examples of unmeasured time involved notated music. The unmeasured preludes of Louis Couperin (fig. 23.1) are a rare example. Although the music is read left to right, as in any other standard notation, there is considerable uncertainty about the sequence of events. The sweeping lines that signify arpeggiations help with phrasing and harmonic analysis, but how to interpret the notation for performance is far from obvious.

Figure 23.2, a schematic diagram of a three-part rondo, shows that a principal theme is followed by a transition, which is followed by a subordinate theme, and so on. We cannot infer anything further about musical time without an instantiation in a musical context, though the rotation of the words hints that we might expect the transition and retransition sections to be shorter than the thematic sections.

Figure 23.3 shows successions of events, whose durations are not known and could (theoretically) be anything. The image differs from the purely schematic figure 23.2 in that it represents actual music: the first three bars of Anton Webern's *Piano Variations*, op. 27, movement 2. The upper part of the image depicts the music's melodic line as it is heard; below that, the pitches are interpreted as a two-voice inversional canon, with the voices aligned to make their correspondence

23.1 Bauyn, II, f. 12v, reduced, Prelude de Monseiur Couperin, in Paris, Bibliothèque nationale de France, Rés. Vm7 674–675, the Bauyn Manuscript, ed. Bruce Gustafson. New York: Broude Trust, 2014, 12v.

THE FIRST RONDO FORM.

71. **The diagram of the First Rondo form is as follows:**

PRINCIPAL THEME	Transition	SUBORDINATE THEME	Re-transition	PRINCIPAL THEME	CODA
Any Part-form.		Different key, usually (next)related. (Codetta).		As before, possibly modified.	

23.2 Percy Goetschius, *The Larger Forms of Musical Composition: An Exhaustive Explanation of the Variations, Rondos, and Sonata Designs, for the General Student of Musical Analysis, and for the Special Student of Structural Composition* (New York: G. Schirmer, 1915), 94.

clearer. The author notes that the contextual inversion operator (*I*) "represents" not only pitch-class inversion about A but also the eighth-note rhythmic relationship between *dux* (the line starting with B♭) and *comes* (the line from G♯). The image offers a characteristically elegant theoretical explication of the passage, one whose clarity benefits from the omission of temporal information. Because the nodes map one-to-one onto events in the score, correspondences to the notation are easy to make. One might question the need for circles around the note names because they do not convey information. The light re-treatment in online figure 23.4 further invigorates the image.

23.3 David Lewin, *Generalized Musical Intervals and Transformations* (New Haven, CT: Yale University Press, 1987), 191.

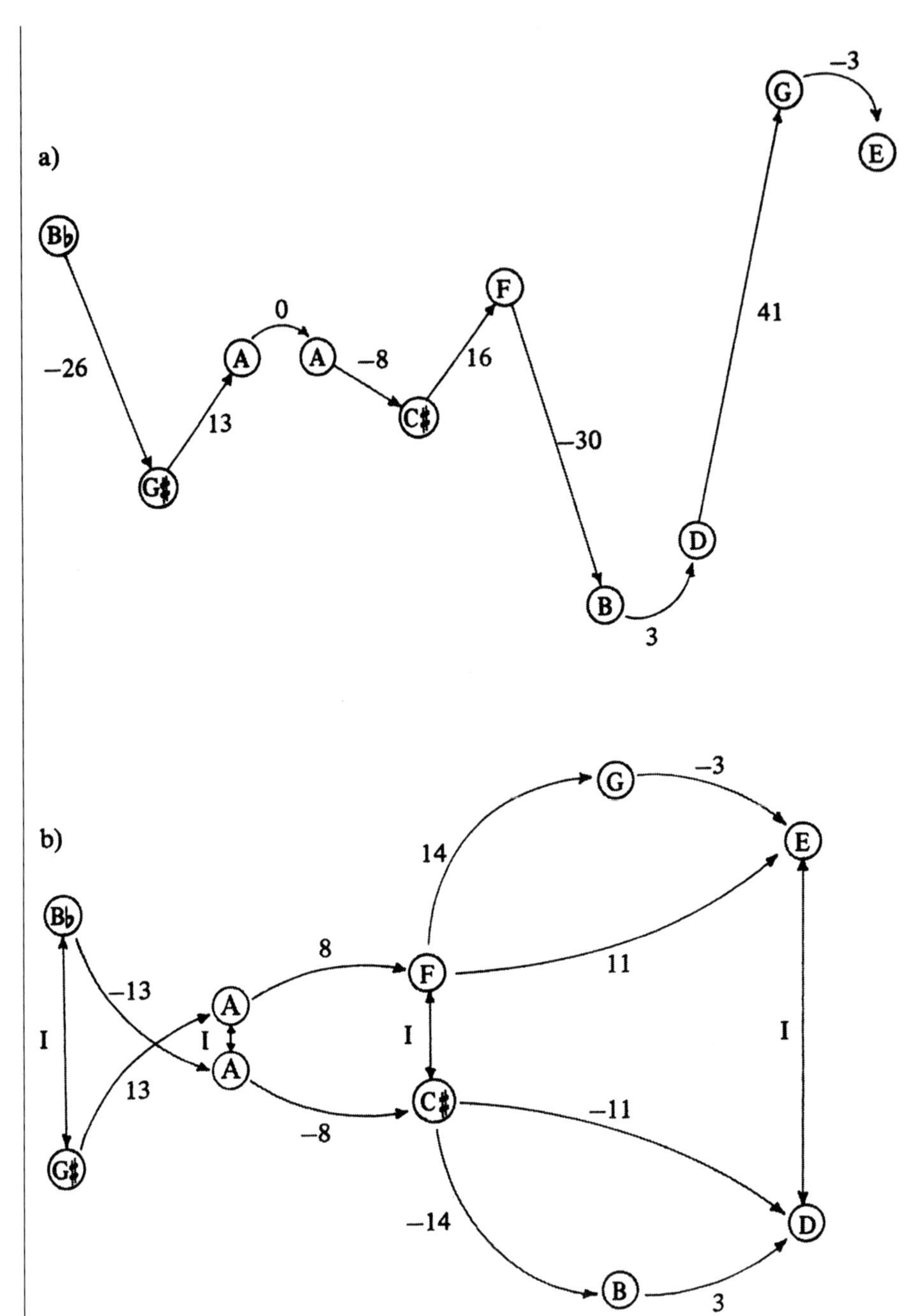

Note the care with which the arrows have been drawn in the lower half of figure 23.3 in particular. The gentle S curves that trace the 13–8–11 path and its inverse are particularly striking. The choice of curved lines for transpositions and straight lines for the *I* operation effectively marks the two types of transformation.

Voice-leading reductions, of which one can find many varieties, are often unmeasured as well. Indeed, some have criticized Schenkerian analyses for omitting an accounting of musical time, though others have disputed that omission. Online figures 23.5 and 23.6 include other types of unmeasured voice-leading reductions.

A different kind of unmeasured chronology appears in images of style successions, such as figure 23.7. While there are some references to specific years, overall the points in the image have no anchors, contextualized only by one another. I will come back to images of this kind in chapter 51.

It is reasonable to omit temporal information from a music visualization when that information does not add value (or doesn't actually exist). But since time is such an integral part of music, one should omit it only by intention, not through neglect.

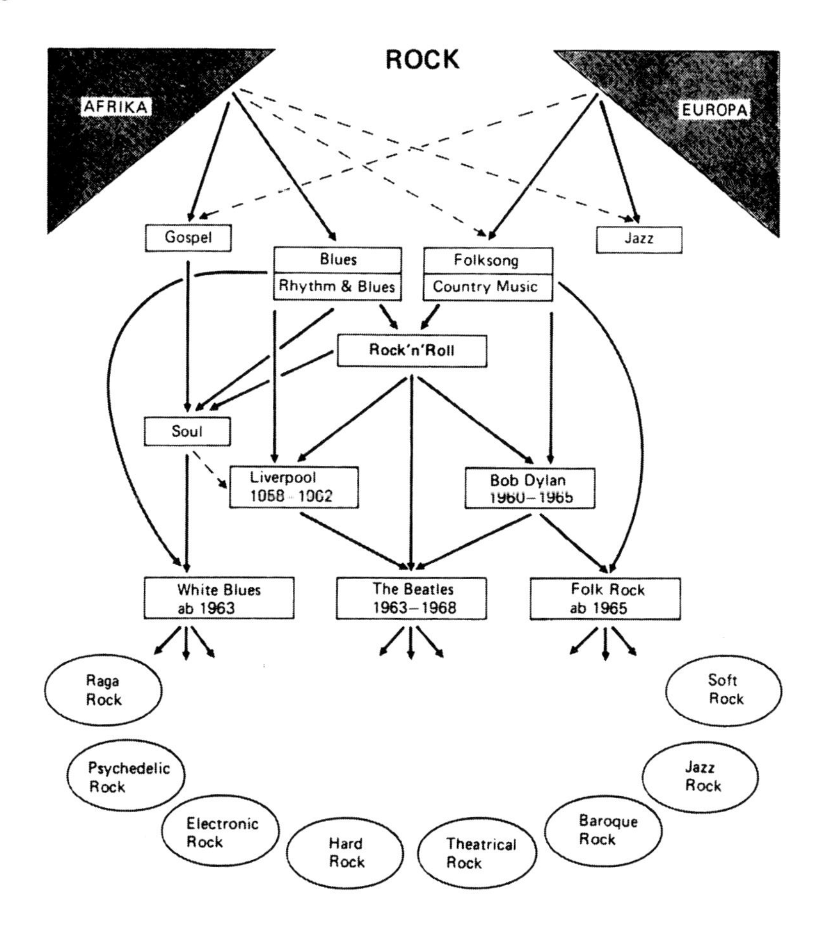

23.7 Siegfried Schmidt-Jones and Barry Graves, *Rock-Lexikon* (Hamburg: Reinbek, 1973), cited in Werner Breckoff et. al., *Musik Aktuell: Informationen, Dokumente, Aufgaben* (Kassel: Bärenreiter, 1971), 242.

CHAPTER 24 Musically Measured Musical Time

We mark the passage of time in many ways, from the extraordinarily precise (atomic clocks won't lose a second for over a million years) to the approximate ("I'll see you next week") and the undefinable ("when pigs fly"). The ubiquity of timepieces in our environment—on kitchen appliances, on our walls, in our cars, and on our computers and cell phones—may make it seem as though our lives are measured precisely by the clock or by its more slowly moving cousin, the calendar. Yet we often mark the passage of time by the moments that make up our lives, even if those moments occur irregularly by clock time. The markers "first frost" or "when the daffodils bloom" may vary by days or even weeks from one year to the next. One's first broken bone, one's first date, or the birth of one's first child occur at different points in people's lives. And although many sports have clocks that govern their internal divisions, in many, the clocks start and stop, and we are more likely to think of a game's structure in terms of its internal organization (halves, quarters, periods). Bowling is measured only by frames, and softball and baseball by outs and innings. We use these internal divisions to measure progress through a game. For instance, baseball's seventh-inning stretch occurs at (precisely!) the end of six and a half innings, whether that comes ninety minutes or three hours after the first pitch.

Likewise, music is frequently measured in terms of things that happen *in* the music. Beats and the aptly named measures are the most obvious of these, but groupings such as phrases, sections, or variations might also be used. This chapter considers the use of music's internal features to measure its unfolding. We will ask what is being measured in the images, by what means, and how this measurement interacts with other kinds of measurement, including physical measurement on the page.

Figure 24.1 helps illustrate the difference between the unmeasured musical time discussed in the previous chapter and this chapter's focus, musically measured time. Above the opening of Schoenberg's *Fourth Quartet* are the tone-row positions of the violin melody, which are numbered 1–12. The tone-row positions for the pitches of the three accompanying lines are below the staff. The image lists order numbers merely in sequence. Since it specifies no time, they are unmeasured. We know only that in the cello, the fifth note of the row is followed by the eighth, then the twelfth, and so on. However, the music itself is measured by the durations specified in the notation. We know that the first violin's opening half notes will each last four times as long as each of the four eighth notes that follow—in musical time. If measured by a clock, the proportions will vary on the basis of both the timing decisions and the physical limitations of the performers.

24.1 Werner Krützfeldt, "Polyphonie in Der Musik Des 20. Jarhhunderts: Die Logik Der Linie," in *Musiktheorie*, ed. Helga De la Motte-Haber and Oliver Schwab-Felisch, 311–34 (Laaber: Laaber, 2005), 323.

Musically measured time is only very crudely suggested by the spacing of notes in the printed score. While typesetters generally leave more space after notes of longer duration, they do not do so proportionally. The primary factor governing horizontal spacing is the need to keep notes "in order" as they are read from left to right. Figure 24.2 illustrates this dramatically, giving the 512th and 1,024th notes at the end of the example (which have one more beam than they should!) roughly the same width as the sixteenth notes in the middle system, even though they last only 1/32 and 1/64 as long.

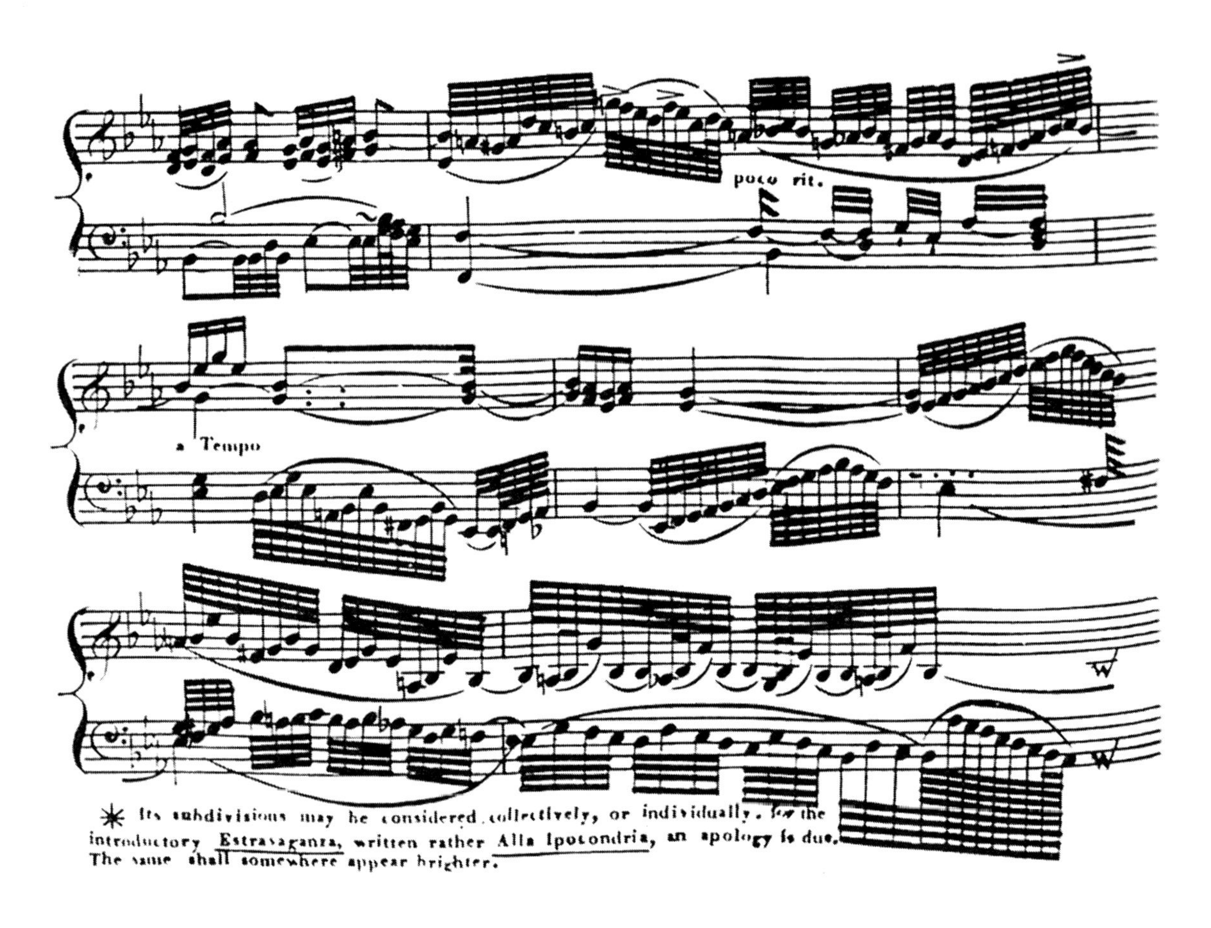
poco rit.
a Tempo
Its subdivisions may be considered collectively, or individually. for the introductory Estravaganza, written rather Alla Ipocondria, an apology is due. The same shall somewhere appear brighter.

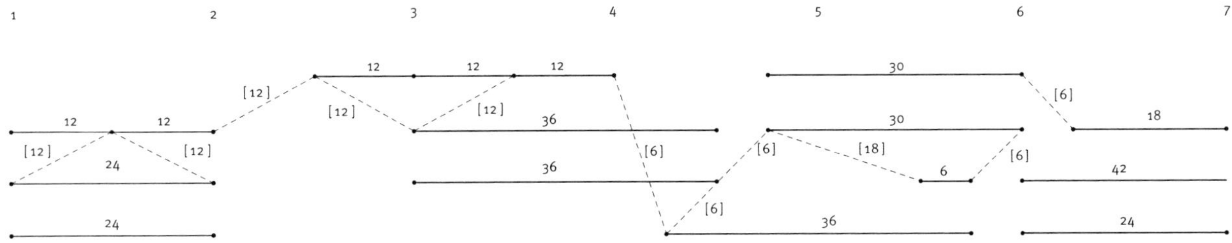
1 2 3 4 5 6 7
12 12 12 30
[12] [6]
[12] [12] 18
12 12 36 30
[12] [12] [6] [6] [18] [6]
24 6
36 42
[6]
24 36 24

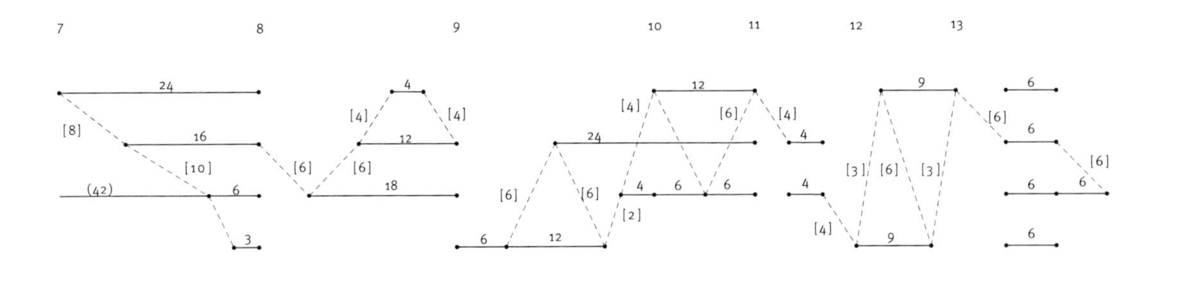
7 8 9 10 11 12 13
24 4 12 9 6
[8] [4] [4] [4] [6] [4] [6]
16 12 24 4 6
[10] [6] [6] [3] [6] [3] [6]
(42) 6 18 [6] [6] 4 6 6 4 6 6
[2]
3 6 12 [4] 9 6

It is not hard to display durations proportionally, however, since simple ratios can represent most durations. Figure 24.3 does this for the fifth of Anton Webern's *Bagatelles for String Quartet*, op. 9. Here, the unit of measure is one-twelfth of a quarter note, the smallest unit that allows all durations in the passage to be represented by integers. The figure includes a line segment representing the duration of each pitch based on that unit, and the line segments are drawn proportionally.

The numbers are essential here because we are not good at judging the relative lengths of the lines, unless they are aligned at one end. We find it easier to tell the difference between, say, the durations 24 and 16 in m. 7 than between 30 and 36 in mm. 4–5. Taken out of their music notational context, the proportions of the durations are easier to see here than in figure 24.1.

Another common unit by which we measure musical time is . . . the measure! In most Western music, measures mark off regularly spaced segments of music that articulate a recurring pattern of accented and unaccented pulses (reinforced by a meter signature). Indeed, if one picks up a piano score from the decades surrounding 1800, one will commonly see the bar lines aligned all the way down the page, with four or five measures on each system.

In analytical images, measure numbers often serve as the link from image to music. While in figure 24.3 the first nine measure numbers are spaced with the same precision as inches on a ruler, more often, the spacing of measure numbers varies. Figure 24.4, like many form diagrams, measures the music's structure with bar numbers. However, the image spaces the bar numbers for purely typographical reasons: see how irregular the distance is between the mathematically equidistant bar numbers 1, 5, 9, 13, and 17 and how similar the spacing is in the sequence 17–19–29–31, which represents spans of two, ten, and two measures. Spacing for typographical convenience distorts the distances involved and provides a skewed sense

***Facing top*, 24.2** Anthony Phillip Heinrich, "Toccata Grande Chromatica," from *The Sylvia*, set 2 (n.p., 1825).

***Facing bottom*, 24.3** Allen Forte, "Aspects of Rhythm in Webern's Atonal Music," *Music Theory Spectrum* 2 (1980): 94.

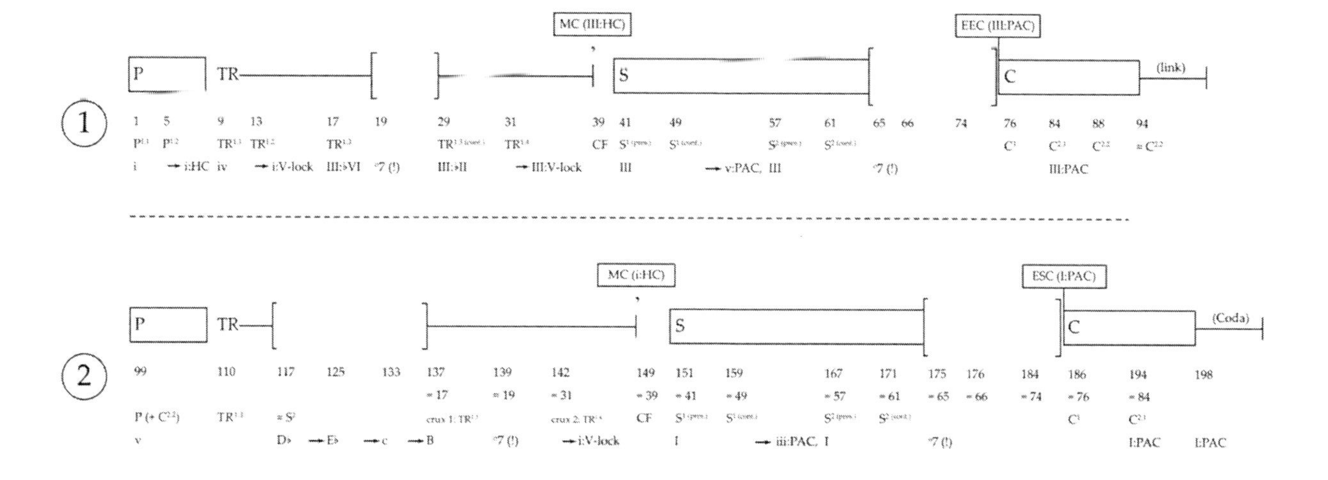

24.4 Andrew Davis, "Chopin and the Romantic Sonata: The First Movement of op. 58," *Music Theory Spectrum* 36, no. 2 (2014): 288.

set	bar	notes
R_0	1–6	
RI_0	7–13	
P_0	13–16	B♭–C is taken by the voice (bs 15–16)
RI_0	17–21	
R_0	22–6	F♯ is taken by the voice (b.24)
RI_0	27–31	
P_0	31–6	
I_0	37–41	D is taken by the voice (b.41)
R_0	42–6	
I_0	46^2–52^1	
R_0	52^2–6	This is interspersed with the residue of RI_0 from the voice:
		R_0: 52^2, G♯–A; 53^1, C; 53^2, B♭–B; 54, D–F–C♯
		RI_0: 52^2, A–C♯–G♯; 53^3, B–B♭–G
RI_0	57–60	
R_0	61–4	F–F♯ is taken by the voice (b.63)
I_0	65–9	
R_0	70–3	C♯ is taken by the voice (bs 71–2)
I_0	74–8	

These sets are grouped to show the periodic return of the opening set in the piano part, R_0. There is a regular alternation of P or R with I or RI sets throughout.

24.6 Christopher Wintle, "Webern's Lyric Character," in *Webern Studies*, ed. Kathryn Bailey (Cambridge: Cambridge University Press, 1996), 255.

of temporal space. (See also online fig. 24.5.) I will look at issues in the depiction of musical proportion in chapter 26 but will note for now one should generally strive for proportional representation, even when that is not an image's explicit aim.

One can also use structural features that are more abstract than beats or measures to measure music. The next two images show ways in which structural elements of a serial work serve as measuring sticks. In figure 24.6, although measure numbers are provided, the completion of row forms in Anton Webern's *Drei Lieder*, op. 25, no. 3, truly measure the music. The note at the bottom of the image points out that the lines are grouped on the basis of the recurrence of row form R_0, though the spacing is not sufficient to make the groups visually obvious. Increasing the space and adding a thin gray line between them would make the point clearer (see the redrawn version in online fig. 24.7). Online figure 24.8 measures time in a movement by Luciano Berio by cycles through the movement's twenty-one-note row.

All kinds of musical units might serve as effective measures of musical time. Spending a little time considering alternatives can yield delightful results. In executing the visualization, ensure that the units are clearly marked and weigh consciously whether correlating those measures with spatial proportion would improve understanding.

Chapter 25 Externally Measured Musical Time (Performance Timing)

When performed, music escapes its own realm and interacts with the physical world. Musical time thus interacts with other times: the time of our lives and the time*keepers* of our lives. We do not lose the measuring sticks of music—we still measure beats, measures, sections, movements—but we can now relate music to other temporal measures, including the "intentional" bodily counters of swaying, walking, marching, and dancing; the involuntary bodily counters, heartbeat and breathing; and the extrabodily marker, the clock. This chapter focuses on the last.

Except for synthesized music, instantiated music is performed music, so the study of externally measured musical time is almost by definition the study of performance timing. The images here blend musical measurements and clock measures in quite varied and sometimes brilliant ways. Most measure time along the horizontal axis, while the vertical axis represents something different. The central decision in any such image is which time scale to prioritize: one in which musical time (beats or measures) is proportionally spaced or one in which clock time (minutes or seconds) governs. Figure 25.1 helps us compare these approaches.

The upper half of figure 25.1 lists features of a performance by Thelonious Monk, measured in *musical* time in the sense the term was used in the previous chapter. It lists measure numbers across the top and indicates the appearance of various features in the place they occur within each measure as transcribed,

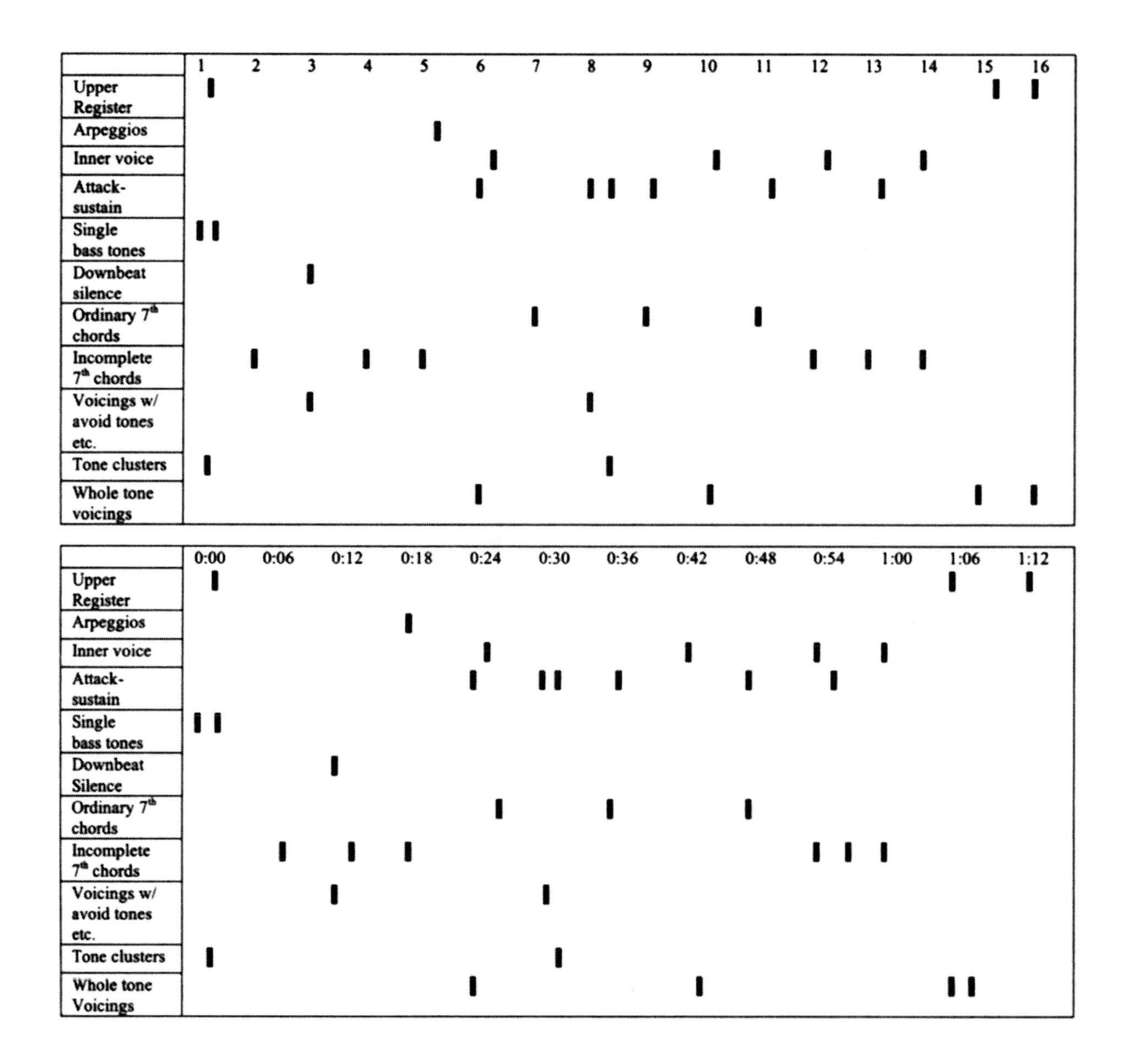

25.1 Evan Ziporyn and Michael Tenzer, "Thelonious Monk's Harmony, Rhythm, and Pianism," in *Analytical and Cross-Cultural Studies in World Music*, ed. Michael Tenzer and John Roeder (New York: Oxford University Press, 2011), 179.

quantized to align with the understood meter. The image's lower half represents the same performance in *clock* time, measured in minutes and seconds. The intent of the two-part image is to show how performance timing deviates from a strict metrical interpretation.

In this layout, however, it is all but impossible to see the interplay between these two measurements since musical and clock time are not dramatically different. To compare the images, it helps to overlay them, as in figure 25.2. Black boxes indicate the measurement of events in musical (notated) time, while white ones show their

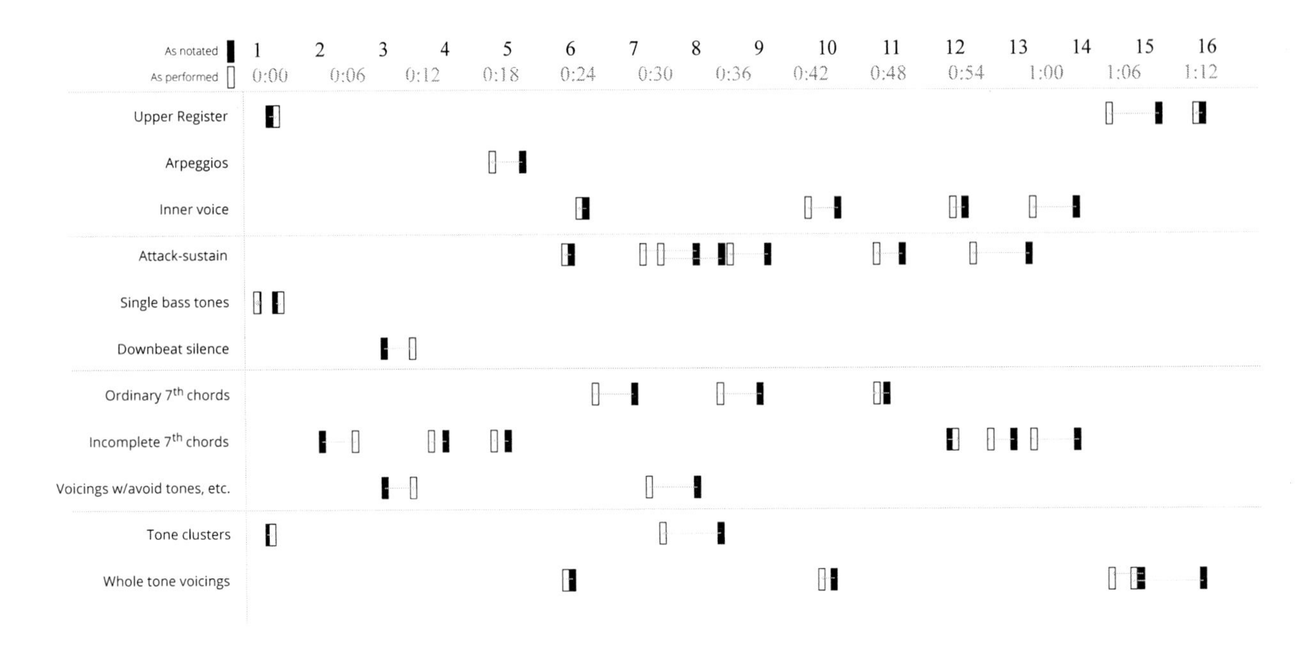

25.2 Redrawing of fig. 25.1.

placement in clock (performance) time. Arrows indicate the direction and distance of Monk's performance deviations from a strictly metronomic performance. The image takes the liberty of removing the outer box, placing the text categories on single lines in a sans serif font, and then spacing the lines evenly. Faint horizontal gridlines make it easier to see which categories the boxes at the right end belong to.

Examined earlier, figure 11.2 (colorplate 3) diagrams the form of the first movement from Beethoven's *Sonata in F Minor*, op. 2, no. 1. The horizontal timescale is based on a performance by Richard Goode (Elektra Nonesuch 9 79328–2). While the figure provides both measure numbers and time indices at the start of each formal division, it is the latter that determines the width of the bubbles.

Figure 25.3 cleverly marries musical and clock time. It interprets an excerpt from a performance of "Desimonae" by saxophonist Wessell Anderson, using a circular representation. In both the left and right images, the twelve o'clock position marks the start of a beat. One counterclockwise trip around the circle traverses one full beat. The bottom of the circle therefore represents a half beat, and the three o'clock and nine o'clock positions are quarter beats. The grid clearly measures musical time. But the performance is mapped using clock time because the time it takes to make a trip around the circle depends on the tempo selected for the analysis. Distance around the circle thus precisely measures fractions of a beat, whatever its length. All events that are part of a given beat have the same shape (open circle for beat 1, closed square for beat 2, etc.). The distance from the center indicates the duration of a note (that is, the time before the following event). In the left image, a note whose duration is precisely a sixteenth note would lie on the solid line, with

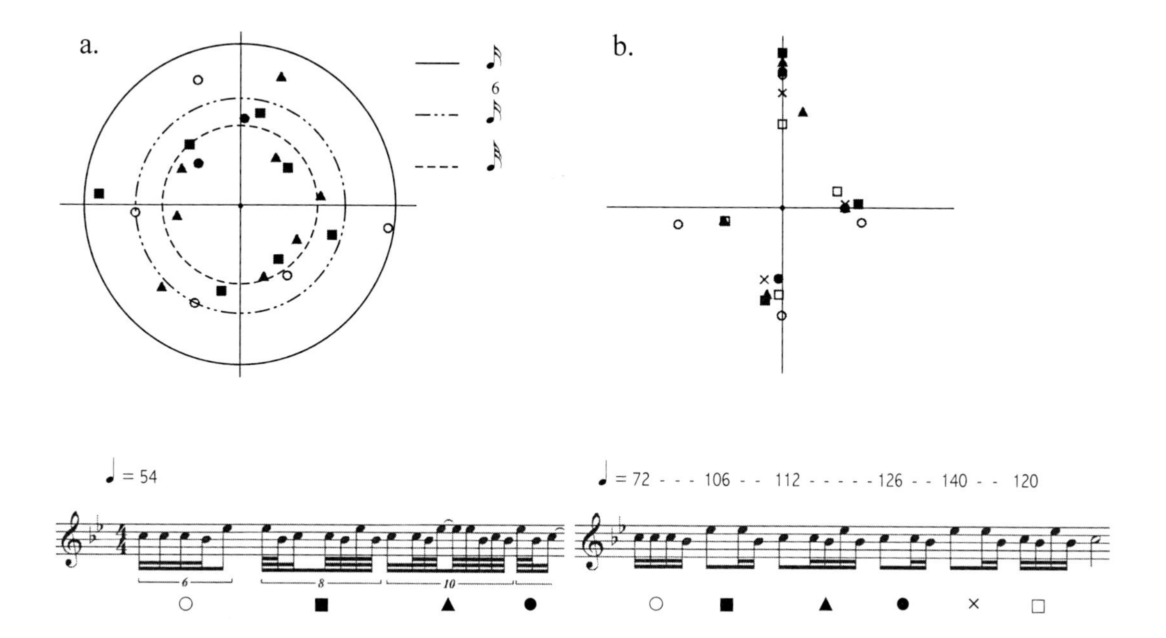

shorter durations plotted closer to the center. As in figure 25.1 (and its redrawing in fig. 25.2), discrepancies between strict musical time and abstract performance time are measured by the discrepancies between the music-measuring grid (the segments of the circle) and the placement of performance events on it. One of the beautiful aspects of this representation is that it allows for an interpretation of the performance using changing tempos. The left image transcribes the passage as consisting of four beats played at a fixed tempo (quarter = 54). The events are distributed scattershot, as if by darts thrown at a board. The right image transcribes the same passage as consisting of six beats with a shifting tempo (transcriptions showing both interpretations lie below the circles). The beat-by-beat tempo changes used in the right image allow an interpretation in which the performance is revealed to be much more regular than it initially appears. Online figures 25.4, 25.5, and 25.6 explore other approaches.

We saw in chapter 12 that graphical representations are often superior to purely numerical representations. We see this again in a performance-timing context. Figure 25.7 traces moment-to-moment tempo changes in an extended recorded "negotiation" between a seller and a would-be buyer among the Yanomami Indians of Venezuela, a custom known as Himou. The rhythmical speech makes it possible to discern a pulse, and therefore pacing—that is, tempo. Throughout the interaction, the tempo changes as the dynamics of the negotiation unfold. From a thirty-eight-minute segment of the recording, tempos at 143 distinct time points were determined, a portion of which is shown in the figure (see especially the third column, "Tempo [actual] in seconds"). The full table provides some 600 data points,

25.3 Fernando Benadon, "A Circular Plot for Rhythm Visualization and Analysis," *Music Theory Online* 13, no. 3 (September 2007): fig. 3, http://www.mtosmt.org/issues/mto.07.13.3/mto.07.13.3.benadon.html.

Running Time	Tempo Points in Sequential Order	Tempo (actual) in seconds	+ Predicted Tempo Value (seconds)	+ Residual Value (Seconds)
Segment A				
	1	.950	.801	.149
	2	.950	.785	.165
	3	.910	.770	.140
	4	.900	.754	.146
	5	.870	.739	.131
	6	.850	.725	.125
	7	.800	.710	.090
	8	.760	.696	.064
	9	.730	.682	.048
	10	.690	.669	.021
	11	.720	.656	.064
	12	.690	.643	.047
	13	.660	.630	.030
	14	.630	.618	.012
	15	.580	.605	-.025
	16	.530	.593	-.063
	17	.510	.582	-.072
	18	.490	.570	-.080
	19	.570	.559	.011
	20	.630	.548	.082
	21	.580	.537	.043
	22	.570	.526	.044
	23	.540	.516	.024
5:59	24	.480	.506	-.026
6:05	25	.500	.496	.004
6:17	26	.490	.486	.004
6:39	27	.480	.476	.004
	28	.440	.467	-.027
	29	.430	.457	-.027
	30	.410	.448	-.038
	31	.430	.440	-.010
	32	.500	.431	.069
	33	.470	.422	.048
	34	.430	.414	.016
	35	.420	.406	.014
	36	.390	.398	-.008
	37	.330	.390	-.060
	38	.310	.382	-.072
	39	.300	.375	-.075
	40	.300	.367	-.067
	41	.300	.360	-.060
	42	.320	.353	-.033
9:50	43	.340	.346	-.006
	44	.300	.339	-.039
	45	.280	.332	-.052
	46	.260	.326	-.066
	47	.240	.319	-.079
	48	.280	.313	-.033
	49	.260	.307	-.047
	50	.250	.301	-.051
	51	.260	.295	-.035
	52	.240	.289	-.049
11:30	53	.290	.283	.007
11:52	54	.280	.277	.003
	55	.300	.272	.028
	56	.300	.267	.033
12:29	57	.260	.261	-.001
12:41	58	.260	.256	.004
12:44	59	.240	.251	-.011
12:52	60	.250	.246	.004
13:13	61	.250	.241	.009
13:23	62	.230	.236	-.006
13:35	63	.240	.232	.008
14:01	64	.230	.227	.003
	65	.210	.223	-.013
14:17	66	.210	.218	-.008
	67	.190	.214	-.024
	68	.190	.210	-.020
	69	.190	.206	-.016
	70	.260	.201	.059
	71	.250	.198	.052
	72	.250	.194	.056
	73	.270	.190	.080
	74	.200	.186	.014
	75	.200	.182	.018
	76	.190	.179	.011
	77	.190	.175	.015

25.7 David Epstein, *Shaping Time: Music, the Brain, and Performance* (New York: Schirmer Books, 1995), 346.

but because printed numbers do not convey magnitude visually, the "story" of the negotiation is entirely invisible.

The data takes on lovely narrative shape when graphed as in figure 25.8, which shows the entire part of the interaction that was timed. The lower section of the figure shows the measured beat durations with dots, with longer durations (that is, slower tempos) higher on the vertical axis. A best-fit curve plots the overall trajectory of the conversation's tempo. The ever-quickening pace of the first third of the negotiation is now apparent. The data, merely listed in figure 25.7, are interpreted here, at the top of the image, where whole-number ratios ("5:3") between notable points in the interaction document an unexpected simplicity in the tempo proportions.

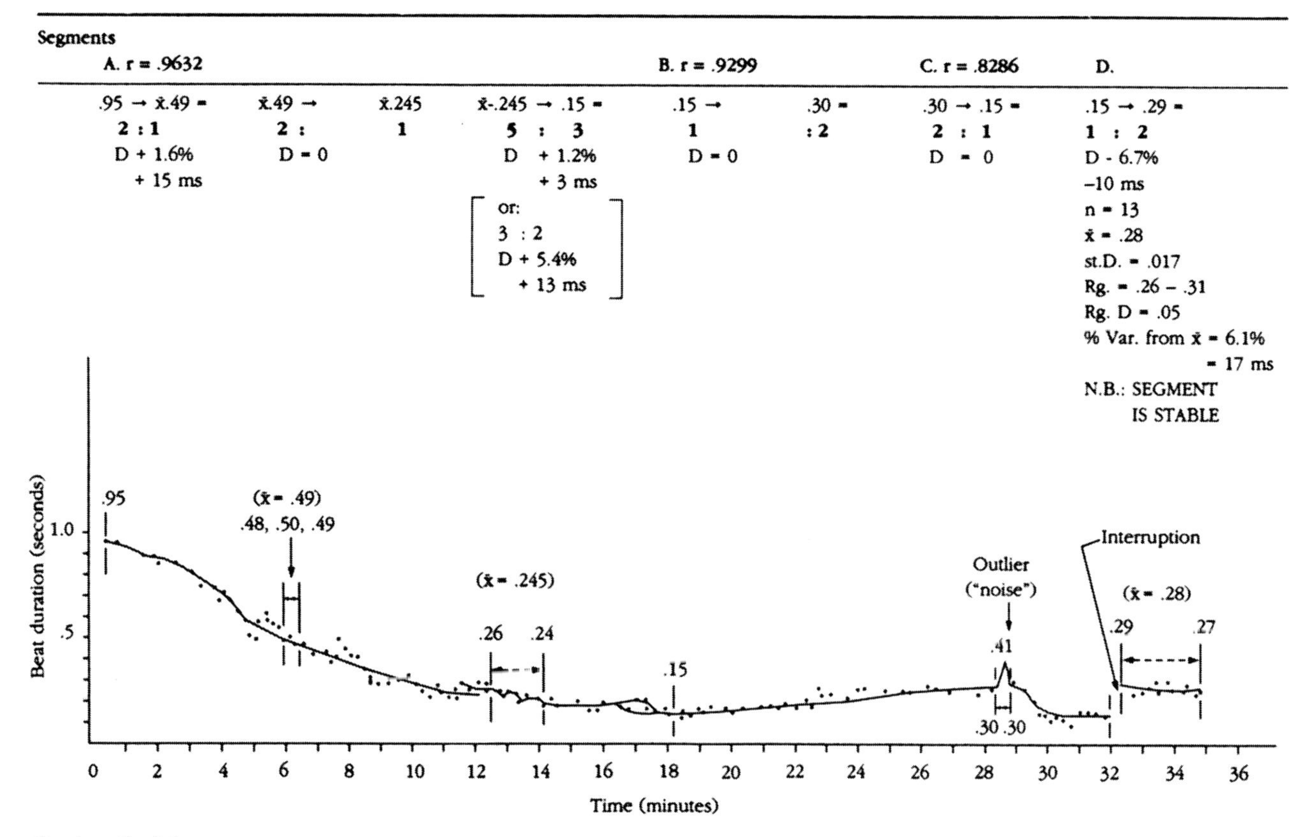

25.8 David Epstein, *Shaping Time: Music, the Brain, and Performance* (New York: Schirmer Books, 1995), 348.

Performance-timing studies are especially effective at comparing performances. For example, figure 25.9 compares recorded performances of Terry Riley's *In C*. In the work, each member of the ensemble (whose makeup is not specified in the score) plays in order through fifty-three short melodic figures, repeating each

25.9 Dora A. Hanninen, *A Theory of Music Analysis: On Segmentation and Associative Organization* (Rochester, NY: University of Rochester Press, 2012), 324, 325.

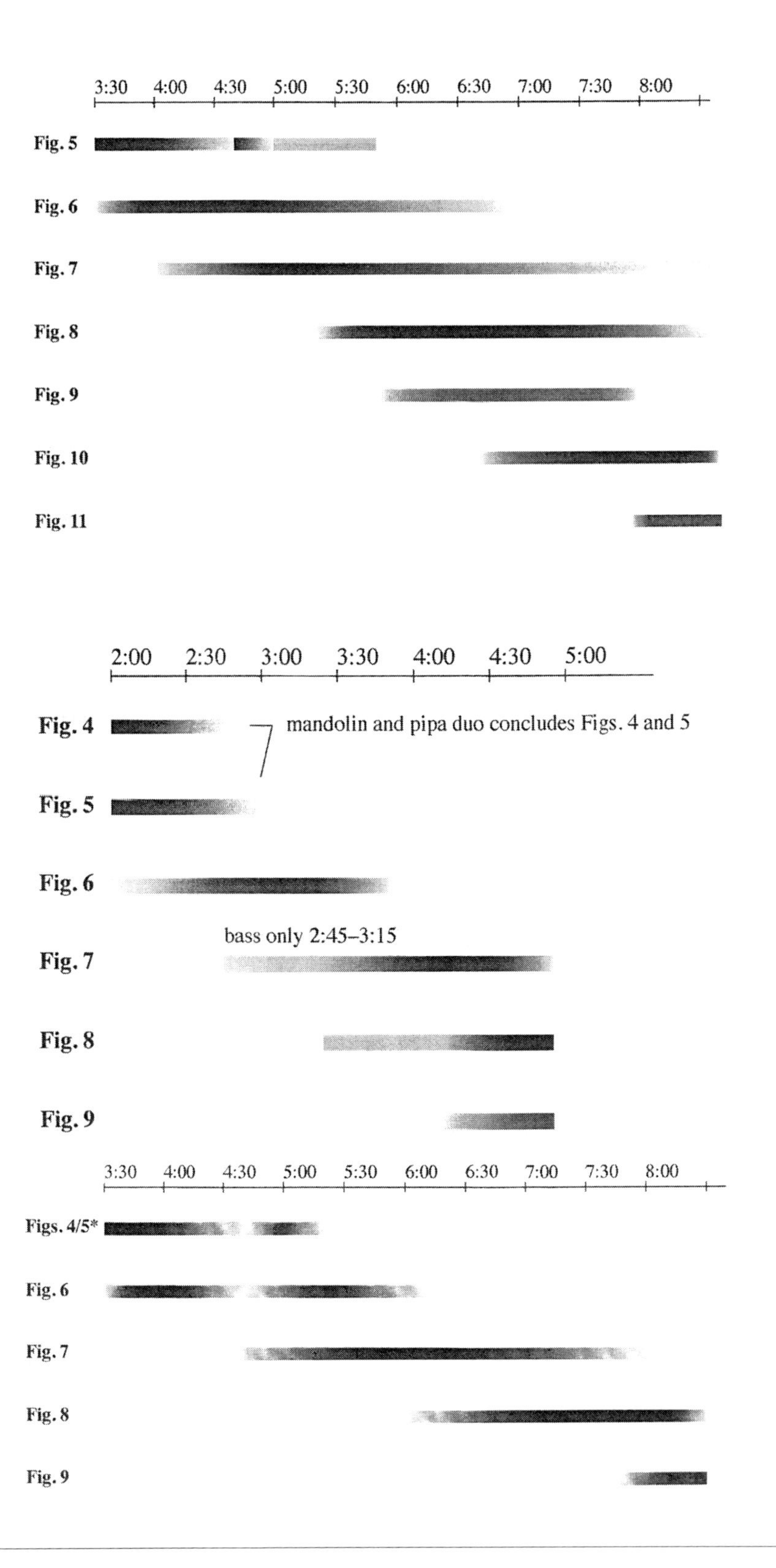

one an unspecified number of times. This leads to a polyphony whose makeup varies depending on the combination of the instrumentation and the number of times each musician decides to repeat each figure. Figure 25.9 includes images for excerpts of three performances, each beginning with the point at which melodic figures 5 and 6 are introduced. As always, combining images facilitates comparison; the larger context promotes greater understanding of each performance. Compressing the vertical space would help the individual performance images integrate more. The image would be even more instructive if it integrated the three performances visually and used equal timescales to make the comparative compression of the second performance more visually apparent. See also online figure 25.10.

Figure 25.11 and online figure 25.12, both involving Ludwig van Beethoven symphonies, provide additional evidence in favor of graphical representations over

Conductor	Basic Tempo	Tempo at m. 196
Norrington	𝅗𝅥=108	𝅗𝅥=104
Gardiner	𝅗𝅥=108	𝅗𝅥=104
Brüggen	𝅗𝅥=106	𝅗𝅥=98
Dohnanyi	𝅗𝅥=104	𝅗𝅥=96
Karajan (1984)	𝅗𝅥=104	𝅗𝅥=92
Toscanini	𝅗𝅥=104	𝅗𝅥=98 (at m.200 𝅗𝅥=90)
Hickox	𝅗𝅥=104	𝅗𝅥=
Giulini	𝅗𝅥=100	𝅗𝅥=82
Steinberg	𝅗𝅥=100	𝅗𝅥=88
Scherchen	𝅗𝅥=100	𝅗𝅥=96
C. Kleiber	𝅗𝅥=100	𝅗𝅥=92
Furtwangler	𝅗𝅥=98–100	𝅗𝅥=86
Karajan (1962)	𝅗𝅥=98	𝅗𝅥=104
Reiner	𝅗𝅥=98	𝅗𝅥=90
...	...	...
Knappertsbusch	𝅗𝅥=84	𝅗𝅥=72
Masur	𝅗𝅥=84	𝅗𝅥=80
Colin Davis	𝅗𝅥=84	𝅗𝅥=82
Ormandy	𝅗𝅥=84	𝅗𝅥=84
Mengelberg	𝅗𝅥=84	𝅗𝅥=84
Bernstein	𝅗𝅥=82	𝅗𝅥=72
Böhm	𝅗𝅥=80	𝅗𝅥=76
Kubelik	𝅗𝅥=80	𝅗𝅥=76
Klemperer	𝅗𝅥=80	𝅗𝅥=70
Leinsdorf	𝅗𝅥=78	𝅗𝅥=84 *(sic)*
Krips	𝅗𝅥=76	𝅗𝅥=76
Boulez	𝅗𝅥=74	𝅗𝅥=72
Stokowski (1940)	𝅗𝅥=86	𝅗𝅥=88

25.11 Gunther Schuller, *The Compleat Conductor* (New York: Oxford University Press, 1998), 148–49, excerpt.

purely numerical ones. Figure 25.11 focuses on a single moment in the first movement of Beethoven's Fifth Symphony. It compares how sixty-six conductors approach measure 196, the moment in the development when the primary motive is first reduced to two notes. The image provides a table showing each performance's overall tempo and the tempo at the moment in question, arranged from the fastest (overall) performance to the slowest. The layout makes analysis difficult. Differences in tempo must be calculated one performance at a time. It is hard to tell at a glance if a performance at that moment is faster, slower, or the same, and it is impossible to discern what is trend and what is outlier. Adding a third column showing the deviation (−4 for Norrington, −18 for Giulini, +8 for Strauss, etc.) and sorting by that column would help. Online figure 25.12 takes a more successful approach.

Representations of performance timing require a special kind of creativity and ingenuity, for they must always convey their information in relation to something more normative, which might be either musically measured time or some kind of clock time. These deviations often tell a musical story, and making that narrative clear is the first goal of such visualizations.

Chapter 26 Proportion

The study of proportion in music brings to mind ancient notions of simple number ratios and the study of proportion in other artistic domains, such as architecture and painting. Studying proportion in music involves the comparison of two or more numerical values, typically measurements of some kind, which might be musical measures or external measures, as described in the previous two chapters. Visualizations involving proportion need to be intentional about the measuring stick chosen to derive those values, since different choices can yield significantly different results.

The famed first movement of Béla Bartók's *Music for Strings, Percussion, and Celesta* illustrates the implications of the choice of measuring unit. Studies of proportion in this movement invariably use the notated measure as the unit of measurement because of its obvious evocation of the Fibonacci sequence (and, by implication, the golden mean). The Fibonacci sequence is a series of numbers in which the next number is the sum of the previous two: 1, 1, 2, 3, 5, 8, 13, 21, 34, 55, 89, and so on. The golden mean is the division of some number into two such that the ratio of the smaller to the larger is the same as the ratio of the larger to the whole—that is, $^{x}/_{y} = {}^{y}/_{(x+y)}$. With a little algebra, the golden mean (y in the equation) can be shown to be approximately 0.618. The ratios between adjacent Fibonacci numbers increasingly approximate this value.

Significant events in the Bartók movement all occur in measures that are part of the Fibonacci sequence, and other divisions in the movement involve Fibonacci numbers. (The movement actually has only eighty-eight measures. The number eighty-nine will be addressed in a moment.) Nevertheless, given that the meter changes sixty-seven times in eighty-eight measures, with meter signatures ranging from $\frac{5}{8}$ to $\frac{12}{8}$, one might imagine that counting eighth notes would be a more sensible basis for proportional analysis. One might also imagine that the proportion in a performance of the piece would be an important factor in the perception of proportion in the piece. Figure 26.1 shows, for all three ways of measuring, the proportions formed by the movement's major formal boundaries, mm. 21, 34, and 55, plus the overall length of eighty-nine measures. With the least precise of these yardsticks, the musical measure (shown in solid black), the proportions are extremely close to the golden mean, 0.6180, to within less than a one-thousandth. Counting eighth notes, however, the proportions (solid gray) are quite distant from the golden mean. Timings from a recording by the Detroit Symphony (conducted by Antal Dorati, recorded in 1983, and released in 1985, shown in dashed gray) are more consistent with the golden mean than counting eighth notes but not nearly as consistent as counting measures. Were it not for the many correspondences with Fibonacci numbers and Bartók's known interest in the sequence, the decision to measure by measures would be suspect. Even counting measures, however, there is some inconsistency. Key events *begin* in measures 21 and 34, and the piece ends at what would be the *start* of measure 89, while a strictly proportional division would put key points *after* the correct number of measures had completed. The dynamic high point does come at the *end* of measure 55, however.

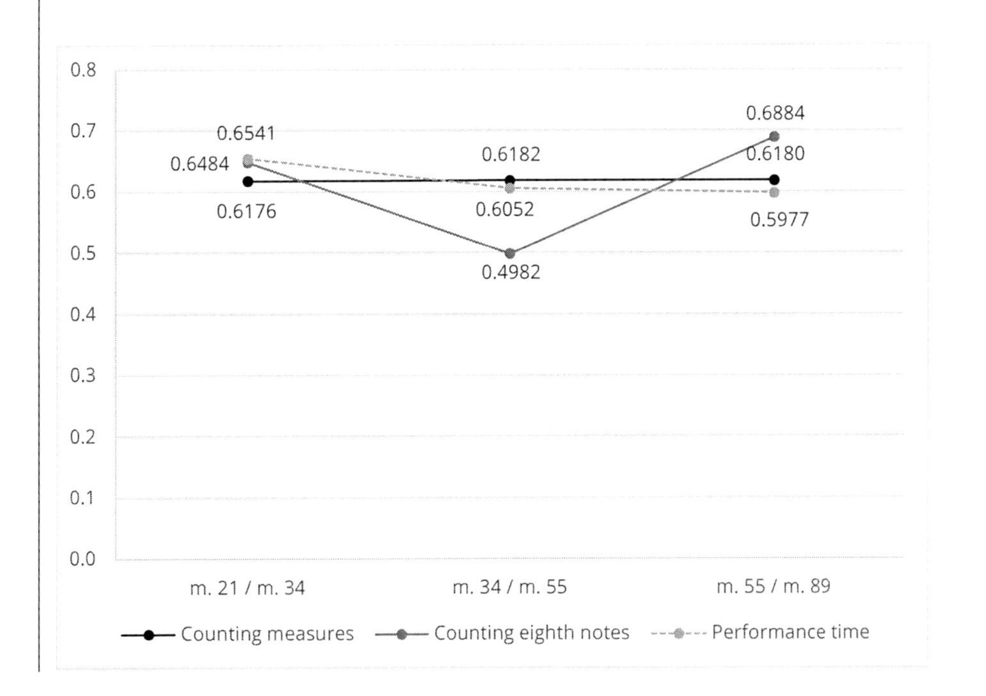

26.1 Proportions between the principal divisions in Béla Bartók, *Music for Strings, Percussion, and Celesta*, movement 1, as calculated by measure, by eighth notes, and in a performance by the Detroit Symphony (Antal Dorati). The golden mean is approximately 0.6180.

If we nevertheless accept that the measure is an appropriate basis for the study of proportion in this work because of the close alignment with the Fibonacci numbers, we can look at attempts to show those proportions visually. The depiction in online figure 26.2, seemingly produced on a typewriter using hyphens and virgules, is not very attractive. The version in figure 26.3 is more attractive and more informative, though it does not include measure numbers. It employs lines of greater height to signify divisions of different scope. The placement of the numbers 55, 89, and the second 34 is deceptive, however. Each appears directly above a vertical line, suggesting that they are markers for that line, when in fact they indicate the durations of the entire span, of which that line is the golden mean divider. They would be better if centered within that span as the first 34 is.

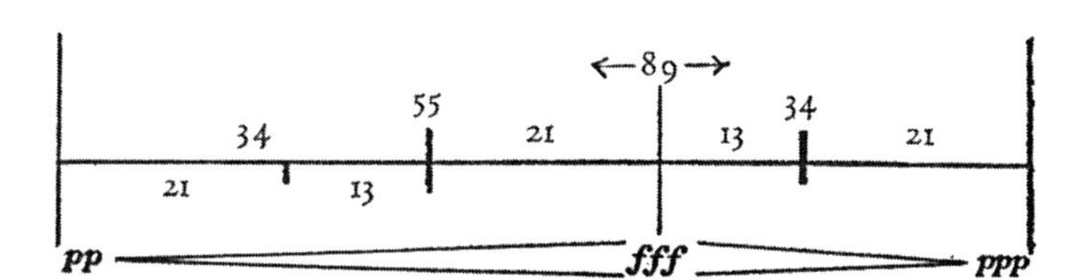

26.3 Ernő Lendvai, *Béla Bartók: An Analysis of His Music* (London: Kahn and Averill, 1971), 28.

Even when the information *is* displayed proportionally, it is important to include the numbers as well. We cannot judge distances visually with any precision, so the numbers provide evidence to support the more holistic impression provided by our vision system. In figure 26.4, the numbers are based on a unit of one-twelfth

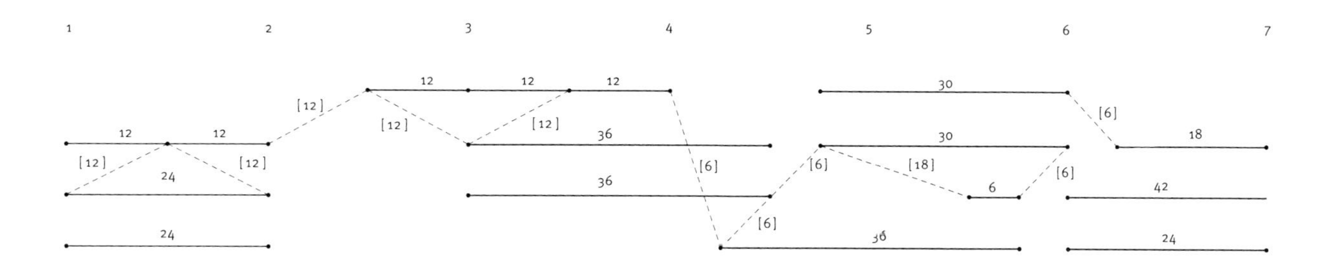

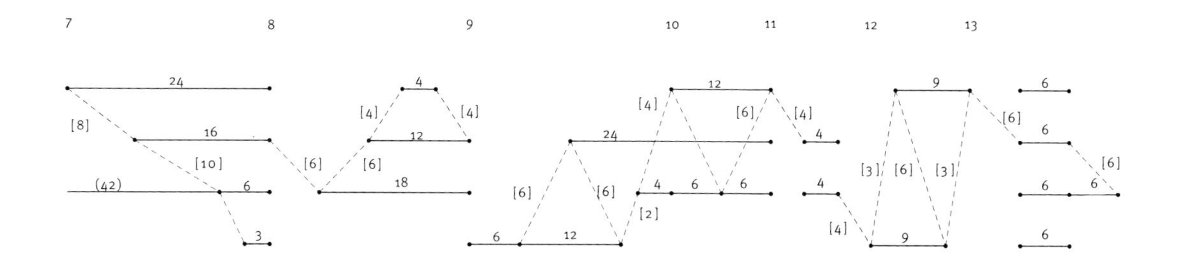

26.4 Allen Forte, "Aspects of Rhythm in Webern's Atonal Music," *Music Theory Spectrum* 2 (1980): 94.

of a beat, the smallest value that allows all durations in the piece to be represented as integers. The image locates events that are not simultaneous with others by connecting them with a dashed line and adding brackets to the duration of the gap between the end of one event and the start of another ("[6]"). The numbers help us confirm, for instance, that the violin 2 pitch in mm. 9–10 is twice the length of the violin 1 pitch in m. 10. Online figures 26.5, 26.6, and 26.7 all fail to make the proportions they represent visually apparent.

Relative to those images, figure 26.8 handles proportion more effectively. It shows the nested proportions of sections using measure numbers, fractions, and a proportionally spaced number line. Unfortunately, the image marks off these proportions with semicircles. Our minds will likely compare the *areas* of the semicircles rather than the *lengths* of their diameters, which will distort the perceived differences. The area under the largest arc is 69 times the area under the smallest, when the scale should be just 8.3.

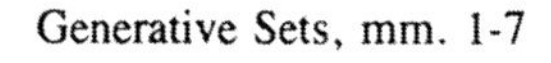

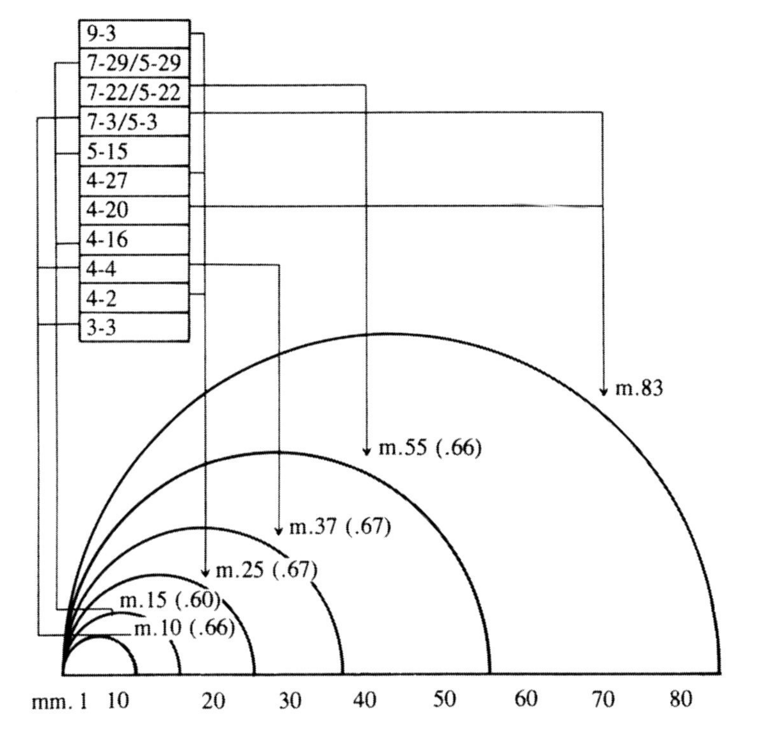

26.8 Charles M. Joseph, "Structural Coherence in Stravinsky's *Piano-Rag-Music*," *Music Theory Spectrum* 4 (1982): 89.

Figure 26.9 depicts durations proportionally. The image shows how three recurring melodic ideas, labeled A, B, and C, each repeat with nearly equal length in a movement by Igor Stravinsky. The boxes are drawn to proportion, with the numbers representing the number of beats per cycle. In the rows labeled AB, BC, and

AC, the image also shows how each idea intersects with the others; for instance, the nineteen-beat cycle of A begins seven beats after B's second twenty-three-beat cycle starts. A couple of minor adjustments help separate the individual lines from their composites: Figure 26.10 uses a gray background to distinguish composite lines, and it shortens the lines that divide segments within these composite rows to make it clearer which melody has contributed a particular divider. The row labels have also been lightly treated and the durations for the composite lines set in a lighter font to distinguish the two types of rows.

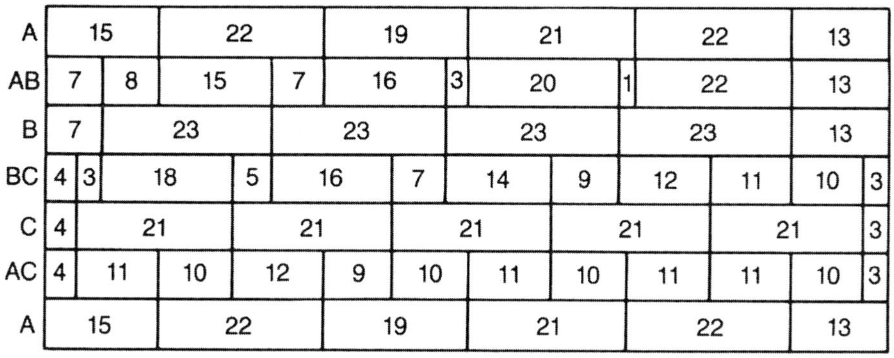

A | 15 | 22 | 19 | 21 | 22 | 13
AB | 7 | 8 | 15 | 7 | 16 | 3 | 20 | 1 | 22 | 13
B | 7 | 23 | 23 | 23 | 23 | 13
BC | 4 | 3 | 18 | 5 | 16 | 7 | 14 | 9 | 12 | 11 | 10 | 3
C | 4 | 21 | 21 | 21 | 21 | 21 | 3
AC | 4 | 11 | 10 | 12 | 9 | 10 | 11 | 10 | 11 | 11 | 10 | 3
A | 15 | 22 | 19 | 21 | 22 | 13

***Left*, 26.9** Jonathan D. Kramer, *The Time of Music: New Meanings, New Temporalities, New Listening Strategies* (New York: Schirmer Books, 1988), 291.

***Below*, 26.10** Redrawing of fig. 26.9.

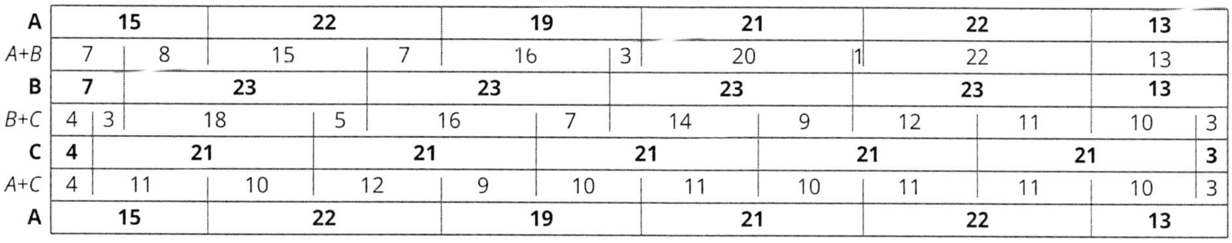

A | **15** | **22** | **19** | **21** | **22** | **13**
A+B | 7 | 8 | 15 | 7 | 16 | 3 | 20 | 1 | 22 | 13
B | **7** | **23** | **23** | **23** | **23** | **13**
B+C | 4 | 3 | 18 | 5 | 16 | 7 | 14 | 9 | 12 | 11 | 10 | 3
C | **4** | **21** | **21** | **21** | **21** | **21** | **3**
A+C | 4 | 11 | 10 | 12 | 9 | 10 | 11 | 10 | 11 | 11 | 10 | 3
A | **15** | **22** | **19** | **21** | **22** | **13**

Figure 26.11 and online figure 26.12 both depict proportion in music as performed, rather than as notated. Figure 26.11 illustrates an effective way of comparing section lengths. It shows each section of a piece on a separate line, with sections graphed proportionally (roughly; see below). Because the lines depicting sections are left-aligned, one can easily compare their relative lengths to reveal the largely symmetrical structure of the work. The image could be improved with a couple of tweaks: It would be better to put the section times at the right ends of each section rather than centering them over each line. The current placement adds clutter and makes it hard to locate the numbers. Additionally, several sections are not drawn to scale. For example, the section lasting 58.12 seconds is shorter than the one lasting 39.4 sections, and the longest section, at 291.09 seconds, should be almost twice as long as the preceding section, 153.71, but it clearly isn't.

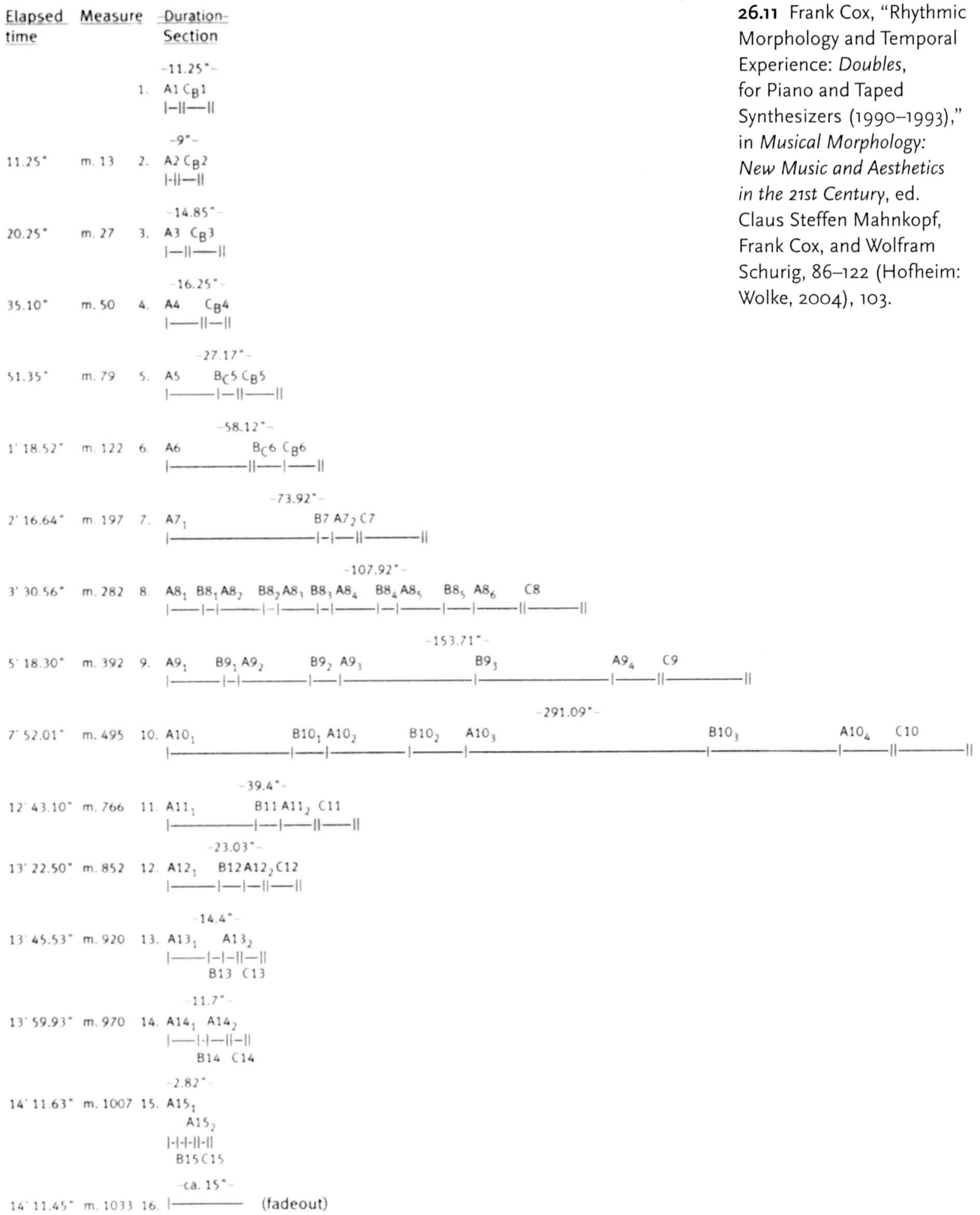

Elapsed time | Measure | Duration / Section

1. –11.25"– A1 C_B1

11.25" m. 13 2. –9"– A2 C_B2

20.25" m. 27 3. –14.85"– A3 C_B3

35.10" m. 50 4. –16.25"– A4 C_B4

51.35" m. 79 5. –27.17"– A5 B_C5 C_B5

1' 18.52" m. 122 6. –58.12"– A6 B_C6 C_B6

2' 16.64" m. 197 7. –73.92"– $A7_1$ B7 $A7_2$ C7

3' 30.56" m. 282 8. –107.92"– $A8_1$ $B8_1$ $A8_2$ $B8_2$ $A8_3$ $B8_3$ $A8_4$ $B8_4$ $A8_5$ $B8_5$ $A8_6$ C8

5' 18.30" m. 392 9. –153.71"– $A9_1$ $B9_1$ $A9_2$ $B9_2$ $A9_3$ $B9_3$ $A9_4$ C9

7' 52.01" m. 495 10. –291.09"– $A10_1$ $B10_1$ $A10_2$ $B10_2$ $A10_3$ $B10_3$ $A10_4$ C10

12' 43.10" m. 766 11. –39.4"– $A11_1$ B11 $A11_2$ C11

13' 22.50" m. 852 12. –23.03"– $A12_1$ B12 $A12_2$ C12

13' 45.53" m. 920 13. –14.4"– $A13_1$ $A13_2$ B13 C13

13' 59.93" m. 970 14. –11.7"– $A14_1$ $A14_2$ B14 C14

14' 11.63" m. 1007 15. –2.82"– $A15_1$ $A15_2$ B15 C15

14' 11.45" m. 1033 16. –ca. 15"– (fadeout)

26.11 Frank Cox, "Rhythmic Morphology and Temporal Experience: *Doubles*, for Piano and Taped Synthesizers (1990–1993)," in *Musical Morphology: New Music and Aesthetics in the 21st Century*, ed. Claus Steffen Mahnkopf, Frank Cox, and Wolfram Schurig, 86–122 (Hofheim: Wolke, 2004), 103.

When representing musical time on one of the primary visual dimensions, one can render that time proportionally. So long as laying out information proportionally does not otherwise hinder the effectiveness of an image, doing so adds a layer of useful information with minimal added effort. Determining whether the measuring stick should be beats, bars, minutes and seconds, or something else is only the first of several decisions to be made.

* * *

While music is experienced only through time, the thoughtful representation of time is sometimes overlooked in music visualizations. The fact that we measure musical time in multiple ways at the same time, by beats, measures, bodily motion, and the clock, complicates the choice of representation. While we have seen that we can sometimes overlook the explicit representation of time, we have also seen the value of representations that more explicitly measure music, whether on its own terms (beats and bars) or in relation to the outside world (minutes and seconds). The latter is essential when discussing time in musical performance. Finally, we have seen the value in creating visual representations of musical time that reflect the same proportions as the music itself.

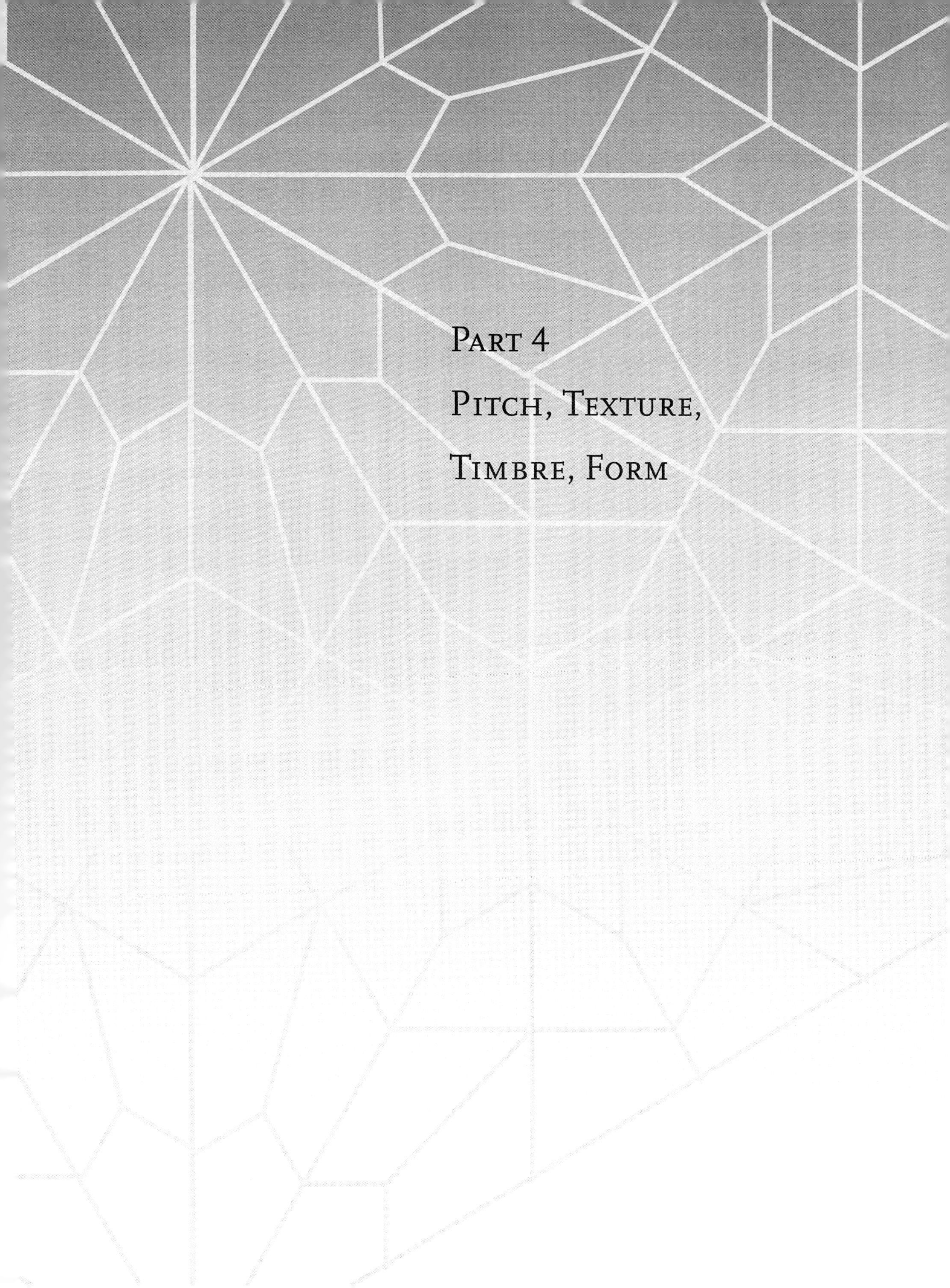

Part 4

Pitch, Texture, Timbre, Form

Now that we have considered the depiction of musical spaces and musical time, we turn our attention to the visualization of other musical parameters. The first three chapters of part 4 will look at representations of pitch that use means other than the five-line staff. Chapters 30–34 explore tuning and temperament (including microtuning), followed by timbre, texture, and the pedagogy of voice leading. After that, in chapters 35 and 36, we consider schematic images and images depicting procedural information, leading to schematic models of musical form. Chapters 37–39 examine the use of tabular information in the form of set-class tables, visualizations of instrument ranges, and the display of translations. We will see that even across such an eclectic range of musical topics, the visualization principles and design techniques discussed thus far remain relevant. These include leveraging pop-out effects, facilitating comparisons, employing appropriate metaphors, eliminating unnecessary ink, and so on.

Chapter 27 Textual Representations of Pitch

One of the challenges in writing about music is that creating musical notation and incorporating it into printed text can require substantial effort. At times, the visual apparatus involved in notation seems to require more effort than necessary. This chapter looks at examples in which music has been stripped of its graphical trappings and rendered simply as text. A textual representation of pitch exchanges the note-as-object metaphor for a note-as-word metaphor because it requires processing by the brain's language center. Sometimes this change is benign, and sometimes it has unfortunate consequences, as we will see.

The tonic sol-fa method shown in figure 27.1 uses solfège syllables, reduced to a single letter, together with a rudimentary rhythmic notation using punctuation marks, much of which could be rendered on a typewriter. Though the system supports only relatively simple music, for those trained in solfège, it would facilitate accurate performance. However, although it would not have been known at the time the system was developed, the textual pitch notation is likely processed by the same language region of the brain that processes the text itself. Unfortunately, the brain cannot effectively do both at the same time. Shape-note notation, described in chapter 29, addresses this problem by replacing the text-based solfège representation with a graphical one.

A number of early analyses of atonal and serial works use numerical representations of pitch to reduce the amount of clutter added by the notational apparatus. Figure 27.2 provides a typical example. Identifying the pitch classes by number

27.1 John Curwen, *Teacher's Manual of the Tonic Sol-Fa Method* (London: Tonic Sol-Fa Agency, 1875), 112.

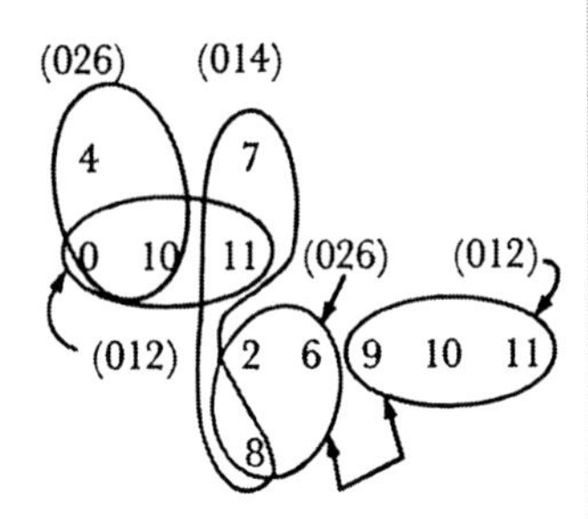

27.2 Gary E. Wittlich, "Sets and Ordering Procedures in Twentieth-Century Music," in *Aspects of Twentieth-Century Music*, ed. Gary E. Wittlich, 388–476 (Englewood-Cliffs, NJ: Prentice-Hall, 1975), 465.

(C=0, B=11) suffices for the analysis, whose primary intention is to identify embedded pitch-class sets. Here, as in online figures 27.3 and 27.4, we see that any time music notation is suppressed, people may forget that the relationships shown refer to music.

Figure 27.5 provides a more detailed image along the same lines. As with online figure 27.4, one will find it most useful with a score handy. The layout is rudimentary, clearly done on a typewriter, yet the information density is high. Note how effectively use of even a regular and italic "font" helps separate the parallel lines of information. The design works in part because the image is concerned with pitch classes, not pitches, and because it treats the pitch information as no more important than that of several other dimensions. The analysis would benefit from additional layering of the information, to help organize it in some fashion. While the image provides enough information for viewers to link it back to the score, a score-based analysis would work better.

There are not many times when a primarily textual visualization of pitch is superior to one involving graphic symbols of some kind.

27.5 Andrew W. Mead, "Detail and the Array in Milton Babbitt's *My Complements to Roger*," *Music Theory Spectrum* 5 (1983): 97.

```
Measure:                1                 2                   3
Number of sixteenths:   3  2  4   3       2    4     1  4  1  3     1  1 4      1 2
Dynamics:               p        (p)          pp     f  pp mf f     p    mf     p
Pitch classes:          5/80---/e3t67/9410-  ----0 /t367e2/925/23e--/83-  9t164--/2e6/8/-----573-/1/50----
Array location (mm.):   R16-15  30   15         32-33   3  6   1  16    28—30 R20    3-4
Secondary details:              mf                      ppp                        pp
                                t2                    --418                  - -  6
                                24                      2                    28

4                       5                  6                   7
3    1 1 3    1  3      1  2   1 1  1 1  1 3     1  1  3    2  2    2  2     2     2  1 2    1  4
p             mp        f      mpmf p f   (pp)   p  p       mf pp     f      p     f  pp     f
--e6t/t3/1/508---/-9/7  -t/07--/-0/t9-/6/61/1/-38e04--/0  -8/104---/925/23e--/--5/6-t7--  23e--/-83/1/e3t67/74/-5970---
6     2-3--4    R33-32  R24-23   R9   10-11-12            5-6    3   6     3-4    6     22  28-30  21 28
          (p)    mf     mf     p pp
          -081--/-2/94---  52  7 5
          R14    31-32  R15    R16
                               p ppp
                               7 8
                               5
```

Chapter 28 Piano Roll Notation

The piano roll, which mechanically controls a player piano (fig. 28.1), also provides a visual representation of the music it encodes. Perforated regions in the paper scroll encode chromatic pitch in one direction and onset time and duration in the other. If we were to try to read a mounted piano roll as notation, we would find that pitch matches the orientation of the piano, lower to the left, higher to the right. Because the roll moves from the top roller to the bottom, time moves from bottom to top. If the scroll were rotated ninety degrees and unfurled so one could read time left to right, then the pitch dimension would be inverted, with low

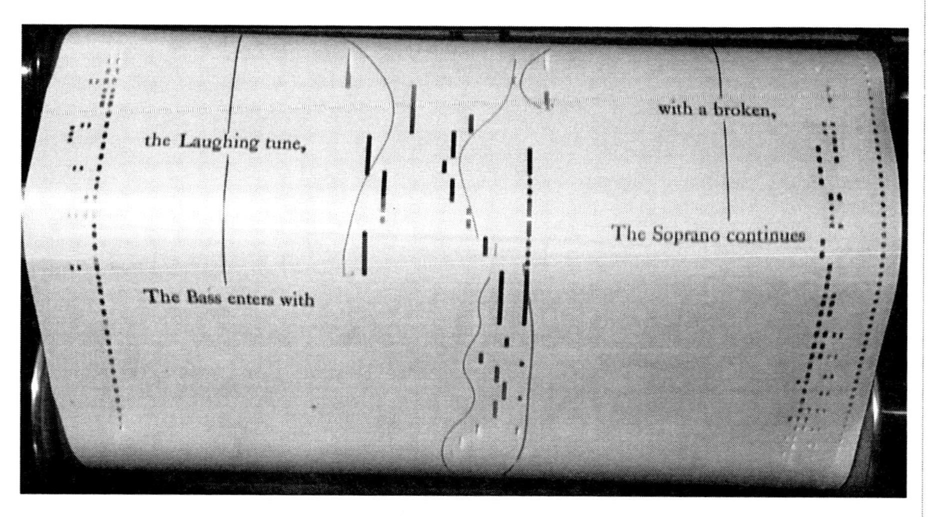

28.1 Screenshot from a video of Percy A. Scholes, *Audiographic Music*, annotated piano roll of J. S. Bach, *The Well-Tempered Clavier*, book 1, Prelude and Fugue in B-flat, as performed by Harold Samuel (1927), accessed October 2, 2021, 4:35, https://www.youtube.com/watch?v=awY2vXG1tOI.

pitches on top, high below. Of course, long scrolls of paper are unwieldy as well, and given the lack of a measuring grid, the piano roll itself is better left to control the piano. However, the roll in figure 28.1 is delightfully annotated, for the education and edification of the family watching the piano do its thing after dinner!

Notwithstanding the limitations of physical piano rolls in communicating about music, notation inspired by the piano roll can provide useful perspectives on music. Piano roll notations typically represent pitch vertically and time left to right, just like standard notation. If we follow the piano roll metaphor strictly, the pitch dimension would map a chromatic space. A diatonic or any other space may also be depicted, however. Piano roll notation differs from traditional notation in two important ways: First, it collapses the texture onto a single grid. This can be especially helpful when reducing scores that otherwise present pitch information out of registral order, as in an orchestral or band score. Second, it represents time more proportionally to the notated rhythms and can thus provide a more visually consistent sense of musically measured time than notated meter does, though one must still decide whether to depict time by musical units (such as beats) or performance time (minutes and seconds).

In figure 28.2, the music of the upper part is converted into a piano roll type of notation in the lower part, which represents the music twice: as written on the left and reduced to within a one-octave register on the right (the final bass C is

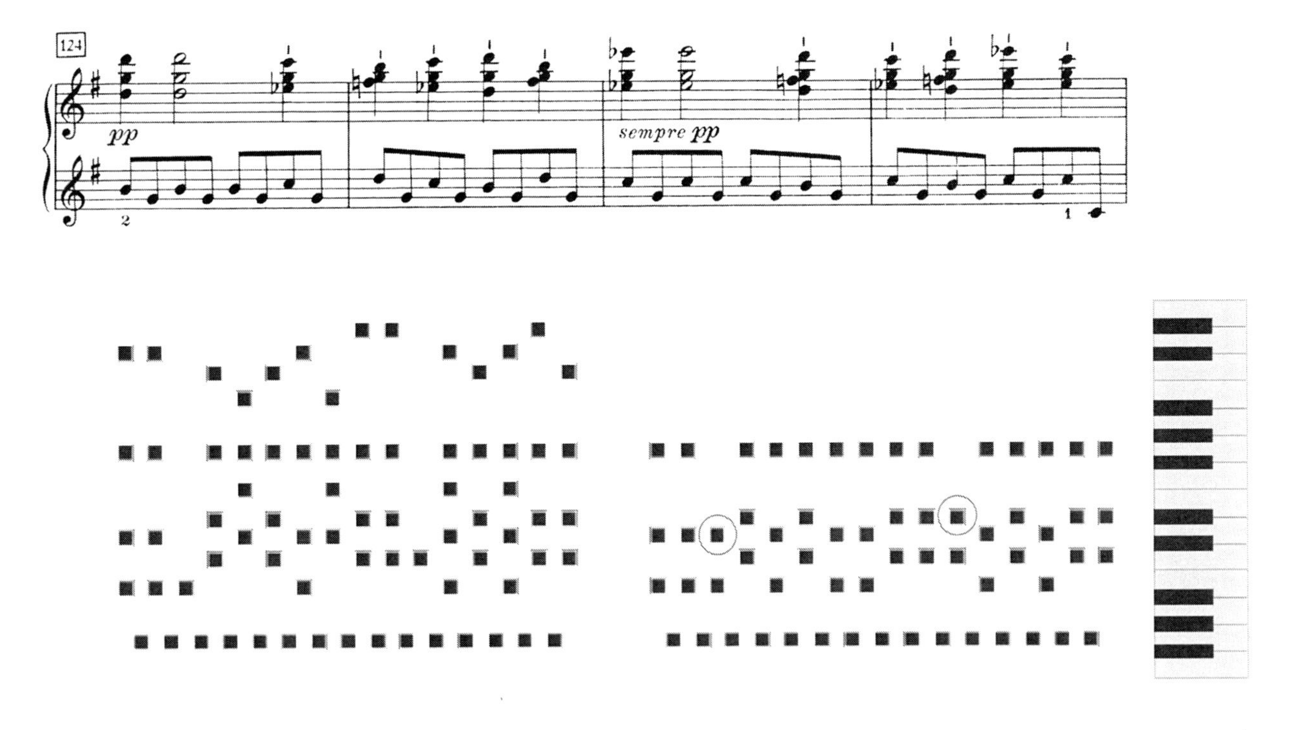

28.2 Guerino Mazzola, Stefan Göller, and Stefan Müller, *The Topos of Music: Geometric Logic of Concepts, Theory, and Performance* (Basel: Birkhauser, 2002), 606 (*above*), 608 (*below*). Circles added.

omitted in both, and the two circled boxes appear to be extraneous). The vertical grid is a diatonic space equaling the spacing of the piano keyboard on the image's far right, except that the E♭s are drawn halfway between where D and E would be. The horizontal grid measures eighth notes. The reduced piano roll notation in the right half makes it easier to see that the second half is an inversion of the first half. However, the addition of a thin dashed line at the inversional axis would make the symmetry much easier to see.

When reducing a multipart texture to piano roll notation, one must decide whether to notate only the pitches or to retain a sense of the texture's individual components. Figure 28.3 does the latter by representing each part of a simple orchestral texture with a different line style. The vertical axis projects diatonic space while the horizontal axis tracks eighth notes in time. Compared to a score, this kind of line drawing more clearly conveys the overall spacing of voices and the relative contour of lines. See chapter 32 for further discussion of this image.

28.3 Robert Cogan and Pozzi Escot, *Sonic Design: The Nature of Sound and Music* (Englewood Cliffs, NJ: Prentice-Hall, 1976), 35.

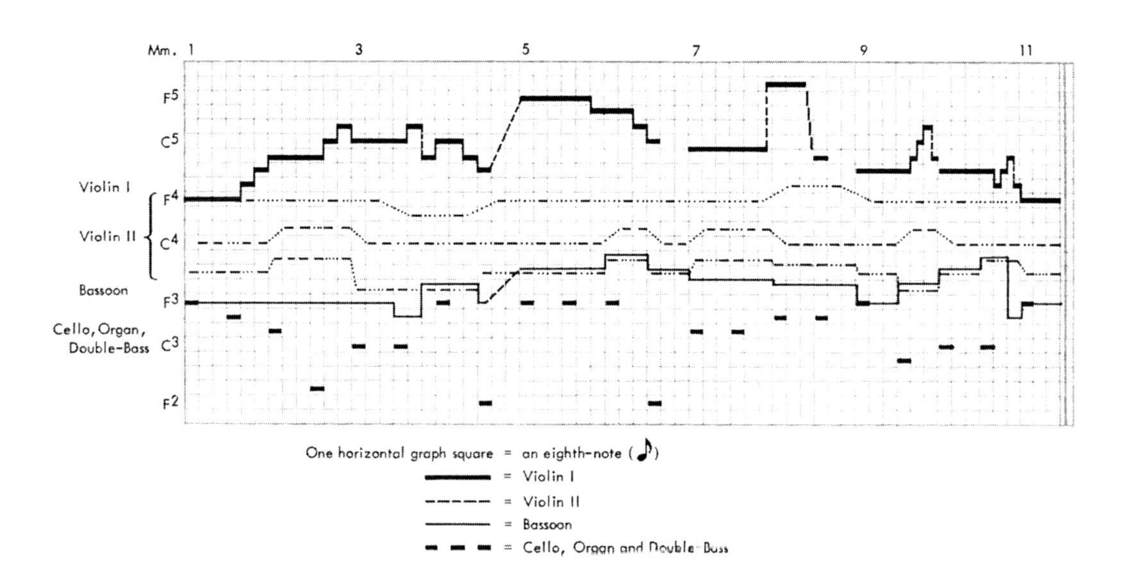

The fingerprints of the piano roll approach are plain in the excellent figure 28.4. The shaded area shows the overall range of a performance of the Indian *Ālāp*, while the solid line shows the focal pitch at each point in the performance. The grid mapping out the pitch space employs different horizontal line styles, which suggest different levels of structural strength. The image groups these further into low, middle, and high octaves, registers that are relevant to the overall trajectory of the performance.

The piano roll informs a series of videos found on the Music Animation Machine website (https://www.musanim.com). In each, a video animation of the score

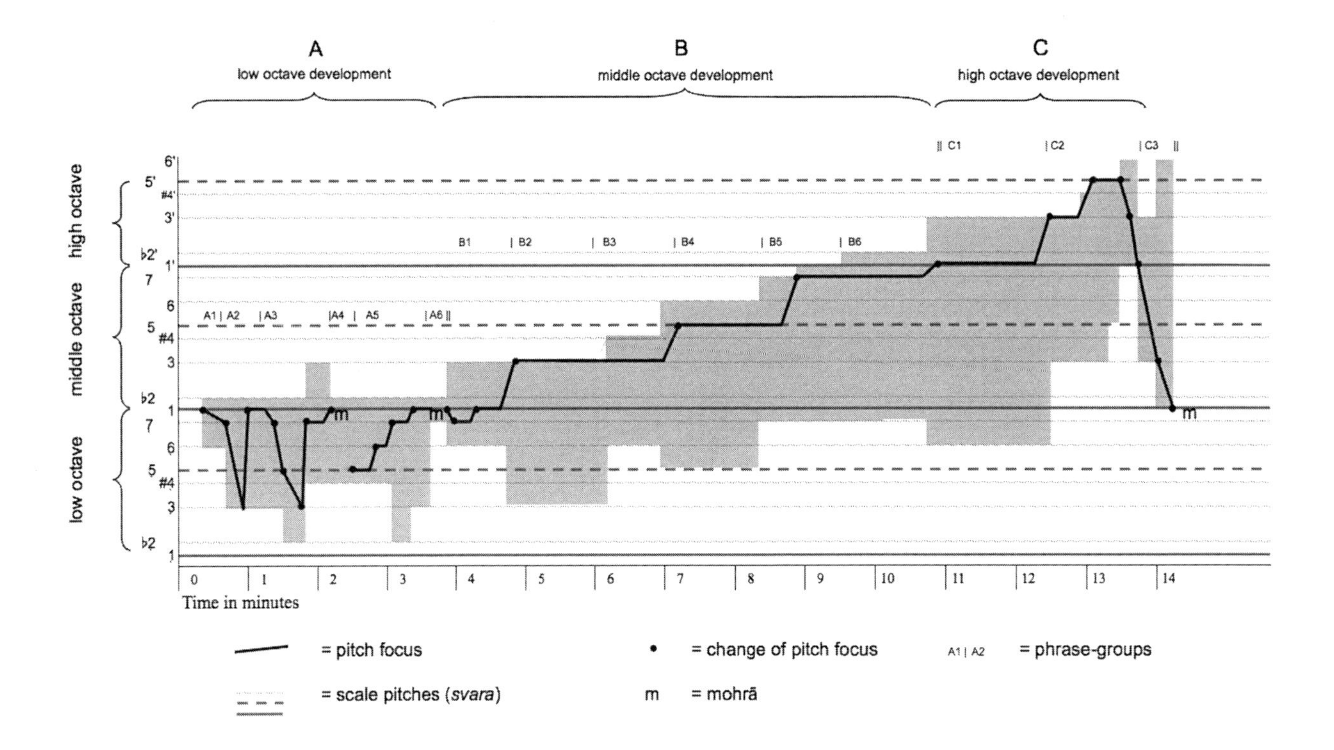

28.4 Richard Widdess, "Dynamics of Melodic Discourse in Indian Music: Budhaditya Mukherjee's *Ālāp* in *Rāg* Pūriyā-Kalyān," in *Analytical and Cross-Cultural Studies in World Music*, ed. Michael Tenzer and John Roeder (New York: Oxford University Press, 2011), 204.

accompanies a synthesized performance of a work. The videos available there have grown in sophistication, and the collection is most impressive. The screenshots in figure 28.5 (colorplate 9) are taken from 3:32 and 10:33 of the remarkable animated score of Stravinsky's *Rite of Spring*. The images represent pitch on a grand staff, which has been adapted to provide more space between lines a major third apart than between those a minor third apart. This makes the vertical grid chromatic (like a real piano roll) rather than diatonic (as it is in fig. 28.2 and fig. 28.3) while preserving the benefits of the five-line staff. The image assigns each pitch class a different color, so notes separated by an octave have the same color. It represents instrument families by different shapes: flutes are oval, double reeds are concave diamonds, clarinets are pointed rectangles, brass are rectangles, and strings are regular diamonds. The fills are translucent, so the bright note outlines show through, making it apparent when two or more instruments are playing the same note, as in the central chord in the upper image and in much of the second image's first half. Percussion hits are fuzzy.

During playback, the image glides right to left, with the center point always representing the present moment. A shape flashes when it is played, and over the duration of the note, the fill shrinks to nothing while the outline remains (in the lower image, traces of this process can be seen in the notes just left of center). Flashes representing an attack sometimes drift to the next note in the melody as they shrink, vanishing just at the point the next note sounds. This can help the listener trace individual parts of the texture. The animation appears to correspond to a "humanized" synthesized version of the score, so attack and release points are rarely perfectly aligned. No horizontal grid appears, nor are rhythmic values shown, so one cannot get more than a gross sense of rhythm, except insofar as one can of course both see and hear the rhythm as the music plays.

The visual-aural analogue is the most powerful element of the animation. Textural elements are easy to see as the animation plays. The upper image, for example, shows the moment when the famous chords from the "Dance of the Adolescents" give way to the triple ostinato at rehearsal 14. Particularly in complex scores, the collapsed view of the piano roll provides a distinct advantage. It appears that vertical thickness indicates dynamics. This contributes to some imprecision in pitch. In the lower image, bars on the left side that represent single pitches nevertheless occupy the vertical space of a fourth. Partly as a result, functional pitch details are hard to discern visually. The lack of ledger lines exacerbates the problem. Without a metrical grid, familiar Stravinskian techniques such as syncopation and polymeter are not visually apparent. Finally, because color recognition is more immediate than shape recognition, the image privileges pitch over timbre. Given the important role of orchestration in the work, it would not have been inappropriate to use color to represent instrument family and perhaps shape for pitch. Doing so would have led to a less varied color experience, however. Academics may quibble about some of the glitz, but the animation is nevertheless impressive, a sophisticated advance on the piano roll. Additional animations from this collection appear in chapter 52.

Piano roll representation excels at tasks such as representing composite textural activity; collapsing complex, multipart scores; and conveying contour. It may be less effective when one needs to display pitch information, and without some adaptation, a piano roll representation tends not to communicate much about rhythm and meter. Nevertheless, it is a useful tool to keep in one's visualization toolbox.

Chapter 29 Alternate Notational Systems

Although chapter 13 touted the value of the standard Western notation system, it is far from the only way composers have tried to communicate with performers. This chapter explores alternatives to the system. These tend to fall into two types: one communicates in some way about the physical act involved in playing an instrument; the other emphasizes structural information about the music.

The first type falls broadly under the term *tablature*, which refers to "a score in which the voice-parts are 'tabulated' or written so that the eye can encompass them" (Dart, Morehen, and Rastall 2001; this source informs the discussion on tablature here). While this description would include such things as reduced scores, the term is generally used more narrowly to describe instrument-specific notation systems. These often differ from common staff notation in significant ways. Tablature systems largely developed in the Renaissance, when instrumental music was becoming increasingly prominent and notational practices had not yet coalesced. Systems existed for many instruments, including the harp and flute; organ tablature is even found in the manuscripts of J. S. Bach. Tablature systems were used most extensively, however, for plucked instruments, notably the lute and guitar.

Lute tablature varies by country and time period, but all such systems show which frets to press on and on which courses of strings. The excerpt in figure 29.1, from a 1601 Dutch collection, is characteristic. A line represents each of the six courses. As in staff notation, the highest line represents the highest-sounding pitch, which is also highest in the visual field of a performer whose head is turned down to look at the instrument's neck, though it is the string closest to the ground. The presence of a letter on a line directs the performer to pluck the strings in that course, while the letter used indicates which fret to press. An *a* indicates an open string, *b* the first fret, and so on. The pitches produced depend on how the strings are tuned. Flagged stems above the staff indicate rhythm, with a duration remaining in effect until another is indicated.

29.1 "Pavana Lachrime," in Joachim van den Hove, *Florida, sive Cantiones* (Utrecht, 1601), 94.

Along with the practice of strumming all the strings of the guitar at once (rather than plucking individual strings) came the development of a chord symbol system called alfabeto. A set of standard chord fingering patterns were arranged in a largely arbitrary order and labeled with letters (see fig. 29.2 for an example involving a guitar with five courses). These symbols were then added to ordinary guitar tablature, as in figure 29.3, which shows the start of a sarabande. The letters do not correspond to our system of labeling on the basis of chord roots, a notion that did not exist yet. A guitarist would memorize the chords assigned to each letter and

29.2 Carlo Calvi, *Intavolatura Di Chitarra E Chitarriglia* (1646; repr., Florence: Studio per Edizioni Scelte, 1980), 5.

29.3 Francesco Corbetta, *Varii Scherzi di Sonata*, book 4 (n.p., 1648), 71.

then play the corresponding fingering, potentially on a different fret, if indicated by a number over the chord symbol (a few examples appear on the left of the lowest system in fig. 29.3).

Tablature for plucked instruments has some advantages. For instance, it does not require the ability to read music and then to translate that into a string-and-fret realization, something that requires more skill. It also accommodates the use of scordatura, in which one or more strings is tuned to a nonstandard pitch. Using tablature, the composer can work out the appropriate scordatura fingering in advance, rather than requiring performers to make the necessary adjustment on the fly, as they must when scordatura is used with standard staff notation. These systems also fulfill music notation's most basic purpose: they describe how to realize a composer's intentions. However, tablature also comes with a disadvantage: musical features such as melodic shape, contrapuntal relationships, and tonal structure are far less apparent than in traditional notation, in which composer and performer are linked at a deeper level. (Kojs [2011] explores tablatures in the context of notation systems that privilege physical gesture.) The previous examples involve historical practice. Online figure 29.4 discusses modern guitar tab.

Klavarscribo, developed in the Netherlands in the 1950s, is a recent attempt to create a piano tablature. It was intended to address primarily the fact that the spatial layout of traditional notation does not align with the spatial layout of the piano keyboard. Figure 29.5 provides a sample in both traditional notation and Klavarscribo. Klavarscribo rotates traditional notation ninety degrees, with what we ordinarily call "low pitches" and "high pitches" becoming in effect "left pitches" and "right pitches," paralleling the left-to-right orientation of the piano. In this sense, it is very much like piano roll notation. Vertical gridlines, grouped in twos and threes, correspond to the black notes of the keyboard. Time is read from top to bottom. Notes played on white keys are open (i.e., they look white) and always appear on spaces. Notes played on black keys are filled (i.e., they look black), always appear on a line, and are drawn on the upper side of the horizontal stems, a subtle indicator that one must reach higher with the finger to play on the black keys. Stem direction indicates which hand to use: leftward stems are played with the left hand, rightward stems with the right hand. Solid horizontal lines serve as bar lines, while dashed lines indicate the location of beats and beat divisions where needed.

29.5 Klavarskribo Institute, *What Is Klavarskribo?*, 2nd ed. (Slikkerveer, Netherlands: Klavarskribo, 1940), 30.

Clearly this system hasn't caught on—and it never will. Adoption of this system would require the republishing of every piece of piano music and the retraining of many millions of pianists and piano teachers. The fact that the system is designed for only a single keyboard instrument (it wouldn't even work for the organ) means that the musicians would need to learn traditional notation, too. But the system is also destined to fail on cognitive grounds. It introduces perception-based problems involving the way our low-level visual processing apparatus works, which undermine it from the start. While the two- and three-line "staves" map neatly onto the keyboard, musicians must traverse an extra layer of visual hierarchy to determine which octave to play a note in, which significantly slows down pitch recognition. The use of stem direction to indicate which hand to play makes the stems informationally significant, but they are not sufficiently visible to carry out this role (beams help, however, when present). Furthermore, the durational information is not bound with the symbols themselves but must be read relative to a grid that is not consistently present.

The second broad category of performer-centric notation includes shorthand notations that convey essential structural information, leaving many details of actual implementation to the performers, who are also informed by professional practices within the style at hand. Baroque figured bass notation (fig. 29.6) directs the performance of two players, who accompany the rest of the ensemble. One plays the notated bass line on a melodic instrument such as a cello, while the other plays a harmonic instrument such as a harpsichord, using the figures to determine which pitches to use in an improvised accompaniment. The harpsichordist has considerable latitude in how this realization might sound, governed by tradition, talent, and taste (or lack thereof, as suggested by certain contemporary treatises).

29.6 George Frederic Handel, "Alleluja, Amen," HWV 272, 18–27.

Lead sheets provide shorthand notation to members of a jazz combo. Consisting of a melody with chord symbols above it ($G\sharp^{o7}$, $E\flat^{7(\flat 9)}$), lead sheet notation functions differently depending on the players. Bass players, guitarist, pianists, drummers, and improvising soloists all read the same set of symbols. What they do with them depends on a combination of the role of the musician in the ensemble, governing stylistic norms, and individual creative impulse. The notation functions in much the same way as figured bass, except that in jazz, the entire ensemble reads from the same notation.

Likewise, the Nashville Number System, first developed in the 1950s and 1960s, efficiently guides many country musicians. Figure 29.7 provides an excerpt of a song. Instead of fixed chord roots as in the jazz lead sheet, the system uses arabic numerals to indicate chord roots relative to whatever key the musicians have chosen. (A minus sign indicates a minor chord.) This basic design is functionally equivalent to roman numeral notation, which makes it easy to perform a song in different keys, and therefore to accommodate singers with different ranges, at the cost of some additional cognitive effort on the part of the performer. Bass notes other than the chord root appear below a line (5|7 would indicate a V chord in first inversion). Nondefault measures (that is, those with more than one chord or a different number of beats) are put in boxes or underlined. If the rhythm is not clear, hash marks (one per beat) depict it. Here is how figure 29.7 would be dictated: "There are 4 beats of pick ups by the guitar. The first line is: One. One. A two-four bar of Six Minor split Five. Four. Four split Two Minor, split One, with two beats on the Four chord. Line two is: Five over Seven. Five split One over Three, with three beats on the Five chord. Four. Five diamond. The last bar is a two-four bar of rest" (C. Williams 2001, 90). The system is efficient and effective.

Shape-note notation, a vocal tablature system developed in the early 1800s, provided the foundation for the influential *Sacred Harp*. It is something of a hybrid of the two types of notations described in this chapter. There are actually two different shape-note systems. One version includes four notehead shapes, which appear on a staff as in standard Western notation, as in figure 29.8. Each symbol represents a pitch in the solfège tetrachord mi-fa-sol-la—that is, a half step followed by two whole steps. In the white-note pitch collection, the pattern can represent both the ranges E–A and B–E. In other words, the notation suffices to represent all diatonic music, except that each symbol functions in two places in the scale. Another version provides a symbol for each of the seven diatonic steps. In either system, those who can read music do not need the symbols, while those who cannot read staff notation can learn the characteristic tonal function of each shape.

As with the plucked-instrument tablature systems, neither this system nor the tonic sol-fa system described earlier requires the ability to read traditional notation beyond a rudimentary level. These vocal systems are intended for amateur

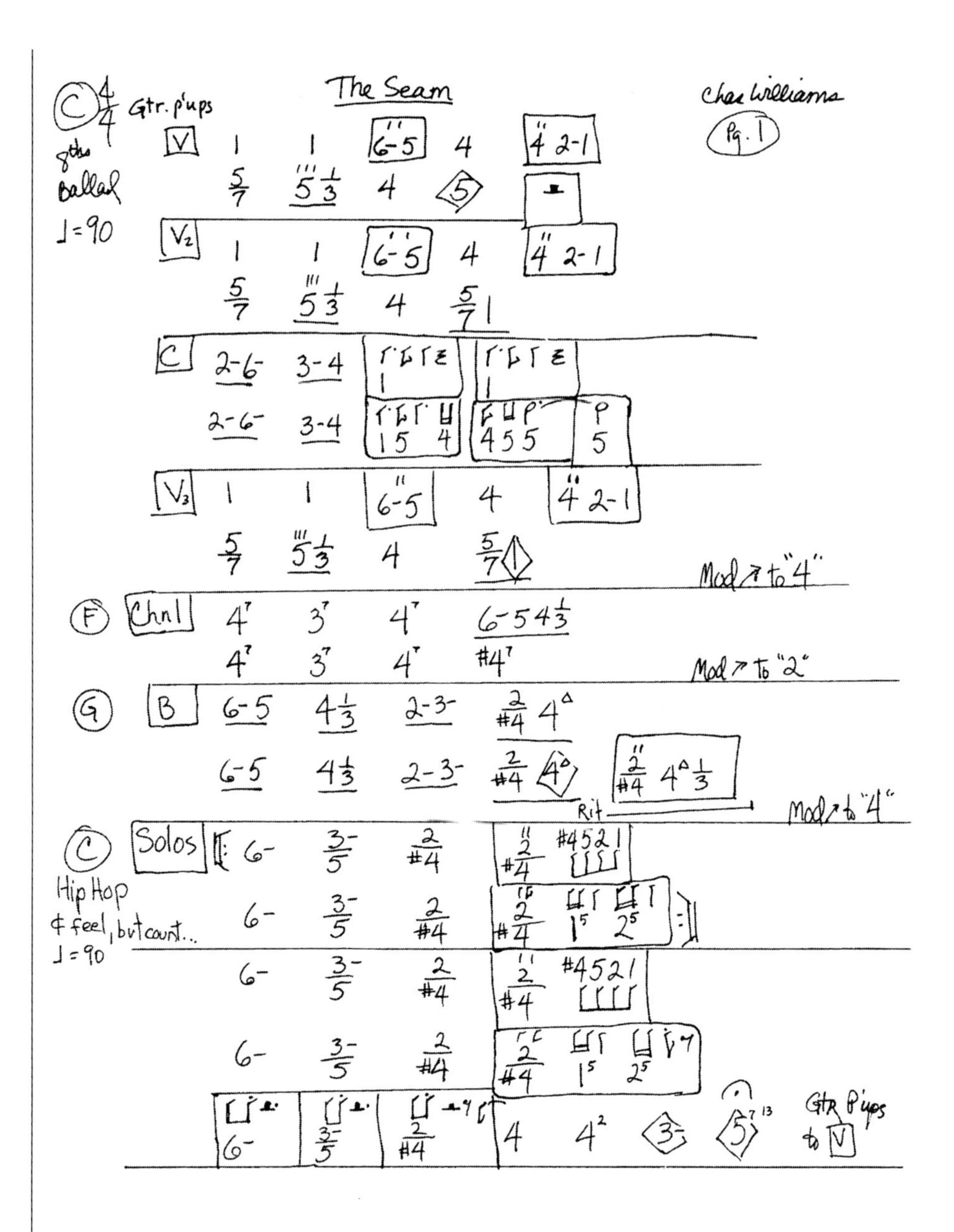

29.7 Chas Williams, *The Nashville Number System*, 10th ed. (Nashville: Big Timbre Music, 2009), 91.

musicians. Their musical scope is limited, compared to the lute and guitar tablature systems, which can be used for music of great complexity.

This chapter has looked at two kinds of alternatives to Western music notation for music that is in some way part of the Western music tradition. Tablature notational systems tell players how to physically play the music on their instruments, or at least where to put their fingers. In some such instances, the notation is aimed at musicians who may not have fully mastered their instruments. (Compare how compilations of popular music include guitar tab but jazz lead sheets do not.) The

spirit of tablature is evident, however, in contemporary compositions in which composers ask woodwind players to execute multiphonics, by specifying unusual fingerings in the score. While tablatures may work for those who use them, they may not convey the linear aspects of music that standard notation does well. The second type, shorthand notations, can arise within musical cultures—such as the late Baroque, mainstream jazz, and country—that allow for some (or a large) degree of performer choice and in which an oral tradition helps shape the realization of the music.

29.8 United Sacred Harp Musical Association, *Original Sacred Harp* (Atlanta: United Sacred Harp Musical Association, 1911), 33.

Chapter 30 Tuning and Temperament

There is an uneasy relationship between acoustically pure intervals and the chromatic pitch-class space on which they are laid. Efforts to represent pitches and intervals in one or both systems, and sometimes to reconcile them with each other, have continued for centuries. The images on these pages represent a few such efforts. The focus here will be more on the presentation than on the content.

Behind a web of lines shooting from the origin, figure 30.1 presents a compendium of pitches produced by frequency ratios relative to the pitch D, which can be measured *x*/*y*, as both *x* and *y* range from 1 to 9. A total of eighty-one nodes and fifty-five lines quarrel for attention, to the detriment of each. The thickness and uniformity of the lines makes the image virtually unreadable. As it turns out, this is a reprint of the original, in which the radiating lines are printed in red. That version, which improves legibility, appears in online figure 30.2, along with a redrawing that makes it yet easier to see the image's elements.

Nineteenth-century German music theorist Joachim von Oettingen created figure 30.3, which is functionally identical to the Tonnetz (see chap. 17). I include it here because it attempts to address tuning issues inherent to such systems. The image shows a tuning system in which all fifths, arranged horizontally, and major thirds, arranged vertically, are pure. One can reach a major third by going four steps to the right (that is, by four 3:2 perfect fifths) or one step up (that is, by a single 5:4 major third). Once brought within the same octave, the ratios between the

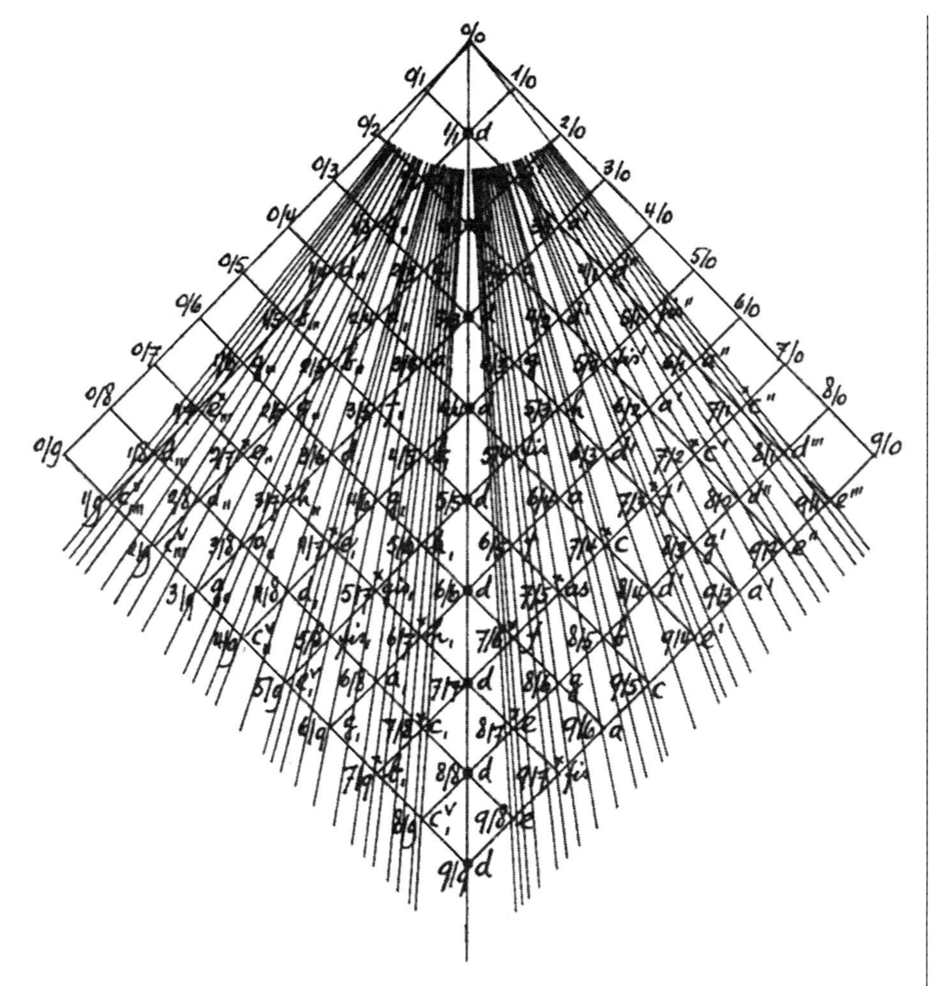

30.1 Reprint of Hans Kayser, *Orphikon: Eine harmonikale Symbolik* (Basel: Schwabe, 1973), 43, from Andreas Holzer, "Das Wiederaufleben Pythagoreischer Tradicionen im 20. Jahrhundert," in *Musiktheorie*, ed. Helga De la Motte-Haber and Oliver Schwab-Felisch, 73–90 (Laaber: Laaber, 2005), 84. The original is in online fig. 30.2.

note reached by different paths will differ by 81/80, known as the syntonic comma. The image inserts a line above pitches that are a comma flat compared to the same pitch without a line. The line might be thought of as a shim that pushes the note flat; likewise, a line below a pitch name indicates that that note is a comma sharp compared to the same pitch without a line. Row and column headers (on the top and right) tell us how many fifths we are from the reference pitch D and how many commas we are from ordinary tuning. See online figure 30.4 for a pair of images that involve enharmonic spelling.

Temperament can also be shown in tabular form, as in figure 30.5. Here, notes of Vicentino's archicembalo, a keyboard instrument with extra keys that employs a nearly equally tempered thirty-one-part division of the octave (some of the divisions are further subdivided here). The table provides the note names, their tuning in cents, and the intervals between them (there are 100 cents in an equal-tempered

30.3 Arthur von Oettingen, *Harmoniesystem in dualer Entwicklung* (Leipzig: Verlag von W. Glaser, 1866), 15.

$5^m\ 3^n$

n:	−8	−7	−6	−5	−4	−3	−2	−1	0	1	2	3	4	5	6	7	8
m																	
2	$\overline{\overline{c}}$	$\overline{\overline{g}}$	$\overline{\overline{d}}$	$\overline{\overline{a}}$	$\overline{\overline{e}}$	$\overline{\overline{h}}$	$\overline{\overline{fis}}$	$\overline{\overline{cis}}$	$\overline{\overline{gis}}$	$\overline{\overline{dis}}$	$\overline{\overline{ais}}$	$\overline{\overline{eis}}$	$\overline{\overline{his}}$	$\overline{\overline{fisis}}$	$\overline{\overline{cisis}}$	$\overline{\overline{gisis}}$	$\overline{\overline{disis}}$
1	$\overline{as}$	$\overline{es}$	$\overline{b}$	$\overline{f}$	$\overline{c}$	$\overline{g}$	$\overline{d}$	$\overline{a}$	$\overline{e}$	$\overline{h}$	$\overline{fis}$	$\overline{cis}$	$\overline{gis}$	$\overline{dis}$	$\overline{ais}$	$\overline{eis}$	$\overline{his}$
0	*fes*	*ces*	*ges*	*des*	*as*	*es*	*b*	*f*	*c*	*g*	*d*	*a*	*e*	*h*	*fis*	*cis*	*gis*
− 1	$\underline{deses}$	$\underline{asas}$	$\underline{eses}$	$\underline{bb}$	$\underline{fes}$	$\underline{ces}$	$\underline{ges}$	$\underline{des}$	$\underline{as}$	$\underline{es}$	$\underline{b}$	$\underline{f}$	$\underline{c}$	$\underline{g}$	$\underline{d}$	$\underline{a}$	$\underline{e}$
− 2	$\underline{\underline{bbb}}$	$\underline{\underline{feses}}$	$\underline{\underline{ceses}}$	$\underline{\underline{geses}}$	$\underline{\underline{deses}}$	$\underline{\underline{asas}}$	$\underline{\underline{eses}}$	$\underline{\underline{bb}}$	$\underline{\underline{fes}}$	$\underline{\underline{ces}}$	$\underline{\underline{ges}}$	$\underline{\underline{des}}$	$\underline{\underline{as}}$	$\underline{\underline{es}}$	$\underline{\underline{b}}$	$\underline{\underline{f}}$	$\underline{\underline{c}}$

30.5 Maria Rika Maniates, "The Cavalier Ercole Bottrigari and His Brickbats: Prolegomena to the Defense of Don Nicola Vicentino against Messer Gandolfo Sigonia," in *Music Theory and the Exploration of the Past*, ed. Christopher Hatch and David W. Bernstein, 137–89 (Chicago: University of Chicago Press, 1993), 171.

1F		1200		1B		580.6	
			38.8				38.7
3E♯		1161.2		5Ḃ♭		541.9	
			38.7				38.7
4Ė		1122.5		2B♭		503.2	
	19.3						38.7
6E̓		1103.2 }	38.7	3A♯		464.5	
	19.4						38.7
1E		1083.8		4Ȧ		425.8	
			38.7		19.3		
5Ė♭		1045.1		6A̓		406.5 }	38.7
			38.7		19.4		
2E♭		1006.4		1A		387.1	
			38.7				38.7
3D♯		967.7		5Ȧ♭		348.4	
			38.7				38.7
4Ḋ		929		3A♭		309.7	
	19.3						38.7
6D̓		909.7 }	38.7	2G♯		271	
	19.4						38.7
1D		890.3		4Ġ		232.3	
			38.7		19.3		
5Ḋ♭		851.6		6G̓		213 }	38.7
			38.7		19.4		
3D♭		812.9		1G		193.6	
			38.7				38.8
2C♯		774.2		5Ġ♭		154.8	
			38.7				38.7
4Ċ		735.5		3G♭		116.1	
	[19.3]						38.7
[6C̓]		[716.2] }	38.7	2F♯		77.4	
	[19.4]						38.7
1C		696.8		4Ḟ		38.7	
			38.8		[19.3]		
3B♯		658		[6F̓]		[19.4] }	38.7
			38.7		[19.4]		
4Ḃ		619.3		1F		0	
	19.3						
6B̓		600	38.7				
	19.4						
1B		580.6					

semitone and therefore 1,200 cents in an octave). Poor column spacing hides the fact that there are two sets of four columns, not eight columns. As we have seen elsewhere, tables of numerals, while offering precision, do not convey meaning as clearly as graphical representations.

Figure 30.6 is a lovely image showing how successions of commonly used, pure intervals align with an equal-tempered pitch-class space. The central image, showing a circle of perfect fifths, aligns dots representing a sequence of acoustically perfect fifths on a spiral divided into equal-tempered semitones. Though the degree of "mistuning" is not quantified, careful study shows the gradual sharpening

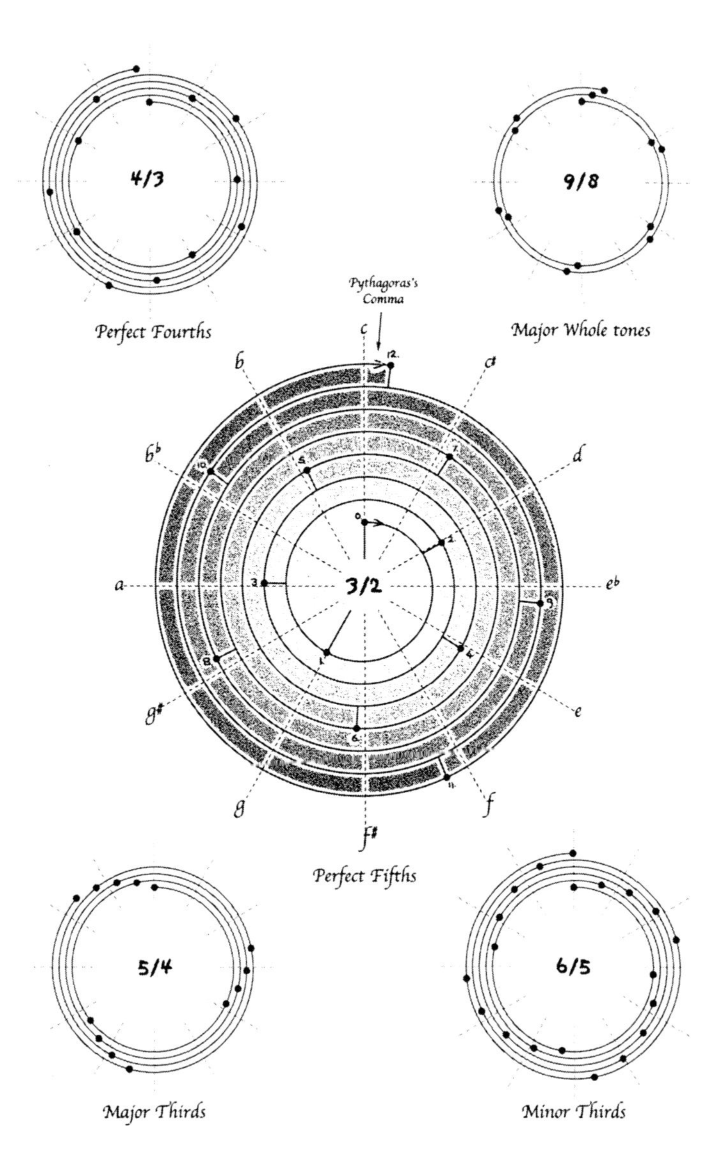

30.6 Anthony Ashton, *Harmonograph: A Visual Guide to the Mathematics of Music* (New York: Walker, 2003), 41.

that results in the Pythagorean comma by the time we come around to C again. The fifths are numbered, which makes them easier to follow as they whip quickly around the circle, though larger numbers in a clearer font would be improve legibility. The other images depict the other common acoustically pure intervals, without the distracting shading but also without the numbers and note names, which reduces the information content slightly.

Similarly, figure 30.7 shows how four tuning systems/scales align with the equal-tempered chromatic collection. The image is rich in information and beautifully executed. The diatonic pitches appear at the top, followed by the intervals above tonic and their equal-tempered size, in cents. For each scale, large fractions show the frequency ratios relative to tonic, and these are translated into cents below the fractions. The ratios between scale members appear above the fractions. Pitches

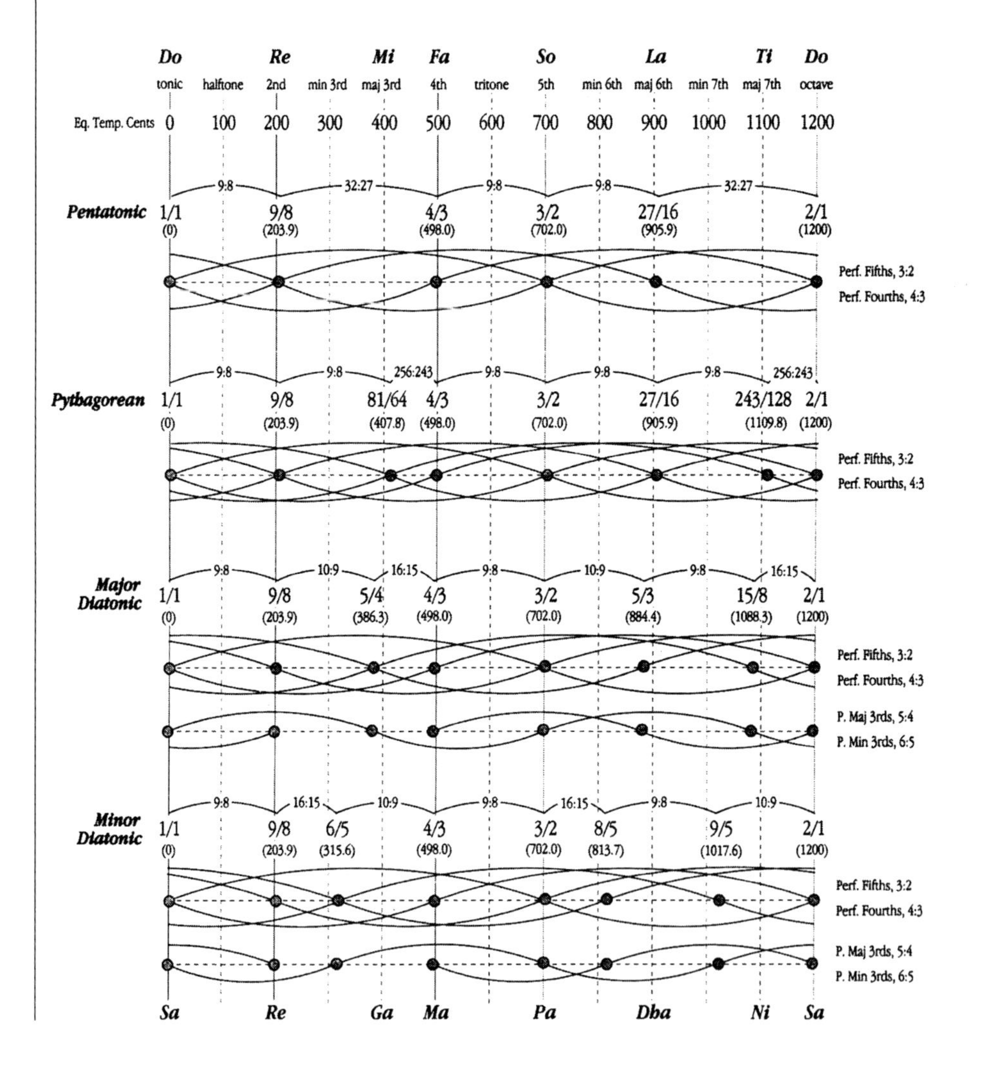

30.7 Anthony Ashton, *Harmonograph: A Visual Guide to the Mathematics of Music* (New York: Walker, 2003), 50.

are plotted on a grid that enables us to see how flat or sharp a note is relative to equal temperament. Finally, arcing lines, indexed at the right of the image, show the specific intervals that form that collection.

Since comparison is such a powerful tool for learning and understanding, the images that best succeed at conveying tuning and temperament directly engage the discrepancies that result from the stacking of different acoustically pure intervals. Such discrepancies might result from a series of 3:2 fifths that don't quite align with 2:1 octaves or 5:4 major thirds, or from tempered pitches (equal or otherwise) that are misaligned with those acoustically pure intervals.

Chapter 31 Microtuning

The scales most commonly found in Western musics rely on the division of the octave into twelve parts, potentially tempered in some way. More complex systems arose around the time chromaticism was burgeoning in the late Renaissance (refer to the discussion around fig. 30.5), and again in the twentieth century, when composers began writing for quarter tones (dividing the octave into twenty-four equal parts) or using more exotic systems, most famously those of Harry Partch. Visual representations of microtonal systems can be divided into two kinds: those designed to represent the system itself and those intended to communicate to performers.

We will start with two examples of the former type. Visualizations of such systems frequently employ tables of ratios and other numerical information, which, as is typical of tables of numbers, can be hard to make sense of. Figure 31.1, one of Partch's representations of his forty-three-tone system, provides a more helpful picture. The depiction begins at the lower left, ascends along the left side to the tritone C-sharp at the top, and continues its ascent down the right side, arriving at G an octave higher. The image marks off intervals proportionally along the way. It provides the locations of equal-tempered pitches within the central column, making clear by how much and in which direction the just and equal-tempered pitches' intervals differ. The design places each pitch across the center axis from its octave-corrected reciprocal, which requires viewers to invert their usual higher-space-equals-higher-pitch mapping. The thick wavy line depicts Partch's theoretical consonance level for each interval on the away-from-center axis, and the intervals are gathered into

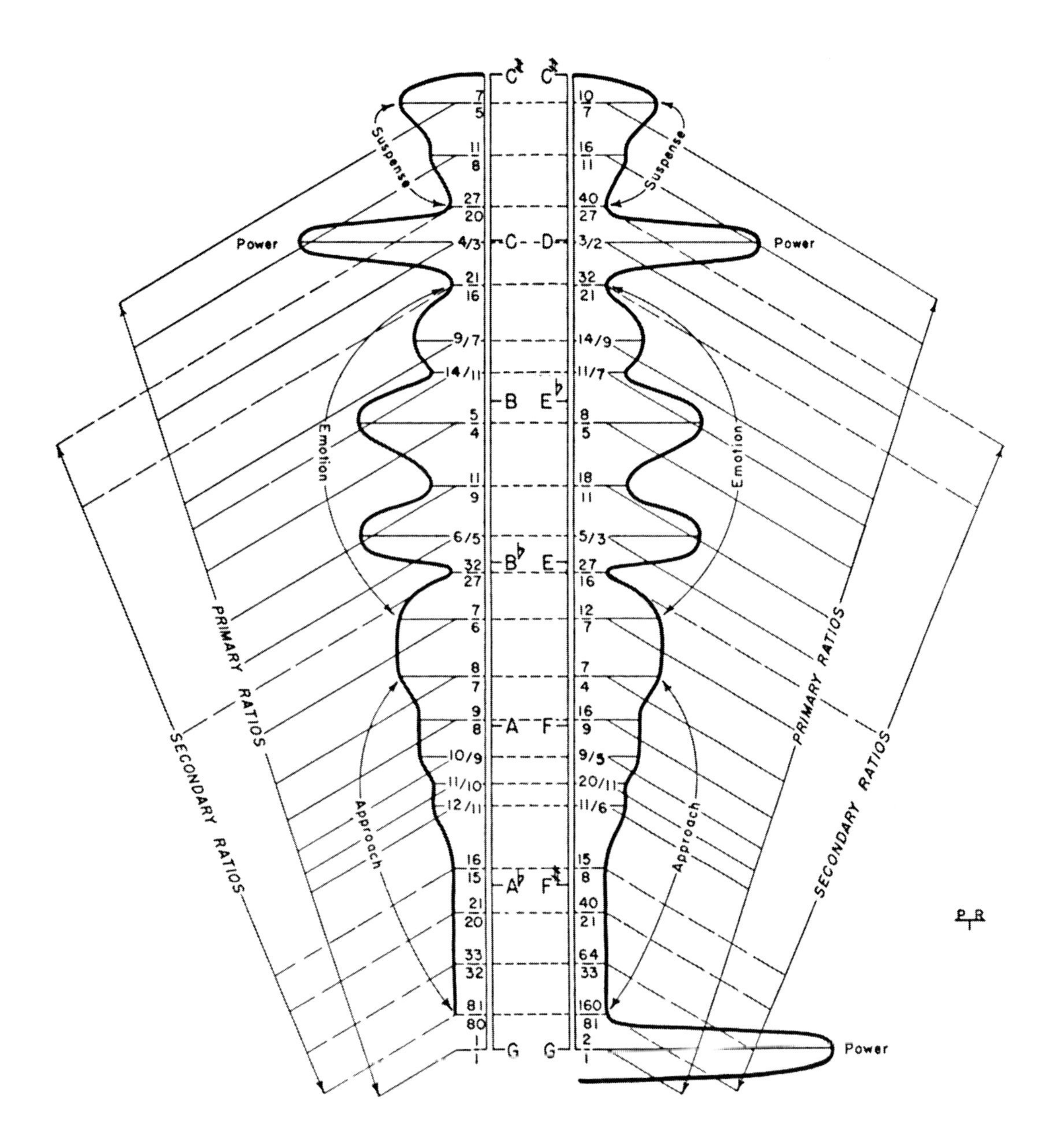

categories (suspense, power, emotion, approach). An ink-heavy means of distinguishing primary from secondary ratios mars what would otherwise be a clean and effective design.

31.1 Harry Partch, *Genesis of a Music*, 2nd ed. (New York: Da Capo, 1974), 155.

In just intonation, all intervals derive from positive and negative powers of 3 (the perfect fifth) and 5 (the major third), plus powers of 2 to adjust the octave. This is implicit in many representations of the Tonnetz (see chap. 17). Partch's system adds powers of 7 and 11. How could one represent a four-dimensional equivalent

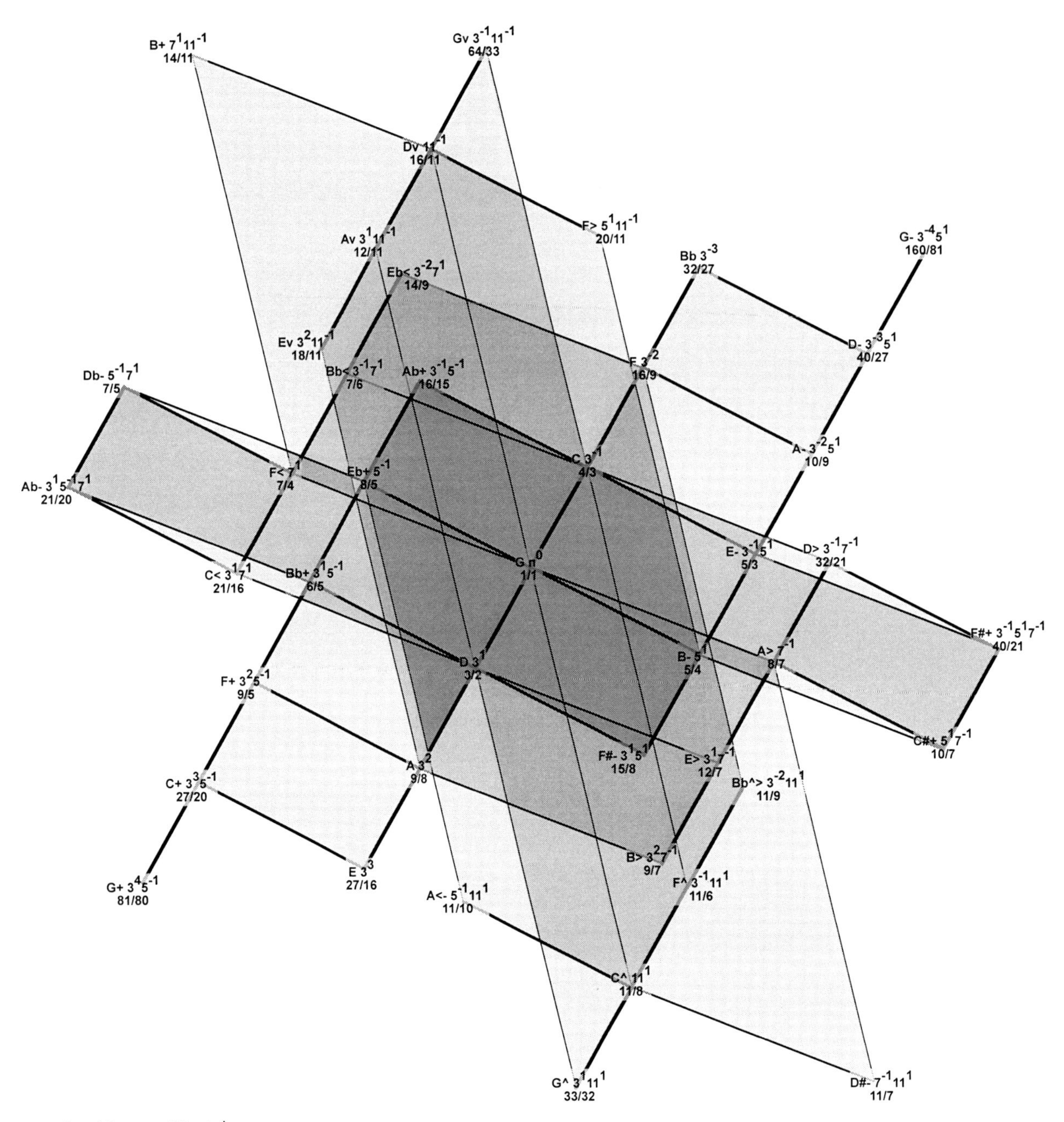

31.2 Joe Monzo, "Partch 43-Tone JI Scale: Shaded Inverted Monzo Lattice," accessed October 25, 2018, http://www.tonalsoft.com/monzo/partch/scale/partch43-lattice.aspx.

of the Tonnetz, as needed to represent this system? Figure 31.2 does a pretty remarkable job of it. From the G at the center of the image, powers of 3 emanate roughly northeast to southwest, powers of 5 perpendicular to that, powers of 7 slightly tilted from the same plane, and powers of 11 vertically. The ratios of figure 31.1 are provided there as well, also represented as powers of the constituent odd numbers (powers of 2 needed to keep the ratios between 1 and 2 are reasonably omitted). Shading gives the impression of panes of glass, which helps group pitches that share two of the four dimensions. While still a bit mind-bending, the representation is impressively clear.

Communicating how to play microtonal pitches to performers, most of whom were trained to play only from traditional notation, requires careful thought. The same demands of traditional notation apply to microtonal notation: it must facilitate fast, accurate reading during a performance. The notation of microtuning inevitably runs into difficulty. By the time microtonality develops, traditional notation is well established, so much so that those developing notational systems employing quarter tones or different divisions of the octave must first decide whether to adapt traditional notation, which performers will understand, or to develop a new system, which might allow for greater precision. Online figure 31.3 proposes a nine-line staff that puts chromatic pitches on lines and quarter tones on spaces.

Figure 31.4 displays three notation systems for quarter tones. Each provides a superset of the accidentals available for normal chromatic music, which therefore can be employed on the familiar five-line staff and come with the cognitive processing benefits associated with that system. The first set is perhaps the most commonly used. The symbols for sharps are quite intuitive: the number of vertical lines indicates the number of quarter steps to raise the pitch. Unfortunately, the symbols for flats do not follow the same convention. Rather, a backward flat symbol represents a quarter flat, which may slow down visual processing. The symbols in the middle row provide the symmetry missing from the upper system and consistently show the number of quarter-step increments. Arrows make the direction of alteration immediately clear, and one can easily see how many quarter steps the accidental refers to: in the symbols with three lines, those lines are more closely spaced than in symbols with two lines, and the four-quarter-sharp/flat symbols change the orientation of the lines altogether, making the symbols more immediately recognizable. The consistency of the system comes with the loss of the familiar single- and double-sharp and flat symbols, however. The third system proposes a hybrid that restores the standard symbols and adds a quarter-tone adjustment arrow, sacrificing internal consistency for potentially greater accessibility for those who don't work consistently in microtonal contexts.

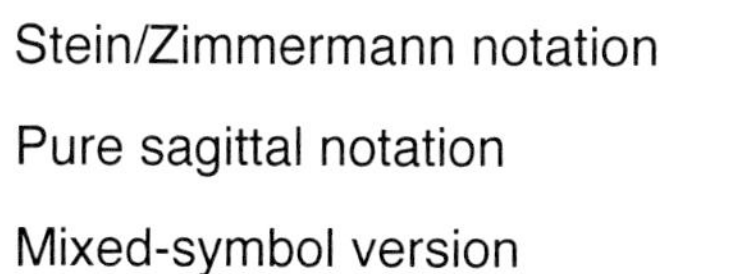

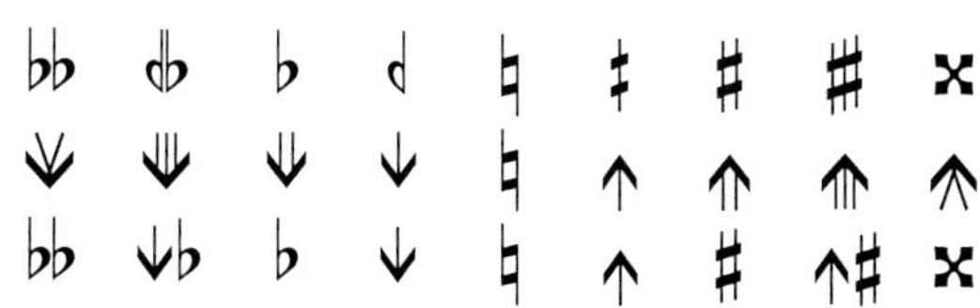

31.4 George D. Secor and David C. Keenan, "Sagittal: A Microtonal Notation System," *Xenharmonikôn: An Informal Journal of Experimental Music* 18 (2012), reprinted (with updates), Sagittal.org, accessed February 22, 2019, 3, http://sagittal.org/sagittal.pdf.

More complex microtonality naturally requires more complex notation. Figure 31.5 proposes a notation system for Partch's forty-three-note system. It employs four different notehead shapes, based on the highest prime number that contributes to the interval: pitches involving a power of 11 are triangles, pitches involving a power no higher than 7 are diamonds, 5s are squares, and 3s (plus the origin note) are regular noteheads. Despite the clear linkage to shape-note notation, which is described in chapter 29, the mapping from shape to pitch is (literally!) exponentially more challenging. Unfortunately, the shapes are too similar to one another to generate the kind of pop-out effects that would allow rapid reading in the manner of traditional Western notation. The system's ultimate efficacy depends considerably on how well it aligns with the design of the instrument on which the music will be performed. Online figure 31.6 shows notation for a more intensely microtonal work.

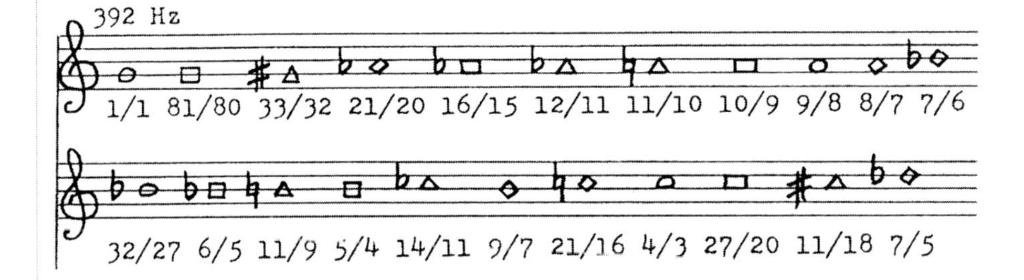

31.5 Partch/Will, cited in Gardner Read, *20th-Century Microtonal Notation* (New York: Greenwood, 1990), 132.

The area of microtonal pitch systems remains ripe for new developments in visual representation. Many design decisions rely on whether the target audience consists of practitioners of microtonality or those who more commonly inhabit the twelve-note universe.

Chapter 32 Timbre

Communicating visually about musical timbre is difficult. The metaphors we use for sound quality are not tied to space and motion in the same way that our metaphors for pitch and time are. Some of our terms for timbre, such as *warm*, *dark*, *bright*, or *birdlike*, come from the natural world. Some, such as *buzzy* or, well, *mechanical*, come from the mechanical world. Others simply refer to the kinds of instruments that produce the sounds themselves: *brassy*, *percussive*, *reedy*.

Our conception of timbre so tightly intertwines with the sources that produce musical sounds that the orchestration of a work serves as a perfectly reasonable representation of timbre. Figure 32.1 and online figure 32.2 do this in works by Björk Guðmundsdóttir and Krzysztof Penderecki. The two images use the same design strategy, though the execution is superior in figure 32.1. There are no extraneous gridlines, black bars clearly indicate start and end points for each instrument, and gray shading helps situate all of this in the song's formal context.

Of course, such images do not show us how timbres are formed, as (usually complex) combinations of simultaneous wave forms at different frequencies and amplitudes. Figure 32.3 explores this process, tracing the sonic shape of the onset (0.2 sec) of pitch D4 as played on a trumpet. It shows the overtone series above, with the customary out-of-tune partials 7, 11, and 13 indicated by closed noteheads. Below, it depicts the changing amplitudes of the various partials in a line graph. These amplitude changes are much easier to read than the tabular representation in online figure 32.4, though the lines get badly tangled and difficult to trace. A redrawing with different characteristics for each line, similar to the way online

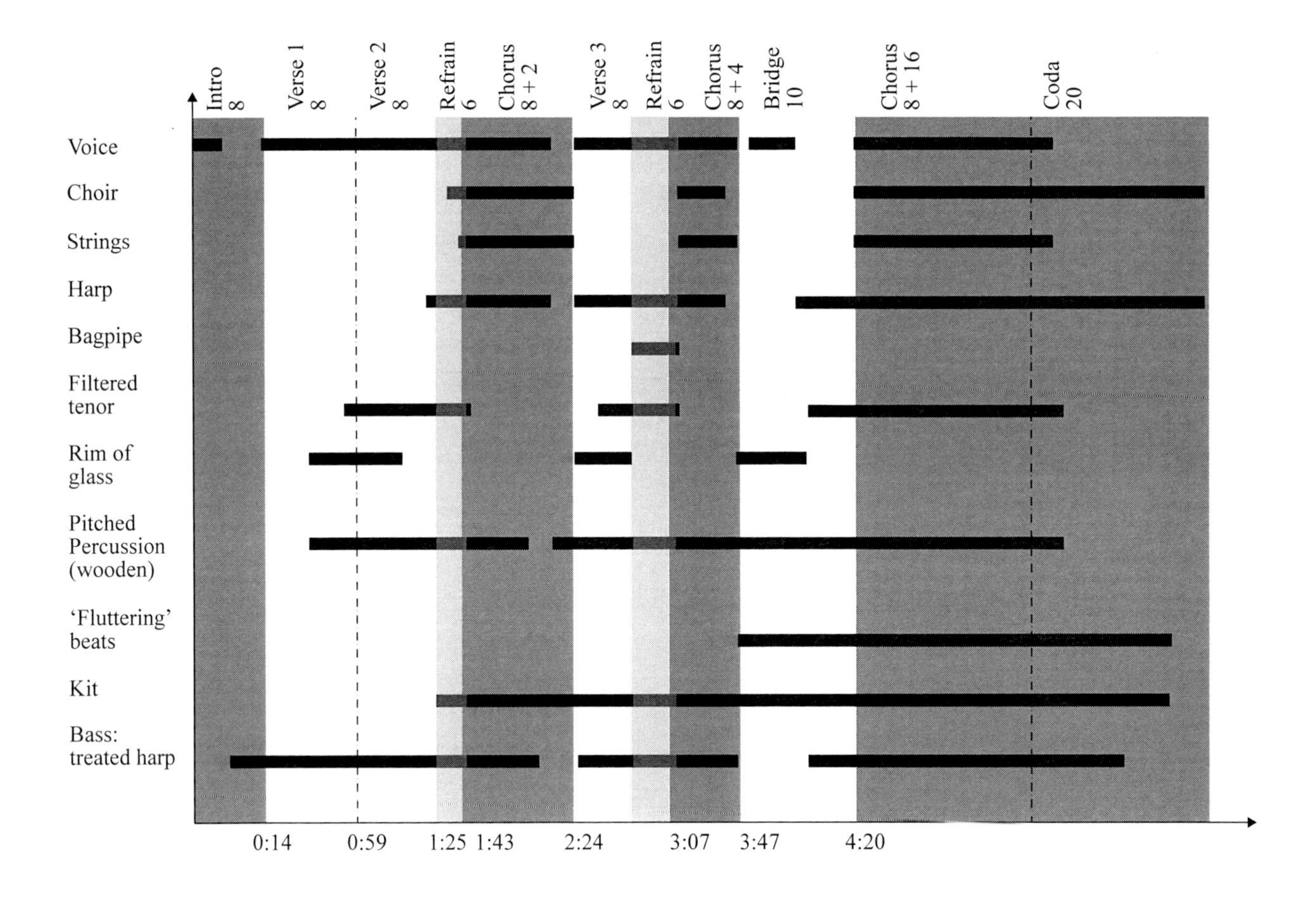

32.1 Nicola Dibben, "Subjectivity and the Construction of Emotion in the Music of Björk," *Music Analysis* 25, no. 1/2 (2006): 179.

figure 12.9 adapts figure 12.8, for example, would be a great benefit. Online figure 32.5 demonstrates the use of color to even more clearly distinguish the lines. Note that all images in which the axes represent numerical values, including the vertical axis here, should display the units.

The voice can produce a wide range of timbres, including the sounds that make up human speech. The International Phonetic Alphabet (IPA) provides symbols (sometimes with diacriticals) to represent these spoken sounds to a reasonable approximation. The simple, but effective, image in figure 32.6 shows how a subset of IPA vowel sounds map onto the placement of the tongue in the mouth. Not labeled in the image, tongue height corresponds to the vertical axis, while the horizontal axis represents the position of the tongue forward or back in the mouth. In this sense, the representation is akin to a vocal tablature, though of course it would be impractical to employ as a musical notation. See also online figure. 32.7.

The images in figure 32.8 come from a remarkable book that, across nearly six hundred pages, draws from a large number of historical sources to summarize and compare descriptions of the timbral properties of orchestral instruments. Blocks or triangles represent the characteristics of the first, second, and third formants of the

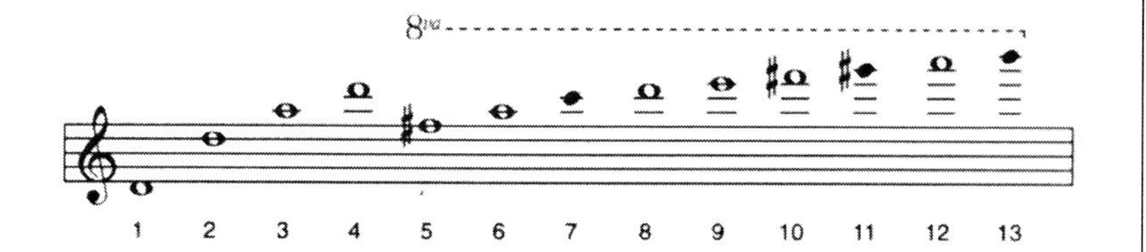

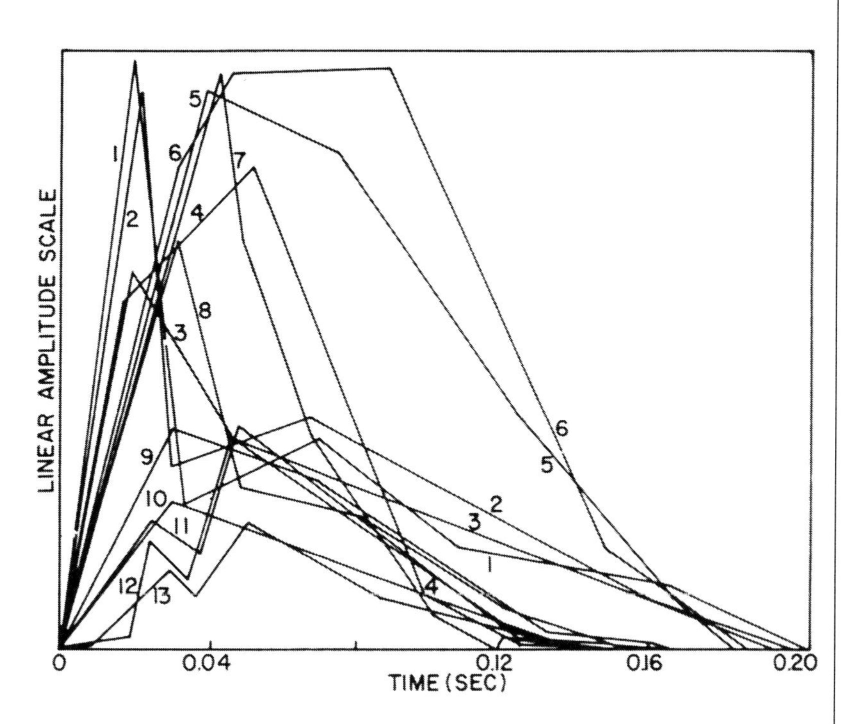

32.3 Jean-Claude Risset and David L. Wessel, "Exploration of Timbre by Analysis and Synthesis," in *The Psychology of Music*, ed. Diana Deutsch, 113–69 (San Diego: Academic, 1982), 31.

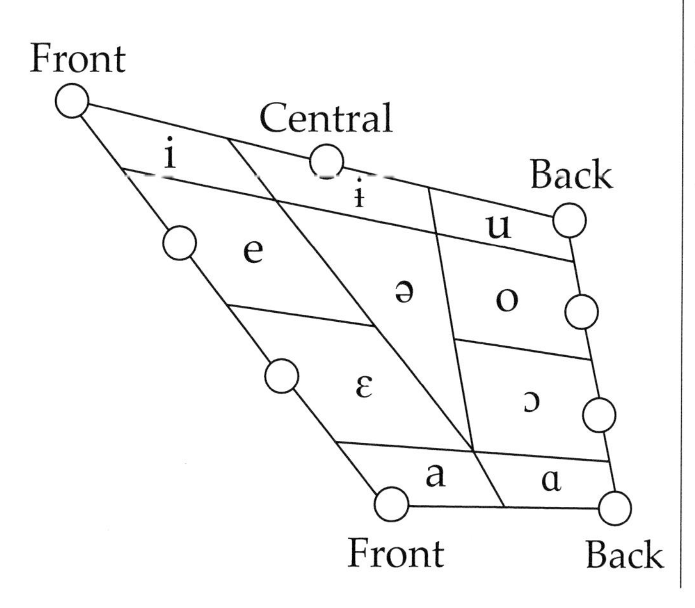

32.6 David Osmond-Smith, *Playing on Words: A Guide to Luciano Berio's "Sinfonia"* (London: Royal Musical Association, 1985), 35.

instrument (the oboe and horn in these cases) as described in each of the sources. Mapping these formants onto vowel sounds provides a reference point that enables one to compare the timbral qualities of different instruments to an instrument that is familiar to each of us: our voice.

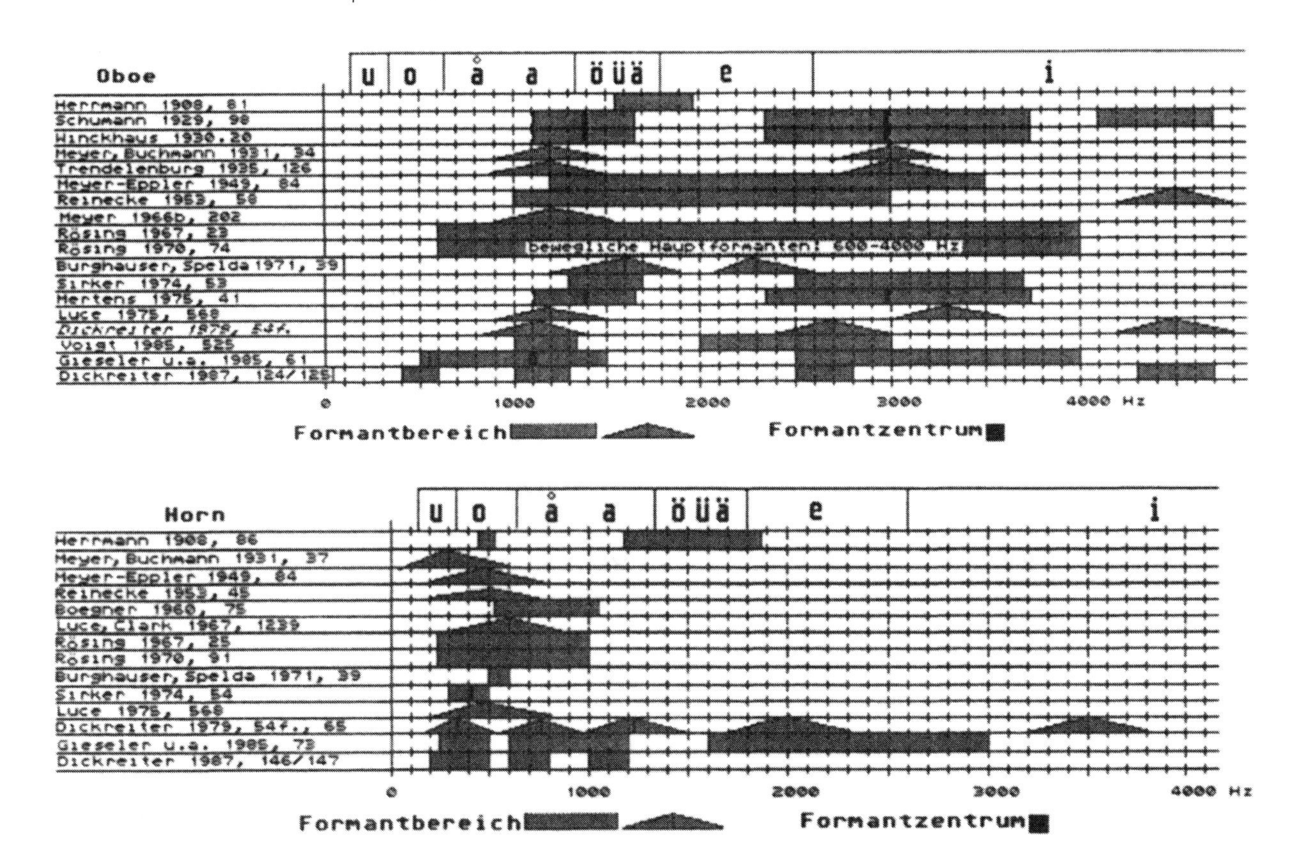

32.8 Christoph Reuter, *Klangfarbe und Instrumentation: Geschichte, Ursachen, Wirkung* (Frankfurt am Main: P. Lang, 2002), 111, 327.

Representing the timbre of tones heard in combination introduces additional challenges. Figure 32.9 shows the first six partials for the pitches constituting four intervals. The use of solid and open bars makes it easy to tell which frequencies derive from which pitch, while a distinctive shading characterizes frequencies associated with both pitches. (Common partials are taller because the vertical axis measures sound pressure level, which will be greater when both pitches contribute to a frequency.) The relatively even spacing of the bars for the perfect fifth compared to the other intervals, particularly the minor second, reflects its sensory perception of consonance.

If we take the harmonic-tones portion of figure 32.9 and rotate it counterclockwise so higher frequencies appear higher in the image, the result is a spectrogram.

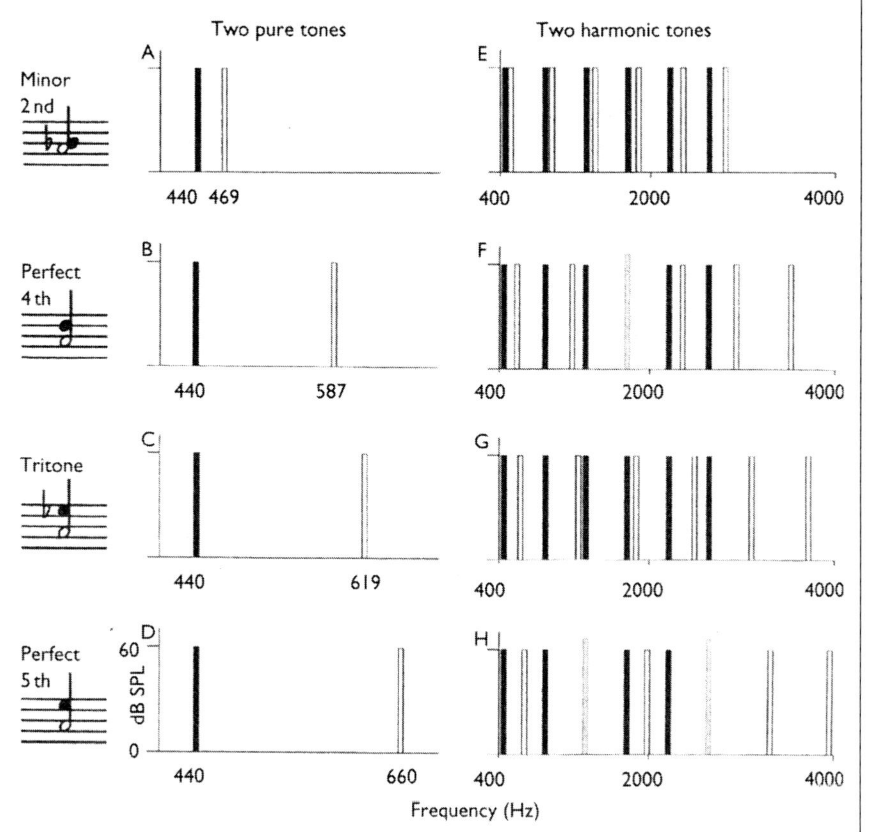

32.9 Mark Jude Tramo, Peter A. Cariani, Bertrund Delgutte, and Louis D. Braida, "Neurobiology of Harmony Perception," in *The Cognitive Neuroscience of Music*, ed. Isabelle Peretz and Robert J. Zatorre, 127–51 (New York: Oxford University Press, 2003), 133.

Spectrograms represent the sound of music as performed (or recorded). The relationship between notated pitch and a spectrogram is similar to the relationship between pure and harmonic tones shown in figure 32.9. Like the piano roll, a spectrogram image provides a more synoptic view of the registral space of a piece. Online figure 25.6 referenced spectrograms, and online figure 32.10 explores them in more depth. See online figure 32.11 for an illustration of the pedagogical utility of spectrograms.

Our conception of timbre resists the kind of spatial representation that our notes-as-objects metaphor allows for in the domain of pitch and rhythm. As a result, the useful visualization of timbre remains a niche endeavor. Nevertheless, we have seen a number of approaches, each of which is useful in some settings.

Chapter 33 Texture

We use the word *texture* to describe the relationships among musical lines. The term has a direct metaphorical link to textiles, which is evident when we talk about interweaving lines, textures that are thick or thin, dense or sparse. Characterizations of musical texture are often quite brief, relying on a small set of standard terms: *monophony*, *homophony* (and the related *melody and accompaniment*), *polyphony* (and sometimes the related *contrapuntal*), and *heterophony*. Figure 33.1 illustrates some of these in simple schematic fashion. (I will not explain the term *holophonic* proposed in this image or critique the image's overgeneralization about the relative historical prevalence of these textures.) As we will see later, visual depictions of musical texture often adopt a quasi–piano roll notation like the one used here, though that is hardly required.

Figure 33.2 and online figure 33.3 represent the basic textures in a more elegant way. They map semblant motion, or the degree to which simultaneously sounding parts move in the *same direction*, against onset synchronization, or the degree to which simultaneously sounding parts move at the *same time*. Figure 33.2 shows the generic qualitative differences for the three primary types of texture. In monophonic music, all simultaneous notes by definition move in the same direction and at the same time, while in polyphonic music the voices are more likely to move at different times and with a mixture of similar and contrary motion. Online figure 33.3 shows how various musical collections map onto this texture space.

Descriptions of texture in individual works of course benefit from more nuance than these broad categories. Since a discussion of texture focuses on musical

Period	Graphic representation	Type
400 - 1450		Monophonic Texture
1450-1750		Polyphonic Texture
1750-1950		Homophonic Texture
1950-		Holophonic Texture

33.1 Panayiotis A. Kokoras, "Towards a Holophonic Musical Texture," *JMM: The Journal of Music and Meaning* 4 (Winter 2007): sec.5.3, http://www.musicandmeaning.net/issues/showArticle.php?artID=4.5.

lines and their interactions, the best images make both clear. Figure 33.4 conveys texture in Gunther Schuller's *Music for Brass Quintet (1961)*, movement 2, through a graphical representation of its different kinds of musical activity. The image distinguishes between sustained and repeating pitches, represented on the horizontal plane, and passages with melodic motion, shown with squiggly lines. This approach is simple and effective. Online figure 33.5 does much the same for a song by Madonna.

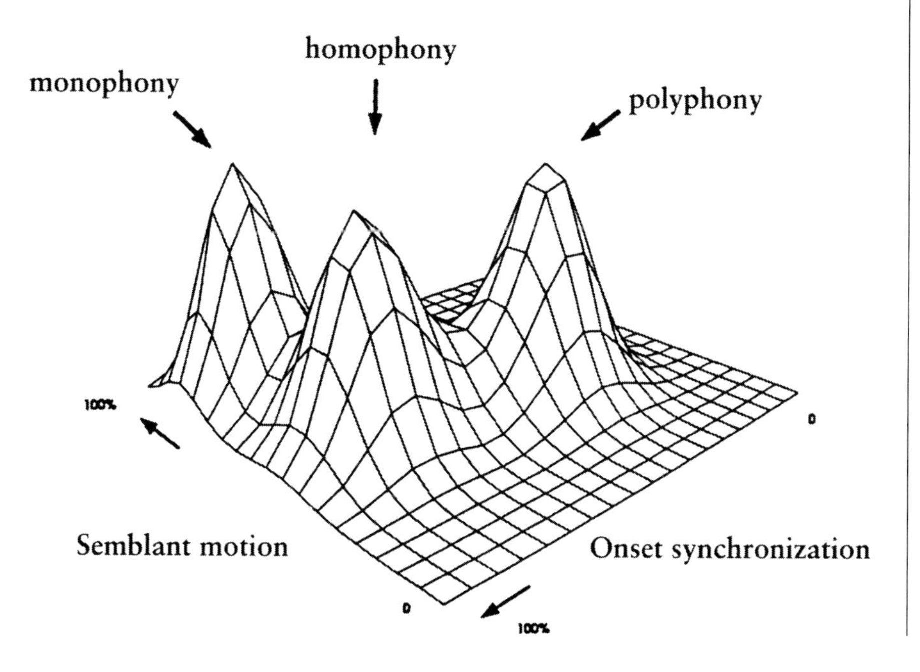

33.2 David Huron, "Tone and Voice: A Derivation of the Rules of Voice-Leading from Perceptual Principles," *Music Perception* 19, no. 1 (2001): 53.

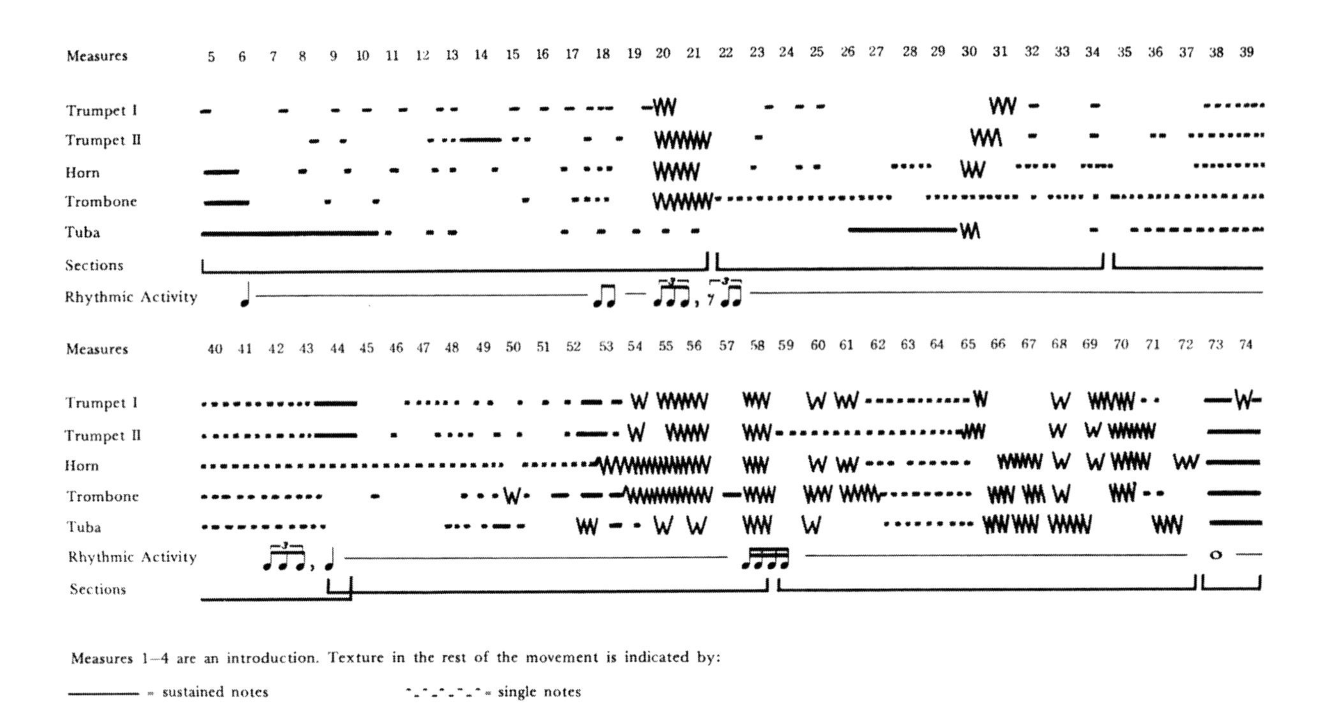

33.4 Mary Wennerstrom, "Form in Twentieth-Century Music," in *Aspects of Twentieth-Century Music*, ed. Gary E. Wittlich, 1–65 (Englewood-Cliffs, NJ: Prentice-Hall, 1975), 55.

Notation inspired by the piano roll can provide different perspectives on musical texture. As noted in chapter 28, such representations normally display pitch on the vertical dimension, and although they would usually display a chromatic pitch space, they may use diatonic or other spaces instead. They usually display time in the customary left-to-right orientation. Piano roll notation differs from traditional notation in two important ways: First, it collapses the parts onto a single grid. This can help when reducing scores that otherwise present pitch information out of order (in an orchestral score, the bassoon part is higher on the page than the trumpet, and the tuba higher than the violins). Second, the temporal tick marks often represent beats or measures, so the representation can provide a more proportionally accurate sense of musically measured time than standard notation does.

When reducing a multipart texture to piano roll notation, one must decide whether to retain a sense of the texture's individual components. The distinction can be seen in online figure 33.6 and figure 33.7, which show a piano roll style of representation of the first twenty-six bars of Béla Bartók's *Fourth String Quartet*. Online figure 33.6, in which the pitches are not connected, clearly shows regions of relative sparseness and denseness, at the expense of conveying how each instrument contributes. In figure 33.7, lines connect the pitches within a part, which makes the shapes of the individual parts clearer. The two sets of tight imitative entrances near the beginning of the excerpt particularly stand out. The image also makes clear the viola's broken octaves at the midpoint and a similar arpeggiated pattern in the cello a little later. On the other hand, it does not differentiate melody and accompaniment, and the vertical lines that serve merely to connect the pitches that are sounding overwhelm the latter. Also, the image does not indicate which instrument is responsible for each line. Finally, while tick marks set off measures horizontally and octaves vertically, because the image uses those tick marks only at the borders and lacks reference labels, it facilitates only a basic understanding of pitch and time in the music.

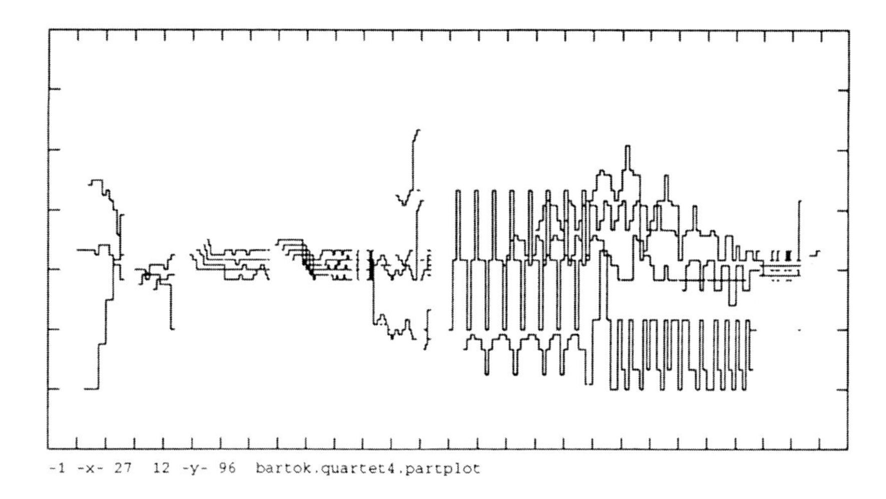

33.7 Alexander R. Brinkman and Martha R. Mesiti, "Computer-Graphic Tools for Music Analysis," *Computers in Music Research* 3 (1991): 8.

Figure 33.8, which we first saw in chapter 28 and which represents the music shown in figure 33.9, addresses some of these issues. The vertical grid projects diatonic space, plotting pitches in the spaces between the lines, while the horizontal grid tracks eighth notes in time. The image represents each part of a simple orchestral texture using a different line style. The violin 1 melody appropriately receives a thicker line. The accompanying bassoon line is thinner. Both parts have a chromatic pitch in m. 7, which is rendered on the line between the spaces representing the adjoining diatonic pitches.

The notation of the violin 2 and bass (cello, organ, double bass) parts merit comment. The legend suggests that both parts are represented by dashed lines, thicker for the bass part (appropriate given the outer voice priority of the style). Yet it appears in the music that the dashes in the bass part result from the rhythm: an eighth note, represented by a solid but short line, with the gap formed by the two eighth rests. The second violin part is made up of an arpeggiated figure, which the image represents as three individual melodic strands. As in the bass part, each pitch is actually a short solid line, not a dashed line. Unlike the bass part, the dashes for the second violin part are linked by dots that connect the pitches in the three melodic strands, at the expense of clarity on where the notes start and stop. (The

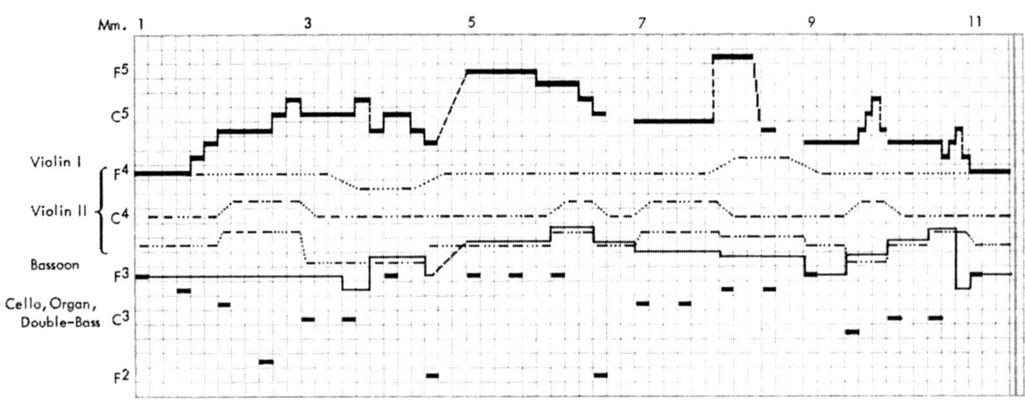

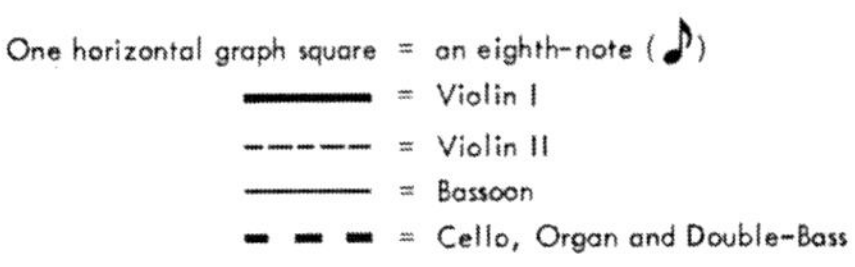

***Above*, 33.8** Robert Cogan and Pozzi Escot, *Sonic Design: The Nature of Sound and Music* (Englewood Cliffs, NJ: Prentice-Hall, 1976), 35.

***Below*, 33.9** Mozart, *Vesperae Solennes de Confessore*, "Laudate Dominum," mm. 1–10.

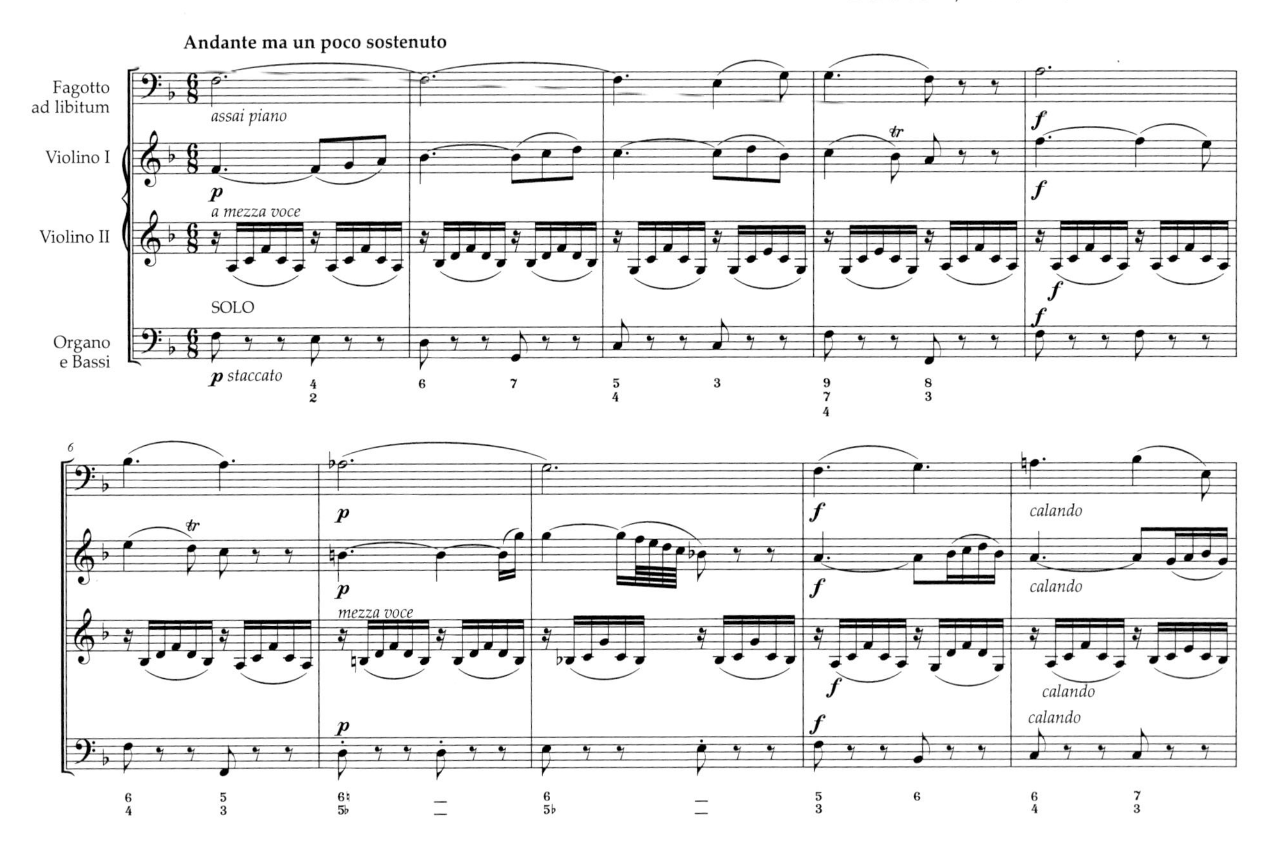

notes are not drawn precisely anyway.) Dashed lines are also used within the first violin part leading into measure 5 and during measures 7, 9, and 10. The longer dashes in the quick descent in m. 8 are actually pitches, however: they represent the sixty-fourth-note descent. Better differentiation of these functions would benefit this image. Reconceiving the grid to show the position of bar lines and octaves more clearly would also help. Though the gridlines are relatively muted, they could be even lighter.

Texture is not exclusive to the pitch domain. Figure 33.10 shows the "texture" of part of the electronic dance track "Sighting" produced by Dave Angel. The image shows when in the track various looped elements occur, including bass drum (BD), hand clap (HC), tom-toms (TT), hi-hat (HH), melodic riffs (R), digital samples (SMP), and synthesized sounds (S). Although the image does not try to convey what is going on within the individual lines, the overall layout is clear. As we have seen elsewhere, the visually noisy fills detract and could be replaced with distinct line styles.

What are the musical lines? How do they interact? These are the questions images of musical texture should aim to answer.

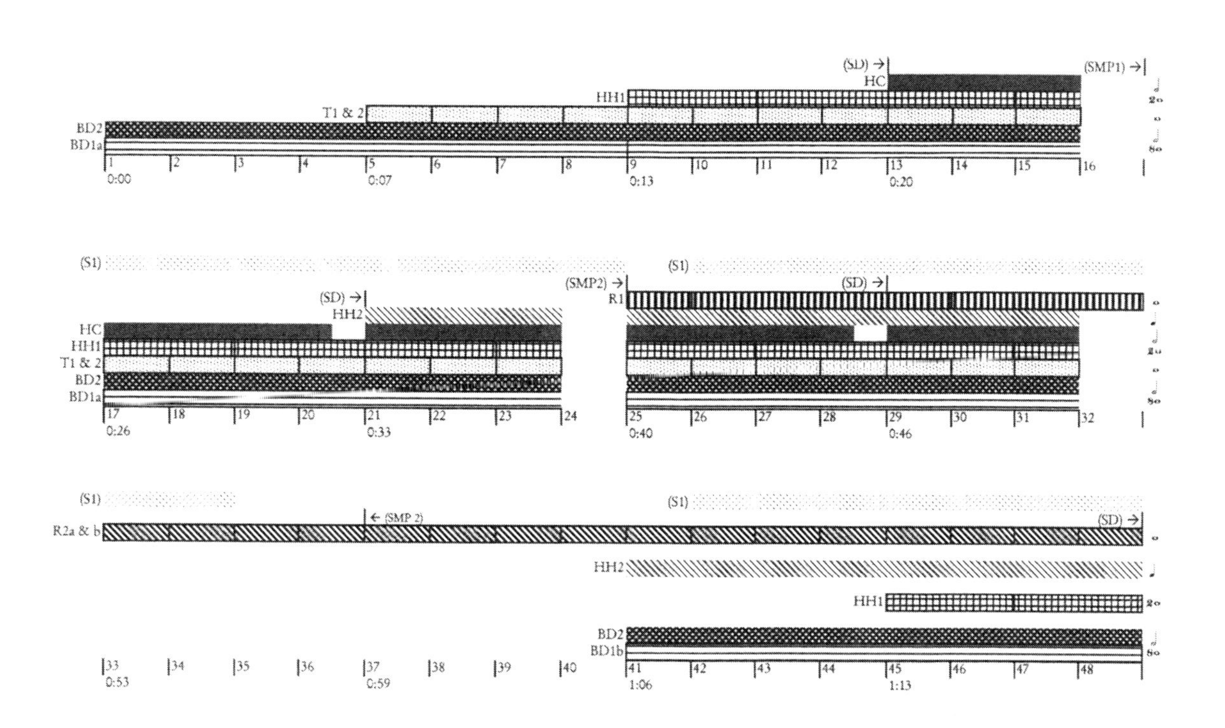

33.10 Mark J. Butler, *Unlocking the Groove: Rhythm, Meter, and Musical Design in Electronic Dance Music* (Bloomington: Indiana University Press, 2005), 282.

Chapter 34 Voice Leading

The study of voice leading remains a central feature of undergraduate music theory textbooks, and the undergraduate core music theory curriculum dedicates a considerable amount of time to helping students master principles of voice leading. When covering this topic, textbook authors must find a balance between *showing* and *explaining*. Thankfully, modern textbooks have found ways to be more "Plaine and Easie" than the 1597 treatise in figure 34.1.

In this chapter, we compare how several textbooks treat the resolution of V^7. Certain aspects of the V^7–I progression are covered almost universally: the treatment of the dominant sevenths' tendency tones (leading tone and chordal seventh), and three options for the resolution—those in which one of the two chords might include an additional root and omit the fifth, plus the one in which the leading tone in an inner voice of V^7 may jump down to the fifth of the tonic triad. Despite this common coverage, the variety of presentations is striking.

Whether in textbooks or classroom handouts, pedagogical visualizations benefit greatly from the principles of information integration found in chapter 7. Since students often use such images not just to learn progressions such as V^7–I for the first time but also to refer back to while trying to apply the principles, the images themselves should contain all relevant information, rather than requiring students to reread the prose when reviewing the subject. As we will see, not all textbooks do so.

Figure 34.2 presents three voicings that omit the fifth in the I chord and three that omit the fifth in the V^7, a small-multiple approach that characterizes each of

OF THE VNISON.

If the treble be	an vniſon with the tenor
and the baſe	a third vnder the tenor
your *Alto* or meane ſhal be	a fifth or ſixth aboue the baſe.
but if the baſe be	a fifth vnder the tenor
the *Alto* ſhal be	a third or tenth aboue the baſe.
Likewiſe if the baſe be	a ſixt vnder the tenor,
then the Alto may be	a 3 or tenth aboue the baſe
And if the baſe be	an eight vnder the tenor,
the other parts may bee	a 3.5 .6 10.or 12.aboue the baſe.
But if the baſe be	a tenth vnder the tenor,
the meane ſhal be	a fift or twelfth aboue the baſe.
But if the baſe be	a twelfth vnder the tenor,
the Alto may be made	a 3. or 10. aboue the baſe.
Alſo the baſe being a	fifteenth vnder the tenor,
the other parts may be	a 3. 5. 6 .10. 12.and 13. aboue the baſe.

OF THE THIRD.

If the treble be	a third with the tenor
and the baſe	a third vnder it
the Alto may be	an vniſon or 8.with the parts.
If the baſe be	a ſixt vnder the tenor,
the *Altus* may be	a third or tenth aboue the baſe.
But if the baſe be	an eight vnder the tenor,
then the *Altus* ſhall be	a fift or ſixt aboue the baſe.
And the baſe being	a tenth vnder the tenor,
then the parts may be	in the vniſon or eight to the tenor or baſe.

OF THE FOVRTH.

When the treble ſhalbe	a fourth to the tenor
and the baſſe	a fifth vnder the tenor
then the meane ſhall be	a 3,or 10,aboue the baſe
But if the baſe be	a 12.vnder the tenor
the *Altus* ſhal be	a 10.aboue the baſe

OF THE FIFTH.

But if the treble ſhal be	a fifth aboue the tenor
and the baſe	an eight vnder it
the *Alto* may be	a 3 or tenth aboue the baſe
And if the baſe be	a ſixt vnder the tenor,
the *Altus* ſhal be	an vniſon or 8 with the parts

OF THE SIXTH.

If the treble be	a ſixt with the tenor
and the baſe	a fift vnder the tenor,
the *Altus* may be	an vniſon or eight with the partes
But if the baſe be	a third vnder the tenor,
the *Altus* ſhal be	a fifth aboue be baſe.
Likewiſe if the baſe be	a tenth vnder the tenor,
the meane likewiſe ſhalbe	a fifth or 12.aboue the baſe.

OF THE EIGHT.

If the treble be	an 8. with the tenor.
and the baſe	a 3.vnder the tenor
the other parts ſhal be	a 3.5.6.10. 12.13.aboue the baſe
So alſo when the baſe ſhal be	a 5.vnder the tenor
the other parts may bee	a 3. aboue the baſe.
And if the baſe be	an eight vnder the tenor
the other parts ſhall bee	a 3 5 10.12.aboue the baſe.
Laſtly if the baſe be	a 12.vnder the tenor
the parts ſhal make	a 10. or 17. aboue the baſe.

34.1 Thomas Morley, *A Plaine and Easie Introduction to Practicall Musicke* (London: Peter Short, 1597), 129–30.

34.2 Walter Piston, *Harmony*, 3rd ed. (New York: W. W. Norton, 1962), 155, 156.

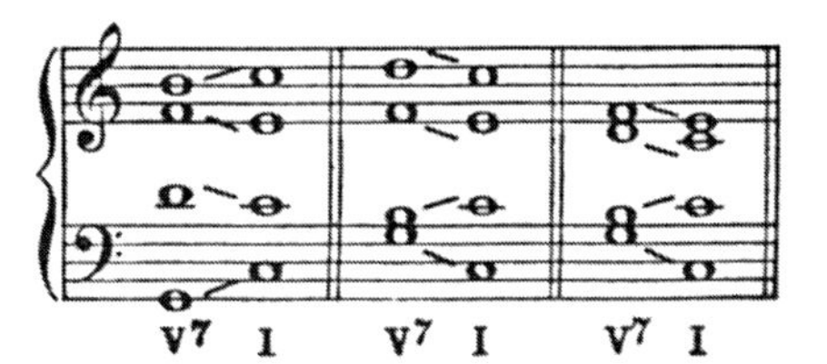

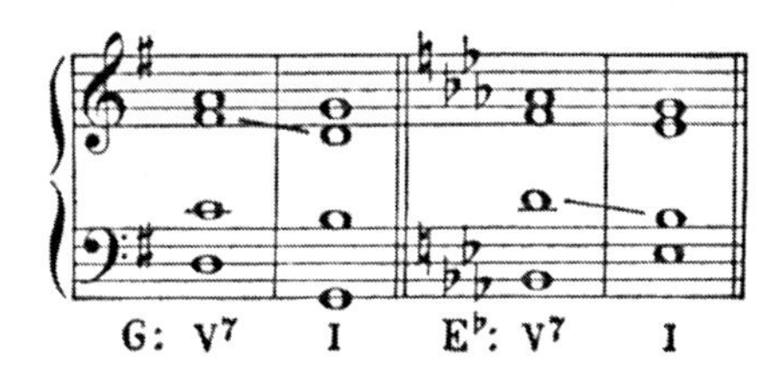

the examples we will examine. Lines show the direction of each voice's movement, though in the examples in the second line, the Gs that are not in the bass look as though they descend to the same G. The roman numerals are omitted in the second line, which is unfortunate. The third line shows but does not explain the option involving the deflected leading tone. Because details are found only in the text, the images cannot stand on their own.

Online figure 34.3 puts V^7 into a three-chord context to convey the different embellishing functions the chordal seventh can serve, a nice touch. The voice-leading examples in figure 34.4 omit the approach to V^7, however, and focus only on the spelling of the two chords and the resolution of the first to the second. The image provides three standard resolutions in A, D, and E. It would be better to lead with the standard resolutions or to otherwise align them. In addition, labels for these are inconsistent: D says which chord member is omitted, while E says which chord member has an extra instance. Also, the image does not make clear to which chords "omitted fifth" and "tripled root" refer. The image intersperses typical resolutions with three that present voice-leading problems. Of these, B and F say to "avoid" a particular error, while C only names the error. With a revised layout and more consistent labeling, the image could stand on its own.

Figure 34.5 shows only the usual resolutions, plus one unusual one, omitting common voice-leading errors altogether. The tendency tone resolution is indicated in the first two examples, not in the third, and only the "frustrated leading tone"

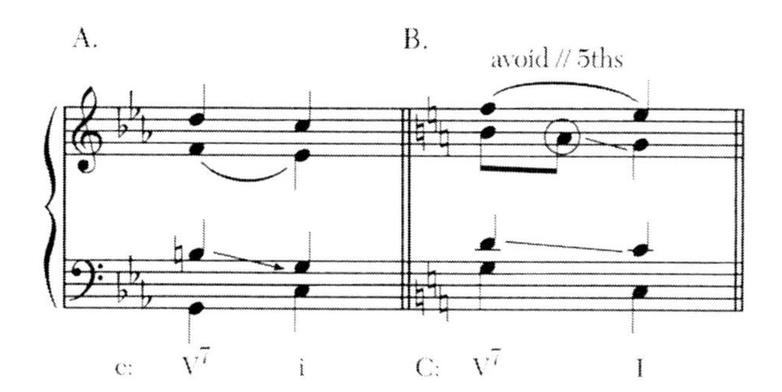

34.4 Robert Gauldin, *Harmonic Practice in Tonal Music*, 2nd ed. (New York: W. W. Norton, 2004), 149.

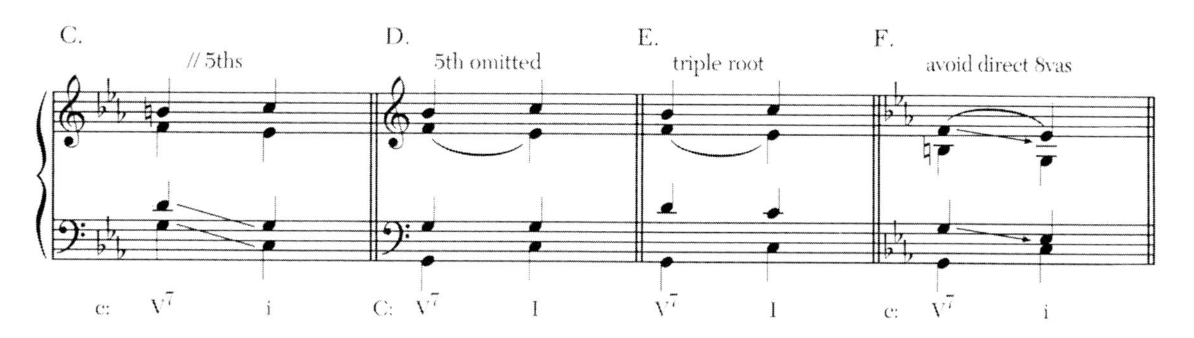

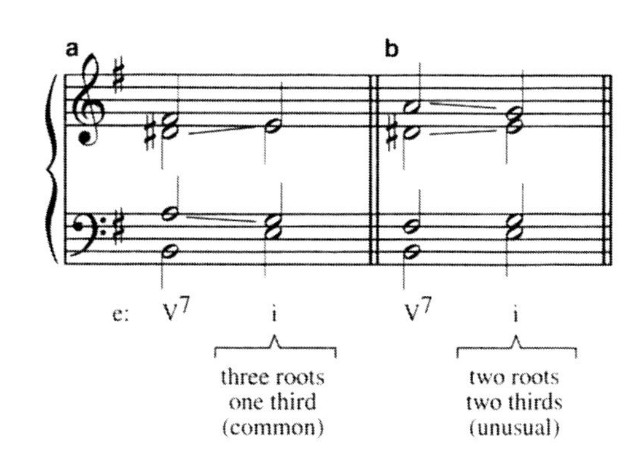

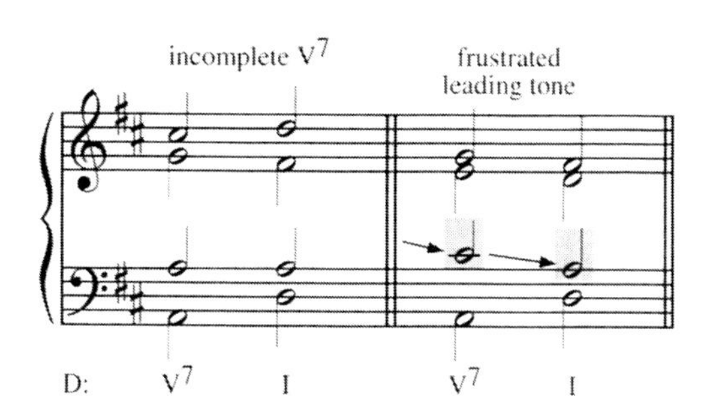

34.5 Stefan M. Kostka and Dorothy Payne, *Tonal Harmony: With an Introduction to Twentieth-Century Music*, 6th ed. (New York: McGraw-Hill, 2009), 213–14.

34.6 Ralph Turek, *The Elements of Music: Concepts and Applications*, 2nd ed. (New York: McGraw-Hill, 1996), 1:327–28.

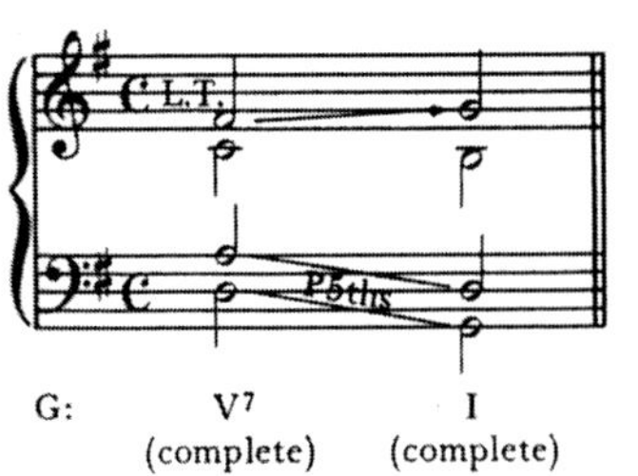

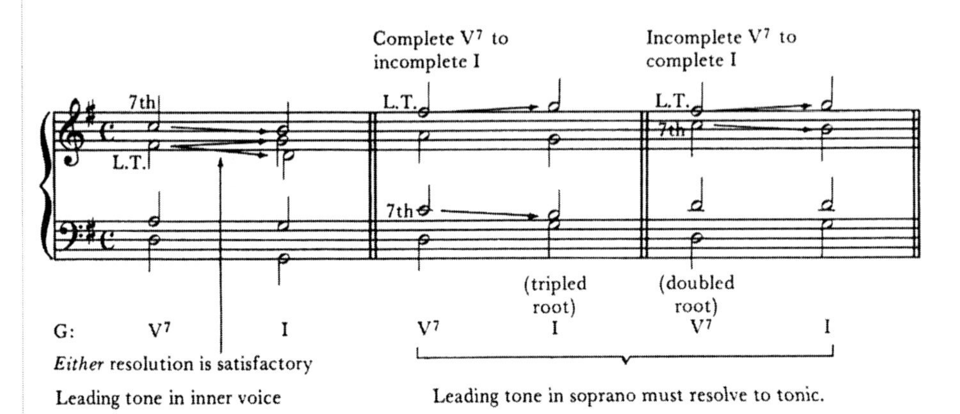

in the last. Consistency would be better. Because these images provide key features of each resolution, they can stand on their own better than the previous examples, however.

Figure 34.6 presents an even more complete picture of the resolution's principles. It starts by outlining the basic "problem" with the progression. Then it presents, with commentary, the possible solutions, starting with the question of whether or not the leading tone is in the soprano. Explaining these options on the image reinforces information presented in accompanying prose and makes it easier to review later. Some typesetting adjustments to hierarchize the information would help.

Online figure 34.7 provides a novel approach to the task, employing schematic paradigms (foreshadowing the next chapter) rather than notated examples. Refer to online figure 34.8 for an image that integrates principles with positive and negative examples and separates layers of information through the use of color (which is why it is included in the online supplement). These features create a highly effective design that all pedagogical visualizations should strive for.

Chapter 35 Schematic and Procedural Representations

In this chapter, we explore two related kinds of visualizations: schematic and procedural representations of musical information. I use the term *schematic representations* to describe images that provide an abstracted view of some phenomenon. Schematic representations often express ways of thinking about something in a generic or conceptual way. They are often reductive, sometimes highly so. In all such images, the layout should accurately reflect the conceptual space. Figure 35.1 expresses the abstract concept of "interval" in a way that beautifully sets the tone for an entire volume that explores the idea. The minimalist schematic image in online figure 35.2 is less successful.

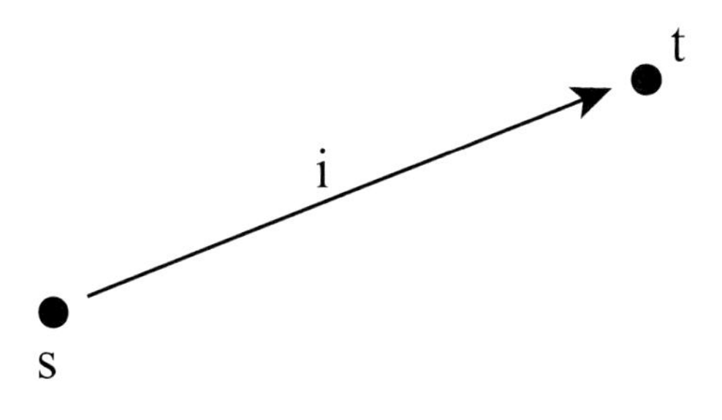

35.1 David Lewin, *Generalized Musical Intervals and Transformations* (New Haven, CT: Yale University Press, 1987), xi.

The next four schematic images are not as minimalist. Figure 35.3 represents the cognitive environment within which the content and structure of a musical event and our aesthetic response to it are understood. The factors are unstructured within the conceptual cognitive space and are thus appropriately unstructured in the image, except that particular aspects under consideration appear in circles whose thickness helpfully indicates their significance. Arrows discreetly direct the motion between domains.

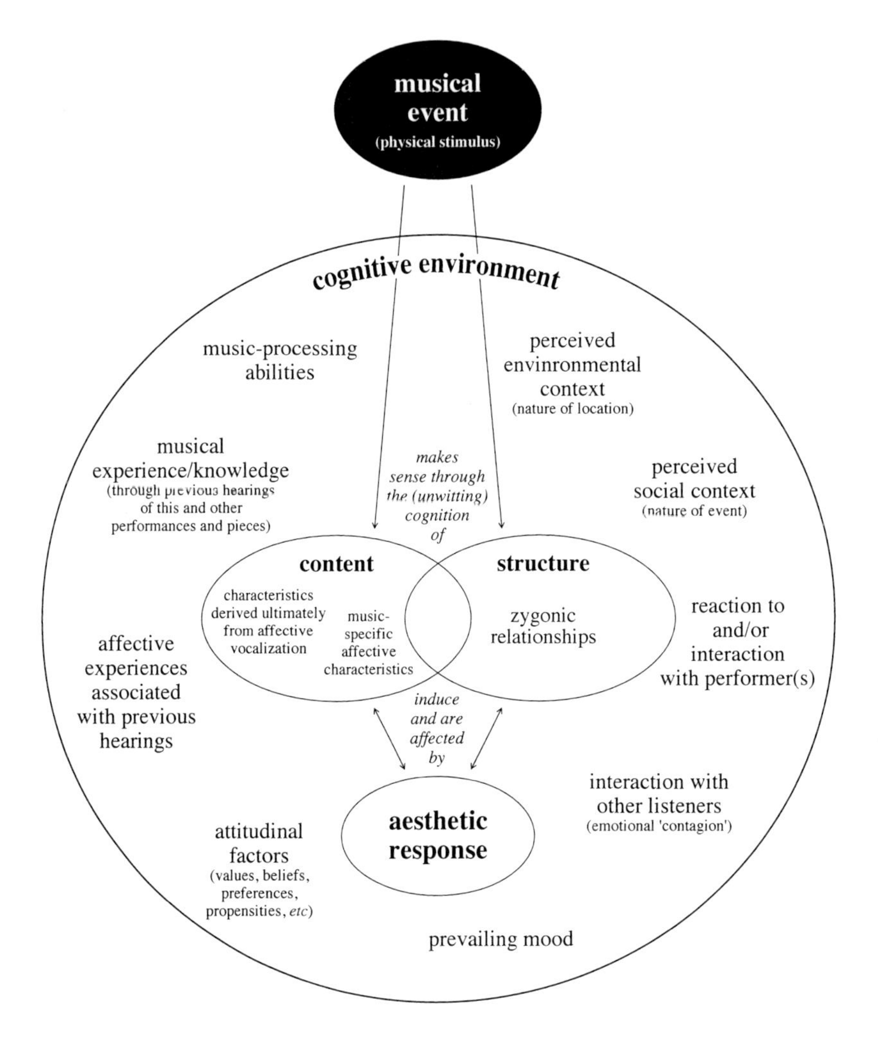

35.3 Adam Ockelford, *Repetition in Music: Theoretical and Metatheoretical Perspectives* (Burlington, VT: Ashgate, 2005), 32.

Figure 35.4 shows a more structured conceptual space. Its Venn-like diagram theorizes how timbre, pitch height, pulse, and movement arise from different combinations of micro- and macroexpressions within the temporal and spatial domains. Gracefully curved lines allow harmony, melody, rhythm, and micromodulation to exist within a single domain (for example, melody exists within the macrospatial

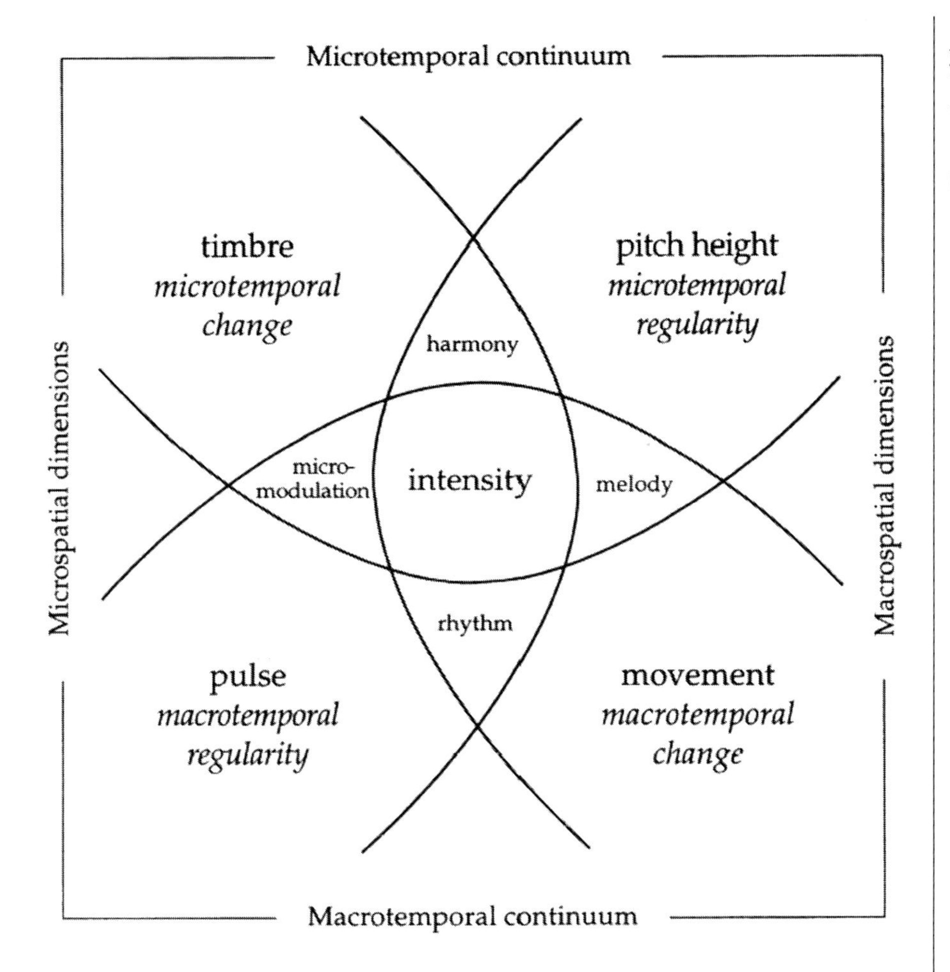

35.4 Erik Christensen, *The Musical Timespace: A Theory of Music Listening* (Aalborg: Aalborg University Press, 1996), 1:153.

dimension but incorporates elements from both micro- and macrotemporal dimensions), while intensity exists along the full range of both domains. The content of the image invites thoughtful consideration, and its attractive layout makes such consideration pleasant.

Figure 35.5 is also a Venn-like diagram, but one rich with layered detail. It proposes a multifaceted model of vocal delivery. Characteristics in boldface are assigned to one or more of three broad categories, while roman text enumerates possible values within some of those characteristics. The light-gray lines that surround the three large ovals (and two subovals) are essential to their demarcation. The shading, while not strictly necessary, helps set off the overlapping regions. See also online figure 35.6.

Figures 35.7 and 35.8 illustrate the difference between schematic and procedural diagrams and will help us transition from one type to the other in this chapter. Paul Hindemith's chord classification system assigns chords to one of six

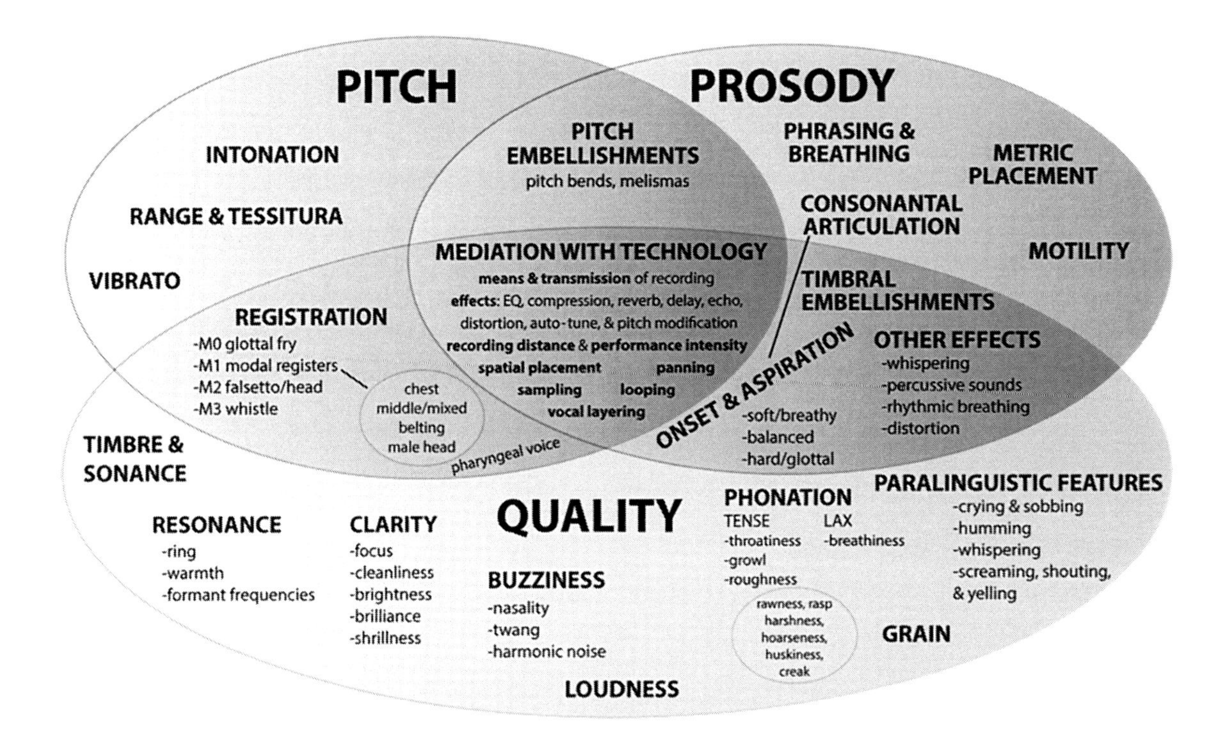

35.5 Victoria Malawey, *A Blaze of Light in Every Word: Analyzing the Popular Singing Voice* (New York: Oxford University Press, 2020), 7.

Facing, 35.7 Paul Hindemith, *The Craft of Musical Composition*, ed. Arthur Mendel and Otto Ortmann (New York: Associated Music, 1945), foldout.

categories based on intervallic and rootedness characteristics. Hindemith's own representation puts the six types in a table that requires an initial determination based on the presence or absence of tritones, then a downward scan to match other characteristics. The flowchart representation in figure 35.8 puts the focus on the classification process itself. Although it allows one to reach the correct chord type faster, this approach does not offer examples of each type or provide a sense of how the categories relate to one another. The emphasis on process over content makes flowcharts generally inferior to other representations. There is almost always a better way to express the information, and in this case, a resetting of figure 35.7 would be that better way.

We continue exploring the representation of procedural information by comparing three descriptions of how to determine the prime form of a pitch-class set. (The prime form is a canonical version of a group of pitch-class sets that are related by transposition or inversion. It is essentially a prototype used to represent all members of the set, in the way that "C–E–G" might be considered the prime form of the set class we call "major triads" in tonal music.). Several sources (Forte 1973, Rahn 1987, Roig-Francolí 2007) spread the prime-form procedure out in prose across multiple pages. I am interested instead in instances in which a single figure illustrates the steps.

The representation in figure 35.9 spells out the prime-form procedure as an algorithm in pseudocode—that is, language that could be translated directly into

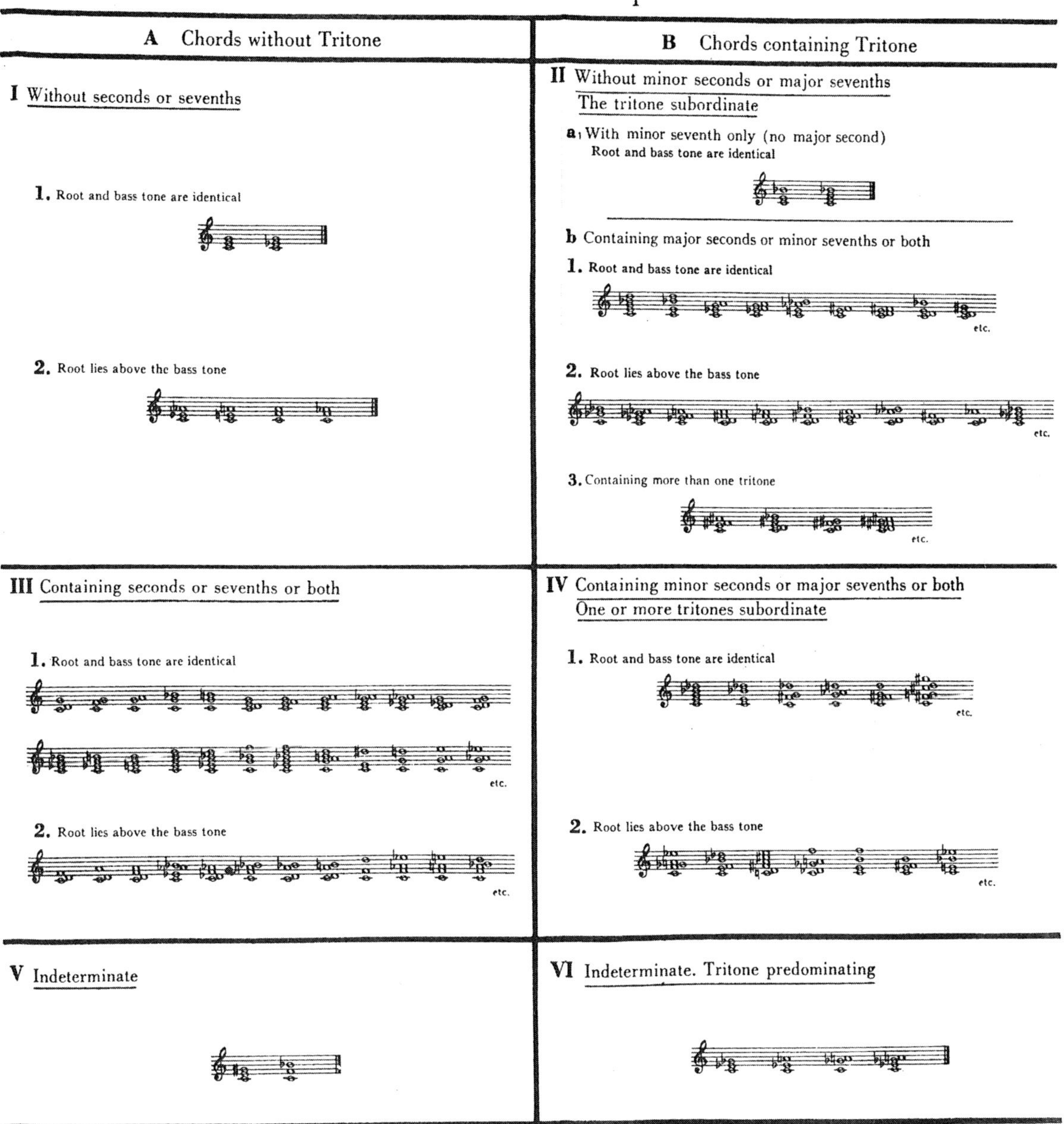

a computer language. Those with programming backgrounds will understand this approach and appreciate its terseness. Others may find it less congenial. Nevertheless, numbering the steps helps avoid the mistake of skipping or repeating a step and is always a good idea when representing procedures of more than three or four steps.

35.8 J. Kent Williams, *Theories and Analyses of Twentieth-Century Music* (Fort Worth: Harcourt Brace, 1997), 230.

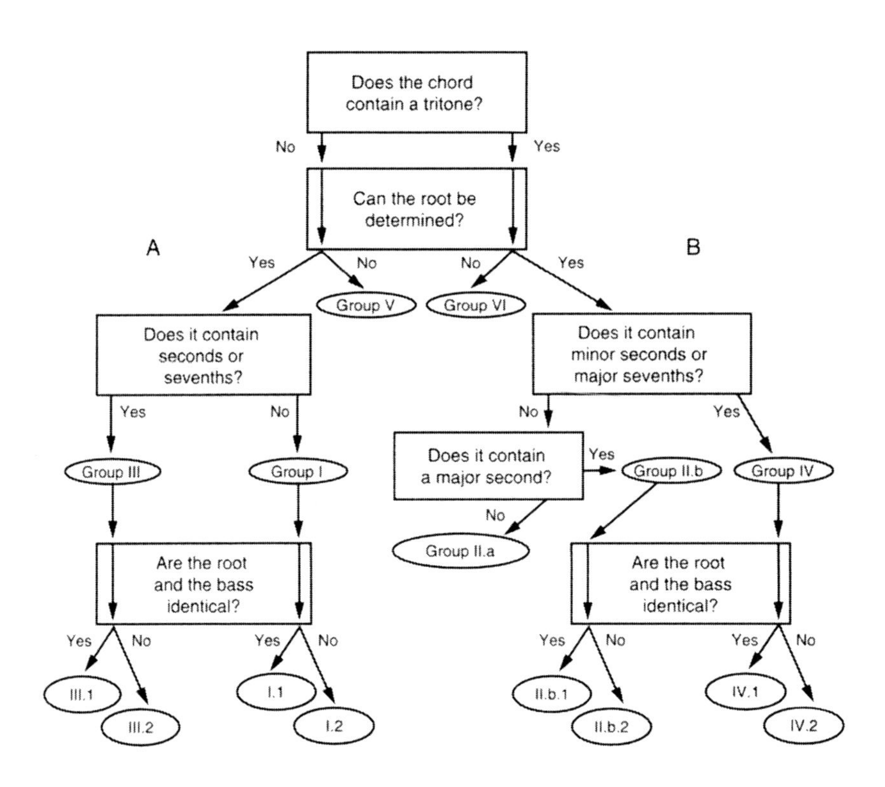

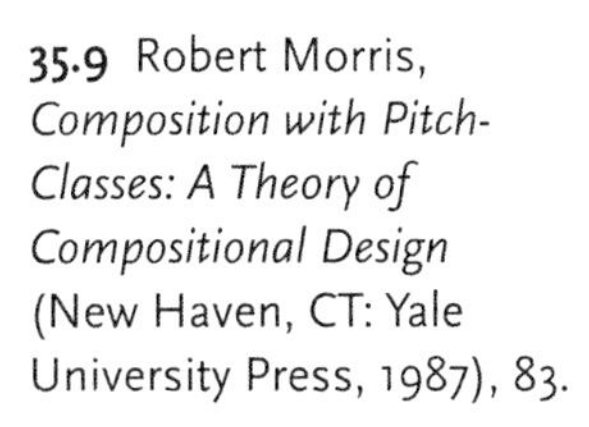

35.9 Robert Morris, *Composition with Pitch-Classes: A Theory of Compositional Design* (New Haven, CT: Yale University Press, 1987), 83.

Prime Form Algorithm

1. START: select P.
2. n = #P.
3. Arrange P and IP in ascending order of pc number.
4. q = 1.
5. Produce the set S of pccycs comprised of all rotations of P and IP (S will have 2 times #P members).
6. Find the subset S′ of all pccycs C in S where $i\langle C_0, C_{(n-q)}\rangle$ is minimal.
7. S ← S′ (delete all members of S not in S′).
8. If S has only one member, go to 12.
9. If q = n, discard all (T_n-related) members of S except one and go to 12.
10. q = q + 1.
11. Go to 6.
12. Transpose the (sole) member of S so it begins on 0.
13. END: the resulting pccyc is called the prime form of P.

Just because something appears in prose rather than pseudocode does not necessarily mean it is easier to understand, however. The prime-form procedure in figure 35.10 contains the same information as figure 35.9, minus "and IP" in step 3 because figure 35.10 does not consider inversionally related sets to be equivalent. While the layout feels more user-friendly, the procedure described has exactly the same steps (find the most compact rotation of pitches, deal with ties, and transpose the final set to start on zero), and these discrete steps are linked together within a paragraph. Carrying out each step requires rescanning the paragraph to find where the last one ended and the next begins. However, the sample problem in the second column enhances the image's pedagogy.

Write the pc's of set **S** in ascending order so that the interval from first to last pc is less than 12. Consider all rotations $r_n(\mathbf{S})$, *n* ranging from 0 to |**S**|-1. Determine which rotation has the smallest value for the directed interval from the first pc to the last. This set is in *normal order*.	*pc's* 0 1 5 6 10 1 5 6 10 0 5 6 10 0 1 6 10 0 1 5 10 0 1 5 6	*ipc(first, last)* ipc(0,10)=10 ipc(1,0)=11 ipc(5,1)=8 ✓ ipc(6,5)=11 ipc(10,6)=8 ✓
There may be more than one rotation that fulfills the requirements of normal order. In any case, the inversions of all sets selected in the first step fulfill the requirement of normal order and must be considered.[70] Now calculate the ordered interval between the first and second pc's and select the smallest entry. Again, there may be several orderings that pass this test. Continue with the interval between second and third, third and fourth, until only one ordered set remains. This set is in *best normal order*.	*pc's* 5 6 10 0 1 ↓I 11 0 2 6 7 10 0 1 5 6 ↓I 6 7 11 0 2 *pc's* 5 6 10 0 1 11 0 2 6 7 6 7 11 0 2	*ipc(first, second)* ipc(5,6)=1 ✓ ipc(11,0)=1 ✓ ipc(10,0)=2 ipc(6,7)=1 ✓ *ipc(second, third)* ipc(6,10)=4 ipc(0,2)=2 ✓ ipc(7,11)=4
Best normal order, transposed to start on 0, is the *prime form* (also called *normal order*) of the set.	T_1((11, 0, 2, 6, 7)) = **(0, 1, 3, 7, 8)**	

35.10 Peter Castine, *Set Theory Objects: Abstractions for Computer-Aided Analysis and Composition of Serial and Atonal Music* (New York: P. Lang, 1994), 66.

The two parts of figure 35.11 combine the best of these two approaches. The upper image outlines the steps for finding a set's normal form, while the lower part describes how to find the prime form. The figure presents the steps discretely, but in prose form. Because a handful of situations require special treatment, the inclusion of three examples is helpful.

In sum, descriptions of multistep procedures should list the steps one at a time. Numbering the steps reinforces the importance of taking them in order and also makes it easier to come back to the place where one left off. Including one or more examples to illustrate the procedure in action can help reinforce how it works.

Procedure	Example 1		Example 2		Example 3	
Excluding doublings, write the pitch classes as though they were a scale, ascending within an octave. (There will be as many different ways of doing this as there are pitch classes in the set, since an ordering can begin on any of them.) Calculate the interval span of each.	A–B♭–F B♭–F–A F–A–B♭	A→F = 8 B♭→A = 11 F→B♭ = 5	F–A♭–A–C♯ A♭–A–C♯–F A–C♯–F–A♭ C♯–F–A♭–A	F→C♯ = 8 A♭→F = 9 A♭→A = 11 C♯→A = 8	C–E–G♯–A–B E–G♯–A–B–C G♯–A–B–C–E A–B–C–E–G♯ B–C–E–G♯–A	C→B = 11 E→C = 8 G♯→E = 8 A→G♯ = 11 B→A = 10
Rule 1: The ordering that has the smallest interval from first to last . (from lowest to highest) is the normal form.	F–A–B♭ F→B♭ = 5 [F, A, B♭] is the normal form.		F–A♭–A–C♯ C♯–F–A♭–A	F→C♯ = 8 C♯→A = 8	E–G♯–A–B–C G♯–A–B–C–E	E→C = 8 G♯→E = 8
Rule 2: If there is a tie under Rule 1, the normal form is the ordering that packs the pitch classes most closely *to one end or the other*. It will have a relatively large concentration of big intervals *at the top or at the bottom.*			F A♭ A C♯ 3 1 4 C♯ F A♭ A 4 3 1 [C♯, F, A♭, A] is more packed to the top than its competitor is to either the top or the bottom, so it is the normal form.		E G♯ A B C 4 1 2 1 G♯ A B C E 1 2 1 4 The two versions are identically arranged, one packed to the top and the other packed to the bottom, so both [E, G♯, A, B, C] and [G♯, A, B, C, E] are normal forms.	
It there is more than one normal form (as there will be for inversionally symmetrical sets, to be discussed later), prefer the one that is most packed to the bottom.					[G♯, A, B, C, E] is more packed to the bottom (bigger intervals at the top), so it is preferred.	

Procedure	Example 1	Example 2	Example 3
1. Start with a set in normal form.	[C♯, F, F♯, G]	[B♭, D, F, F♯]	[F, F♯, A]
2. Extract the succession of intervals.	C♯ F F♯ G 4 1 1	B♭ D F F♯ 4 3 1	F F♯ A 1 3
3. Choose the interval succession reading from left to right or right to left, whichever has the smallest intervals toward the left.	1-1-4 is more packed to the left than 4-1-1.	1-3-4 is more packed to the left than 4-3-1.	1-3 is more packed to the left than 3-1.
4. Replicate that interval succession starting on 0.	0 1 2 6 1 1 4 Prime form = (0126)	0 1 4 8 1 3 4 Prime form = (0148)	0 1 4 1 3 Prime form = (014)

35.11 Joseph N. Straus, *Introduction to Post-tonal Theory*, 4th ed. (New York: W. W. Norton, 2016), 45 (*above*), 67 (*below*).

Chapter 36 Formal Models

The basic elements of an analysis of musical form include the division of a piece into sections and then an exploration of the relationships between those sections. Chapter 47 explores analysis of form in actual works. This chapter presents abstract formal models of sonata-allegro form, continuing the previous chapter's focus on schematic images. Sonata form provides a useful vehicle in which to explore schematic representations, since the issues are so well known: Should an image represent the two-part origins from the Baroque binary form, or the three-part division into exposition, development, and recapitulation? How should one represent the range of tonal models? Which method will best show the form's range of thematic options? And so on.

As noted in the previous chapter, sometimes a highly reductive schematic representation can be powerful. Figure 36.1 lays out the basic structure of sonata form in the starkest possible way. It represents the structure of classical sonata form at its most fundamental, conveying clearly the two-part/three-part dichotomy that characterizes theories of sonata form, but nothing else.

‖: Exposition :‖‖: Development—Recapitulation:‖

36.1 William E. Caplin, *Analyzing Classical Form: An Approach for the Classroom* (New York: Oxford University Press, 2013), 262.

The more fleshed-out model of sonata form in online figure 36.2 fully embraces a three-part view of the form, including each section's essential characteristics. The version in figure 36.3 shows three parts visually, but the image's text suggests a two-part structure—kind of. The development and retransition are labeled as the "First section of the second part," while the recapitulation is called simply the "second section." The graphical nesting of smaller sections into larger sections keeps the hierarchy clear. Dashed lines add visual noise, however, and they are not needed here.

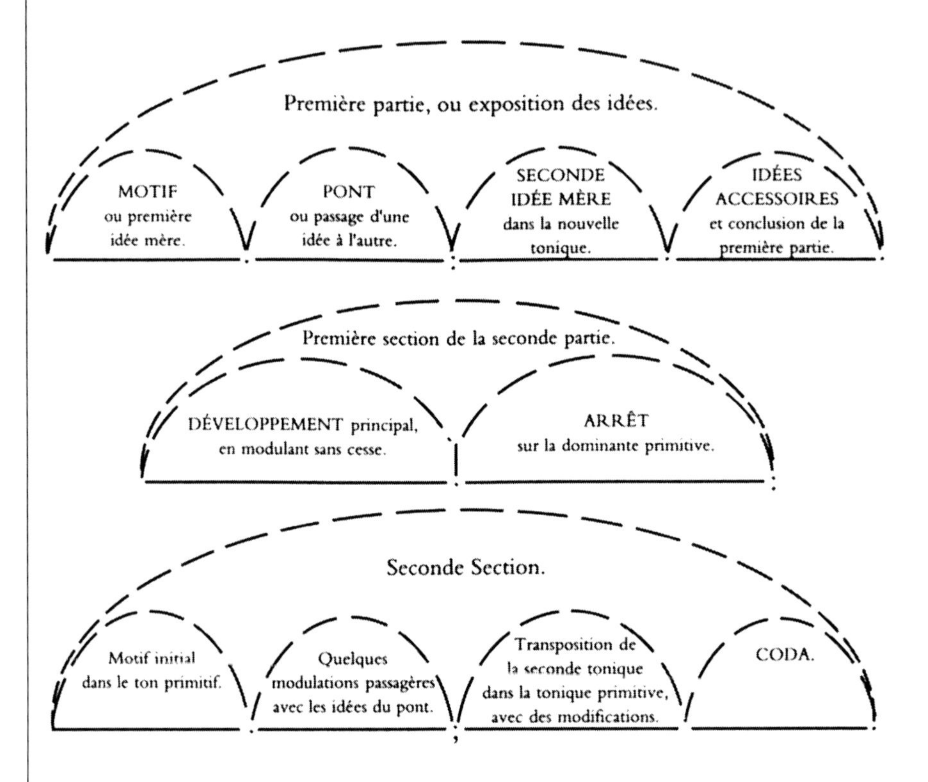

36.3 Anton Reicha, *Traité de haute composition musicale* (Paris: Zetter, 1824–25), 2:300.

Figure 36.4 provides evidence in favor of graphical representations. The prose helpfully explains some of the form's subtleties (even while it offers an unusual characterization of the form), but without a drawing, one must labor to construct a mental model of the form.

36.4 Hugo Leichtentritt, *Musical Form* (Cambridge, MA: Harvard University Press, 1951), 123.

First Part. Exposition of the themes: First or principal theme in the tonic. Transition to the dominant. Second theme, in the dominant. Third or closing theme in the dominant.

Second Part. Development or working-out section: Free fantasy on motifs from the first part, chosen at will. Much more elaborate modulation, final return to the main key, the tonic, and to the principal theme.

Third Part. Reprise: Repetition of the entire first part, this time, however, the second and third themes are not written in the dominant key, but in the tonic. The little appendage at the very close of the movement is called a coda (in Italian, a tail). In case the coda is considerably expanded, as often happens in Beethoven's larger works, it is considered a special fourth part.

Fourth Part: Often corresponding to the second part, as a second development section. In this case the parallelism between the first and third parts and the second and fourth parts is apparent.

We conclude with a side-by-side comparison of two graphical depictions of the sonata model that bear a strong resemblance to each other, despite their different audiences (fig. 36.5 is from a music appreciation text, while fig 36.6 is from a professional treatise on sonata theory). Both are excellent examples of clarity and efficient design. A close analysis of some of their details will clarify how powerful even small design decisions can be at shaping meaning.

36.5 Joseph Kerman, *Listen*, 3rd brief ed. (New York: Worth, 1996), 166.

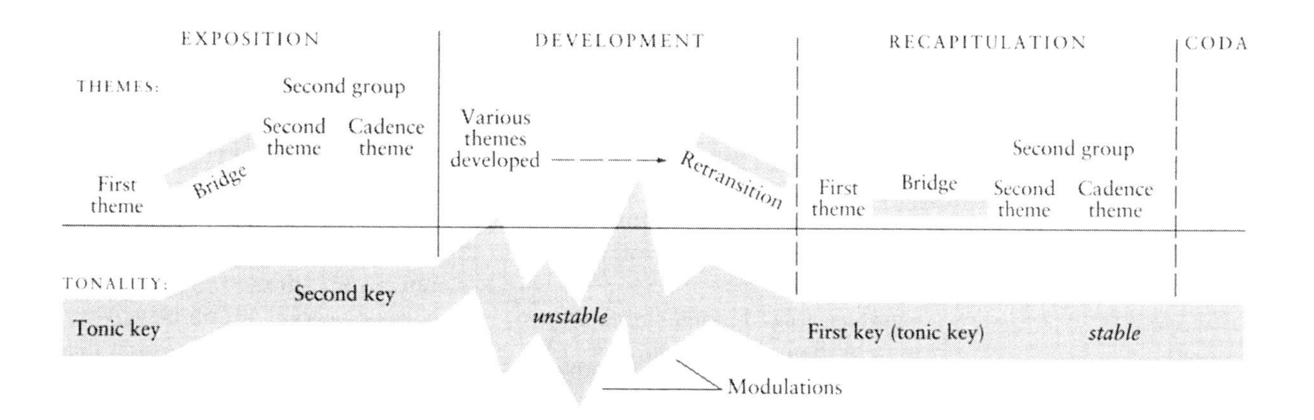

The images represent the form at essentially the same level of magnification. Both images show the tonal contrast that begins during the link to the second theme area by drawing a diagonal line that implies an infusion of energy and then showing formal elements on a higher level until the tonic is restored at the recapitulation. This infuses the images with a dynamism that is nicely suggestive of the experience of listening to a sonata form. There are two notable differences in how the images treat the sonata scheme during the nontonic part of the form. First, figure 36.5 conveys explicitly the harmonic wanderings that characterize the development section, whereas figure 36.6 is curiously silent on the tonal plan of the development, except to note that it ends on (not in) the dominant. Second, the first image suggests that the energy infusion that occurs during the bridge between the first and second themes is reversed during the retransition at the end of the development, whereas the second represents it (more accurately) as a sudden drop with the arrival on tonic at the recapitulation.

Another, more significant difference between the images lies in how, through subtle means, they position the role of the development section within the form. Figure 36.5 preserves the binary origins of sonata form by dividing exposition and development with a solid line while using dashed lines between development, recapitulation, and coda, suggesting that they are part of a single larger section. In figure 36.6, brackets at the bottom suggest a three-part division, but the line

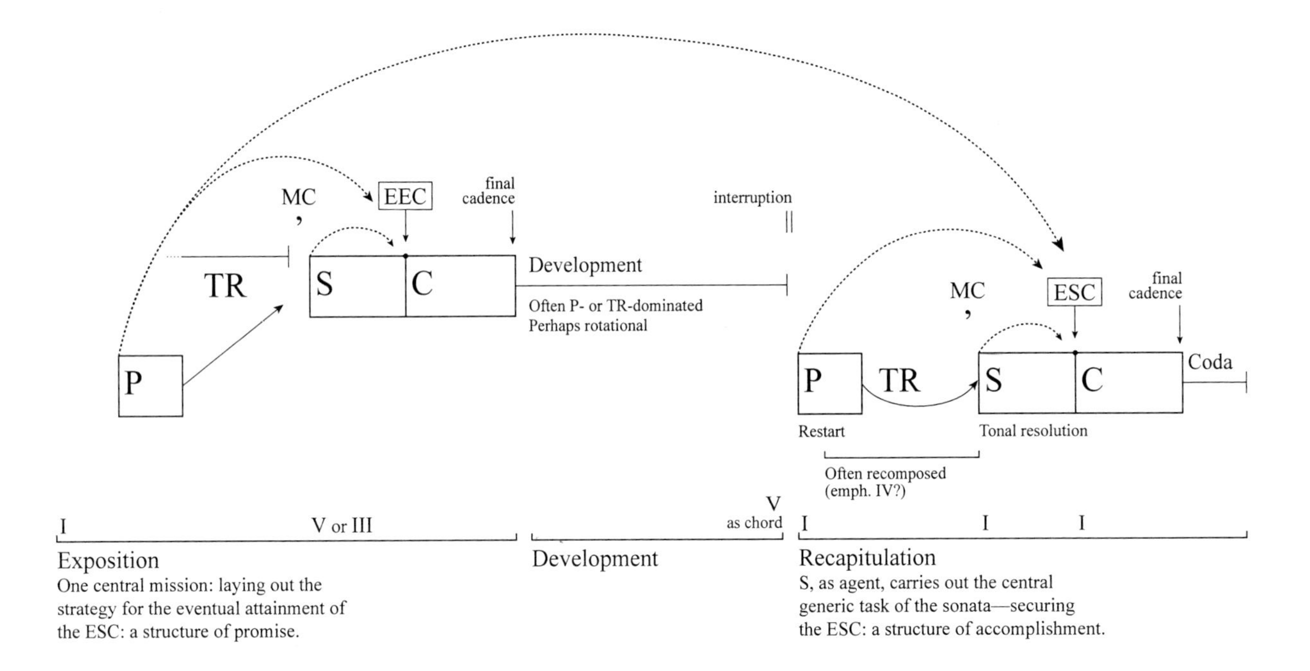

36.6 James Hepokoski and Warren Darcy, *Elements of Sonata Theory: Norms, Types, and Deformations in the Late-Eighteenth-Century Sonata* (New York: Oxford University Press, 2011), 17.

extending directly from the exposition's C block and terminating at the interruption that precedes the recapitulation unmistakably links the development to the exposition, as an extension of the tonal contrast introduced during the transition.

I will close with a handful of additional comparisons. Given our tendency to read top to bottom, the exposition-development-recapitulation structure is the lede in figure 36.5, while it is secondary in figure 36.6. In figure 36.5, next most prominent is the tonal scheme, which is dominated by the thick, shaded area (which is green in the original). In figure 36.6, first billing goes to the thematic areas and the story told with and around them. Its placement of P, S, and C in boxes conveys that these are *sections*, elevated in significance over the other parts of the form, and helps draw attention to the fact that the alignment of P vis-à-vis S + C is quite different in exposition versus recapitulation. Figure 36.6 provides some additional detail about the role of exposition and recapitulation through the use of small-font text. It should be noted that figure 36.6 is accompanied by figure 6.3 (see chap. 6), which provides a more detailed look at the sonata exposition. Providing information with the detail of figure 6.3 in the synoptic image of figure 36.6 would create a more powerful figure. Indeed, a review of schematic diagrams of musical form suggests that the addition of more information layers would enrich many of them.

CHAPTER 37 Pitch-Class Set Tables

Tables are multivariate, inherently organizing information in a way that facilitates discovery and supports comparison (see chap. 9). The list of set classes is a classic table-based music visualization. Many such tables have been published. They differ in how they are laid out, how much information they include, and how they sort the sets. In general, a richer set of information is preferable, so long as that information provides value to the intended audience. We will look at excerpts of several set tables and discuss their individual strengths. Since set tables typically occupy several pages, I have chosen to compare the treatment of four- and eight-note sets. (Throughout this chapter, I use the more wieldy word *set* instead of *set class* to refer to a collection of pitch-class sets related by transposition and inversion.)

Figure 37.1 displays two groups of three columns. Sets in the right group are the complements of those in the left. Within each group, the first column provides Forte's label for each set, plus a parenthetical indication of the number of distinct members for set classes that are symmetrical under transposition, inversion, or both (the default 24 is not listed for other sets), though curiously, it lists this only for the smaller of the complementary set pair, almost as an afterthought rather than an independent bit of information meriting its own column. The second column lists the prime form, and the last provides the interval-class vector. The information is concise; the only extraneous ink appears in the commas separating prime-form members.

37.1 Allen Forte, *The Structure of Atonal Music* (New Haven, CT: Yale University Press, 1973), 179.

4-1(12)	0,1,2,3	321000	8-1	0,1,2,3,4,5,6,7	765442
4-2	0,1,2,4	221100	8-2	0,1,2,3,4,5,6,8	665542
4-3	0,1,3,4	212100	8-3	0,1,2,3,4,5,6,9	656542
4-4	0,1,2,5	211110	8-4	0,1,2,3,4,5,7,8	655552
4-5	0,1,2,6	210111	8-5	0,1,2,3,4,6,7,8	654553
4-6(12)	0,1,2,7	210021	8-6	0,1,2,3,5,6,7,8	654463
4-7(12)	0,1,4,5	201210	8-7	0,1,2,3,4,5,8,9	645652
4-8(12)	0,1,5,6	200121	8-8	0,1,2,3,4,7,8,9	644563
4-9(6)	0,1,6,7	200022	8-9	0,1,2,3,6,7,8,9	644464
4-10(12)	0,2,3,5	122010	8-10	0,2,3,4,5,6,7,9	566452
4-11	0,1,3,5	121110	8-11	0,1,2,3,4,5,7,9	565552
4-12	0,2,3,6	112101	8-12	0,1,3,4,5,6,7,9	556543
4-13	0,1,3,6	112011	8-13	0,1,2,3,4,6,7,9	556453
4-14	0,2,3,7	111120	8-14	0,1,2,4,5,6,7,9	555562
4-Z15	0,1,4,6	111111	8-Z15	0,1,2,3,4,6,8,9	555553
4-16	0,1,5,7	110121	8-16	0,1,2,3,5,7,8,9	554563
4-17(12)	0,3,4,7	102210	8-17	0,1,3,4,5,6,8,9	546652
4-18	0,1,4,7	102111	8-18	0,1,2,3,5,6,8,9	546553
4-19	0,1,4,8	101310	8-19	0,1,2,4,5,6,8,9	545752
4-20(12)	0,1,5,8	101220	8-20	0,1,2,4,5,7,8,9	545662
4-21(12)	0,2,4,6	030201	8-21	0,1,2,3,4,6,8,10	474643
4-22	0,2,4,7	021120	8-22	0,1,2,3,5,6,8,10	465562
4-23(12)	0,2,5,7	021030	8-23	0,1,2,3,5,7,8,10	465472
4-24(12)	0,2,4,8	020301	8-24	0,1,2,4,5,6,8,10	464743
4-25(6)	0,2,6,8	020202	8-25	0,1,2,4,6,7,8,10	464644
4-26(12)	0,3,5,8	012120	8-26	0,1,2,4,5,7,9,10	456562
4-27	0,2,5,8	012111	8-27	0,1,2,4,5,7,8,10	456553
4-28(3)	0,3,6,9	004002	8-28	0,1,3,4,6,7,9,10	448444
4-Z29	0,1,3,7	111111	8-Z29	0,1,2,3,5,6,7,9	555553

The sorting of the list, according to Forte's set names, almost corresponds to a sort by interval-class (ic) vector. In the author's original published list (Forte 1964), the ic vector determined set classes. After transposition and inversion replaced the ic vector as the basis for set equivalence (as in Forte 1973), sets that had the same ic vectors but were not related under T_nI had to be added to the list. Unfortunately, since the original set labels were already in use, Forte decided to append the newly minted sets to the end of the numbering system, where they fell outside the vector-based sorting. This list thus sorts by Forte number.

A set's prime form would provide a better basis for sorting a set table, and this approach appears in figure 37.2, which contains exactly the same information as 37.1, though the columns are reordered. The curious use of em dashes in the Forte numbers adds unnecessary ink, and center-aligning them has made it more difficult than desirable to scan for a particular tetrachord, since the names jump left and right (see the discussion of online fig. 12.2). The number of distinct members is replaced with a degree of symmetry, while interval-class vectors have added unnecessary punctuation.

TETRACHORDS						OCTACHORDS
[0,1,2,3]	2	<3,2,1,0,0,0>	4—1/8—1	<7,6,5,4,4,2>	2	[0,1,2,3,4,5,6,7]
[0,1,2,4]	1	<2,2,1,1,0,0>	4—2/8—2	<6,6,5,5,4,2>	1	[0,1,2,3,4,5,6,8]
[0,1,2,5]	1	<2,1,1,1,1,0>	4—4/8—4	<6,5,5,5,5,2>	1	[0,1,2,3,4,5,7,8]
[0,1,2,6]	1	<2,1,0,1,1,1>	4—5/8—5	<6,5,4,5,5,3>	1	[0,1,2,3,4,6,7,8]
[0,1,2,7]	2	<2,1,0,0,2,1>	4—6/8—6	<6,5,4,4,6,3>	2	[0,1,2,3,5,6,7,8]
[0,1,3,4]	2	<2,1,2,1,0,0>	4—3/8—3	<6,5,6,5,4,2>	2	[0,1,2,3,4,5,6,9]
[0,1,3,5]	1	<1,2,1,1,1,0>	4—11/8—11	<5,6,5,5,5,2>	1	[0,1,2,3,4,5,7,9]
[0,1,3,6]	1	<1,1,2,0,1,1>	4—13/8—13	<5,5,6,4,5,3>	1	[0,1,2,3,4,6,7,9]
[0,1,3,7]	1	<1,1,1,1,1,1>	4—Z29/8—Z29	<5,5,5,5,5,3>	1	[0,1,2,3,5,6,7,9]
[0,1,4,5]	2	<2,0,1,2,1,0>	4—7/8—7	<6,4,5,6,5,2>	2	[0,1,2,3,4,5,8,9]
[0,1,4,6]	1	<1,1,1,1,1,1>	4—Z15/8—Z15	<5,5,5,5,5,3>	1	[0,1,2,3,4,6,8,9]
[0,1,4,7]	1	<1,0,2,1,1,1>	4—18/8—18	<5,4,6,5,5,3>	1	[0,1,2,3,5,6,8,9]
[0,1,4,8]	1	<1,0,1,3,1,0>	4—19/8—19	<5,4,5,7,5,2>	1	[0,1,2,4,5,6,8,9]
[0,1,5,6]	2	<2,0,0,1,2,1>	4—8/8—8	<6,4,4,5,6,3>	2	[0,1,2,3,4,7,8,9]
[0,1,5,7]	1	<1,1,0,1,2,1>	4—16/8—16	<5,5,4,5,6,3>	1	[0,1,2,3,5,7,8,9]
[0,1,5,8]	2	<1,0,1,2,2,0>	4—20/8—20	<5,4,5,6,6,2>	2	[0,1,2,4,5,7,8,9]
[0,1,6,7]	4	<2,0,0,0,2,2>	4—9/8—9	<6,4,4,4,6,4>	4	[0,1,2,3,6,7,8,9]
[0,2,3,5]	2	<1,2,2,0,1,0>	4—10/8—10	<5,6,6,4,5,2>	2	[0,2,3,4,5,6,7,9]
[0,2,3,6]	1	<1,1,2,1,0,1>	4—12/8—12	<5,5,6,5,4,3>	1	[0,1,3,4,5,6,7,9]
[0,2,3,7]	1	<1,1,1,1,2,0>	4—14/8—14	<5,5,5,5,6,2>	1	[0,1,2,4,5,6,7,9]
[0,2,4,6]	2	<0,3,0,2,0,1>	4—21/8—21	<4,7,4,6,4,3>	2	[0,1,2,3,4,6,8,10]
[0,2,4,7]	1	<0,2,1,1,2,0>	4—22/8—22	<4,6,5,5,6,2>	1	[0,1,2,3,5,6,8,10]
[0,2,4,8]	2	<0,2,0,3,0,1>	4—24/8—24	<4,6,4,7,4,3>	2	[0,1,2,4,5,6,8,10]
[0,2,5,7]	2	<0,2,1,0,3,0>	4—23/8—23	<4,6,5,4,7,2>	2	[0,1,2,3,5,7,8,10]
[0,2,5,8]	1	<0,1,2,1,1,1>	4—27/8—27	<4,5,6,5,5,3>	1	[0,1,2,4,5,7,8,10]
[0,2,6,8]	4	<0,2,0,2,0,2>	4—25/8—25	<4,6,4,6,4,4>	4	[0,1,2,4,6,7,8,10]
[0,3,4,7]	2	<1,0,2,2,1,0>	4—17/8—17	<5,4,6,6,5,2>	2	[0,1,3,4,5,6,8,9]
[0,3,5,8]	2	<0,1,2,1,2,0>	4—26/8—26	<4,5,6,5,6,2>	2	[0,1,2,4,5,7,9,10]
[0,3,6,9]	8	<0,0,4,0,0,2>	4—28/8—28	<4,4,8,4,4,4>	8	[0,1,3,4,6,7,9,10]

37.2 John Rahn, *Basic Atonal Theory* (New York: Longman, 1987), 140–41.

The table in figure 37.3 blends features from figures 37.1 and 37.2, placing complementary sets across from each other, adopting and extending the minimal-punctuation approach of figure 37.1, and sorting by prime form rather than Forte number as in figure 37.2. It tweaks the degree of symmetry information by distinguishing the number of T_n operations from the number of T_nI operations under which a set is invariant. Descriptors for sets that have identities in other

	Tetrachords						Octachords	
Chromatic	(0123)	4-1	321000	1, 1	765442	8-1	(01234567)	Chromatic
	(0124)	4-2	221100	1, 0	665542	8-2	(01234568)	
	(0125)	4-4	211110	1, 0	655552	8-4	(01234578)	
	(0126)	4-5	210111	1, 0	654553	8-5	(01234678)	
	(0127)	4-6	210021	1, 1	654463	8-6	(01235678)	
Octatonic	(0134)	4-3	212100	1, 1	656542	8-3	(01234569)	
Major/Phrygian	(0135)	4-11	121110	1, 0	565552	8-11	(01234579)	
	(0136)	4-13	112011	1, 0	556453	8-13	(01234679)	
All interval	(0137)	4-Z29	111111	1, 0	555553	8-Z29	(01235679)	
Hexatonic	(0145)	4-7	201210	1, 1	645652	8-7	(01234589)	
All interval	(0146)	4-Z15	111111	1, 0	555553	8-Z15	(01234689)	
	(0147)	4-18	102111	1, 0	546553	8-18	(01235689)	
	(0148)	4-19	101310	1, 0	545752	8-19	(01245689)	
	(0156)	4-8	200121	1, 1	644563	8-8	(01234789)	
	(0157)	4-16	110121	1, 0	554563	8-16	(01235789)	
Major seventh chord	(0158)	4-20	101220	1, 1	545662	8-20	(01245789)	
	(0167)	4-9	200022	2, 2	644464	8-9	(01236789)	
Dorian/octatonic	(0235)	4-10	122010	1, 1	566452	8-10	(02345679)	
	(0236)	4-12	112101	1, 0	556543	8-12	(01345679)	
	(0237)	4-14	111120	1, 0	555562	8-14	(01245679)	
Whole-tone	(0246)	4-21	030201	1, 1	474643	8-21	(0123468T)	
	(0247)	4-22	021120	1, 0	465562	8-22	(0123568T)	
Whole-tone	(0248)	4-24	020301	1, 1	464743	8-24	(0124568T)	
Stack of fourths/fifths	(0257)	4-23	021030	1, 1	465472	8-23	(0123578T)	Diatonic octad
Dominant/half-diminished seventh	(0258)	4-27	012111	1, 0	456553	8-27	(0124578T)	
French augmented sixth	(0268)	4-25	020202	2, 2	464644	8-25	(0124678T)	
Major/minor triad	(0347)	4-17	102210	1, 1	546652	8-17	(01345689)	
Minor seventh chord	(0358)	4-26	012120	1, 1	456562	8-26	(0134578T)	
Diminished seventh chord; maximally even	(0369)	4-28	004002	4, 4	448444	8-28	(0134679T)	Octatonic; maximally even

37.3 Joseph Straus, *Introduction to Post-tonal Theory*, 4th ed. (New York: W. W. Norton, 2016), 379.

contexts ("Chromatic") are a nice addition, though the boxes around them are unnecessary. Among the tables discussed here, including a fourth that is described in online figure 37.4, this one employs the most effective column spacing.

Set-class tables are a bit like phone books of yore: a handy place to look up information that we haven't quite memorized. A clean, information-rich design, with no wasted ink, always works best.

CHAPTER 38 Instrument Ranges

Composers, budding or experienced, often need a reference showing instrument ranges. The information helps them make sure not to write music beyond where an instrument is capable of playing or (ideally) beyond where it is comfortable doing so. Because of the substantial amount of information and its multivariate nature, designing such a reference requires creative solutions. The thought process involved can extend to other information-rich tasks.

Figure 38.1, from the manual for the music-notation software Finale, shows instrument ranges for nearly every common musical instrument, including ranges for three levels of performer, along with an indication of the transposition level of each and, implicitly, the clef that is ordinarily used for that instrument. Expressing the ranges using notation and using the written pitch rather than the sounding pitch works well for the image's most common audience, composers trying not to exceed an instrument's written range. The image's comprehensiveness makes it a handy resource, although that comprehensiveness is hampered by its visual uniformity. While the instruments are arranged in roughly score order, viewers may find it difficult to tell where one family ends and another starts. The instrument names and other text appear in a font whose color and weight are the same as those of the staff. Even rendering the instrument names in boldface would improve usability. Also, the design does not facilitate easy comparison of ranges between instruments.

Figure 38.2 and online figure 38.3 present information about the ranges of the cornet family and the tenor trombone. Figure 38.2 gives the full list of available

38.1 Finale, "Finale 2012 Instrument Ranges," accessed June 7, 2017, excerpt, https://usermanuals.finalemusic.com/Finale2012Win/Content/Finale/Instrument_Ranges.htm.

pitches for each of three different cornets, with each notated pitch aligned over the sounding pitch (Klang). The ranges are divided into registers; open pitches (those played with no valves depressed) are helpfully shown in open noteheads. Aligning the instruments in this way makes it possible to see where the same sounding pitch falls within the range of a particular instrument. Online figure 38.3 shows similar information for the tenor trombone. A single image that showed similar information for all instruments would of course be unwieldy. Individually, however, each image is quite successful.

38.2 Hermann Erpf, *Lehrbuch der Instrumentation und Instrumentenkunde* (Mainz: Schott, 1959), 349.

38.4 Brian Blood, "Sounding Range of Orchestral Instruments," *Music Theory Online: Score Formats*, last modified November 8, 2018, http://www.dolmetsch.com/musictheory26.htm.

Figure 38.4 tries valiantly to show thirty-two instrument ranges in a single place. Unfortunately, the design fails rather spectacularly. Because all of the lines are undifferentiated, the image looks more like a woven basket than anything else. Since horizontal lines are information-bearing while vertical ones are not, one could improve the image by simply differentiating the horizontal lines from the

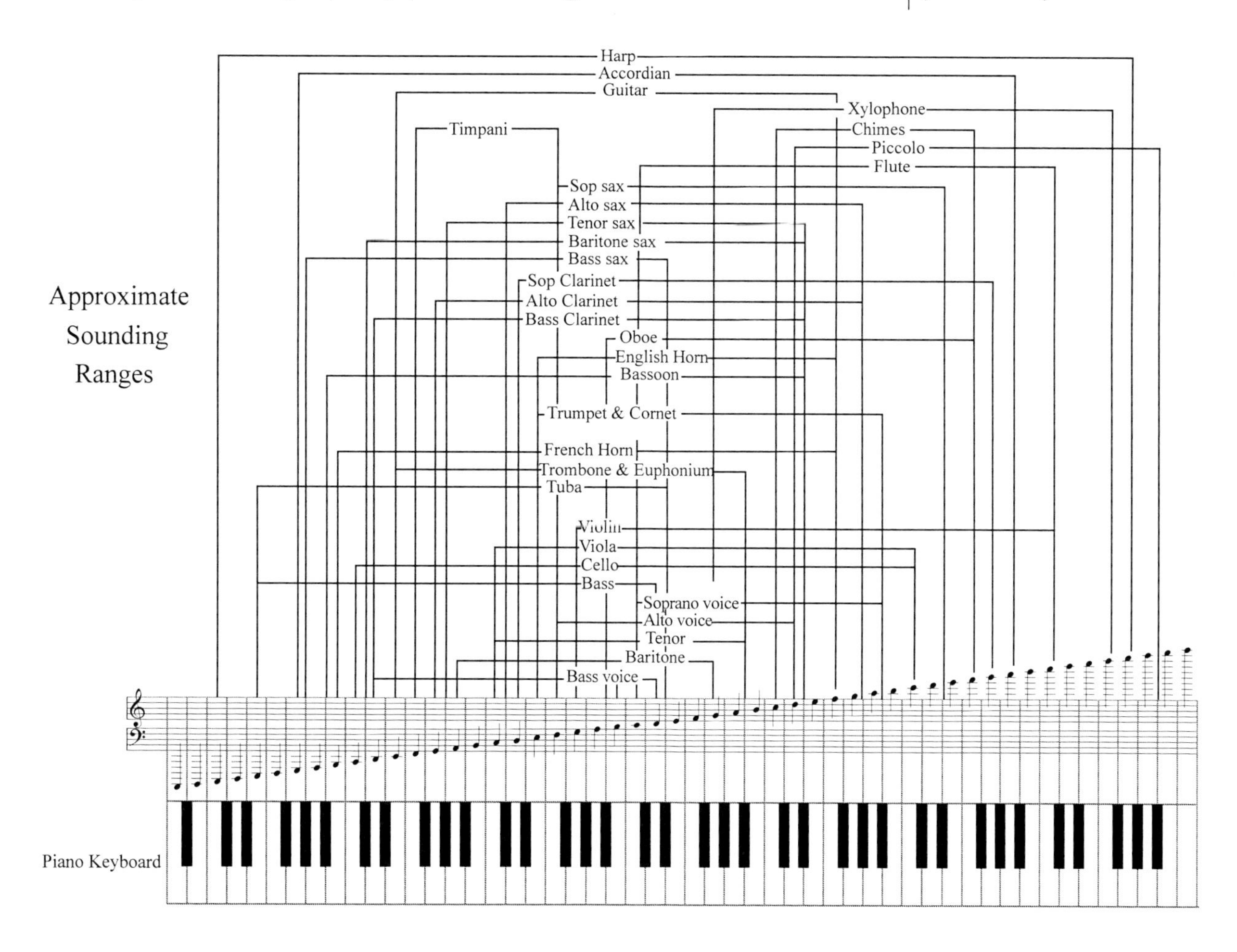

vertical ones. Centering the instrument names within those horizontal ranges would reinforce the relative center of gravity for each instrument, further facilitating comparison. See for instance how the names of the four string instruments are aligned. Since the instrument names pop out more than the lines do, their alignment inadvertently implies something about the instruments' ranges that isn't true. Online figure 38.5 replaces the entangled lines with colored bars, producing a more congenial reference. See also the excellent image in online figure 38.6, which adopts a similar design and adds additional layers of information.

Figure 3.5 (colorplate 2), discussed in part 1, is rich with nicely layered information. It serves as a model of excellence in the representation of instrument ranges and in visual design in general. A careful review of that image and the narrative about it in chapter 3 will be time well spent.

Chapter 39 Translations

Translations, in journal articles, textbooks, and anthologies, sometime seem thrown together in ways that do not consider the viewer. In a layout showing a translation, the corresponding text should align in a way that facilitates comparison of the original and the translation. Figure 39.1 provides an effective model for such a translation. The text structure, down to the indentation of subphrases, is set identically in the two versions. The use of a contrasting color (the Latin text is green in the original) helps distinguish the texts further. The use of all caps to signal passages that are sung homophonically adds a helpful layer of explanatory information. Online figure 39.2 employs a similar approach.

Readers of a translation often want to know not only what the lines mean but what the individual words mean. This issue is alluded to briefly in the commentary in online figure 39.2, but it is addressed more explicitly in anthropological (and hence ethnomusicological) studies than in music theory and musicology contexts. The translation in figure 39.3 includes a gloss on the text, in addition to the original and phrase-level translation. Each appears in a different font: bold for the original, regular text for the gloss, and a slightly larger font for the translation. Because the gloss is word by word, not line by line, the vertical alignment of text and translation better facilitates the comparison. The vertical spacing hampers the image, however. The grouping of the three phrases could certainly be clearer. The translation of the same text in figure 39.4 improves on this spacing. The latter image comes from an article with extensive linguistic interpretations, so the use of

39.1 Joseph Kerman, *Listen*, 3rd brief ed. (New York: Worth, 1996), 75.

LISTEN **Josquin, *Pange lingua* Mass, from the Gloria**

(Capital letters indicate phrases sung in homophony.)

	Qui tollis peccata mundi, MISERERE NOBIS.	You who take away the sins of the world, have mercy upon us.
0:34	Qui tollis peccata mundi, SUSCIPE DEPRECATIONEM NOSTRAM.	You who take away the sins of the world, hear our prayer.
	Qui sedes ad dexteram Patris,	You who sit at the right hand of the Father,
	miserere nobis.	have mercy upon us.
1:18	Quoniam tu solus sanctus, tu solus Dominus, tu solus altissimus, Jesu Christe, cum sancto spiritu, in gloria Dei Patris. AMEN.	For you alone are holy, you alone are the Lord, you alone are the most high, Jesus Christ, With the Holy Spirit, in the glory of God the Father. Amen.

specialized abbreviations is appropriate. For the more general audience addressed in figure 39.3, however, the need to refer to a separate legend to interpret the symbols hampers understanding. The redrawing in figure 39.5 builds on the strengths of both. It addresses the vertical spacing issue of figure 39.3; increases the contrast

39.3 Linda Barwick, "Musical Form and Style in Murriny Patha *Djanba* Songs at Wadeye (Northern Territory, Australia)," in *Analytical and Cross-Cultural Studies in World Music*, ed. Michael Tenzer and John Roeder, 316–54 (New York: Oxford University Press, 2011), 330.

Lirrga PL08

Text phrase A

aa muli kanybubi kanybubi

SW female mermaid mermaid

"Mermaid women, mermaids!"

Text phrase B

wuyi =ga niwiny =ga yi =ngi

country =FOC 3DU.PRO =FOC FAR.DEIC =now

"Their country is there now"

Text phrase C

kangarkirr bugim +mi kwang

water lily white +face 3SG.S.R.stand

"Where the white-faced water lily stands"

Abbreviations for the linguistic gloss:

SW = song word
FOC = focus marker
3DU.PRO = third-person dual pronoun
FAR.DEIC = deictic meaning "far"
3SG.S.R = third-person singular subject, realis mood

Text 15. PL08 Kanybubi (mermaid) #2

Marett DAT1998/12 items xxiv, xxv

Composed by Pius Luckan

a	*muli*	*kanybubi*	*kanybubi*	*muli*	*kanybubi*	*kanybubi*
SW	female	mermaid	mermaid	female	mermaid	mermaid

Mermaid women, mermaids! mermaid women, mermaids!

wuyi	*=ga*	*niwiny*	*=ga*	*yi*	*=ngi*
country	=FOC	3DU.PRO	=FOC	FAR.DEIC	=now

Their country is there now.

kangarkirr	*bugim*	*+mi*	*kang*
water lily	white	+face	3SG.S.R.stand

Where the white-faced waterlily stands.

Lirrga PL08

Phrase A

aa	**muli**	**kanybubi**	**kanybubi**
song word	female	mermaid	mermaid

"Mermaid women, mermaids!"

Phrase B

wuji	**= ga**	**niwiny**	**= ga**	**yi**	**= ngi**
country	*focus marker*	*third-person dual pronoun*	*focus marker*	*deictic meaning "far"*	= now

"Their country is there now"

Phrase C

kangarkirr	**bugim**	**+ mi**	**Kwang**
water lily	white	+ face	*third-person singular subject, realis mood: stand*

"Where the water-faced water lily stands"

39.4 Lysbeth Ford, "Marri Ngarr Lirrga Songs: A Linguistic Analysis," *Musicology Australia* 28, no. 1 (2011): 40, https://doi.org/10.1080/08145857.2005.10415277

39.5 Redrawing of figs. 39.3 and 39.4.

between the original text (bold, serifed font) and the gloss (smaller, light sans serif font); and replaces the linguistic markers in both with their own translations, which in figure 39.3 had been relegated to a legend.

From the perspective of visual design, the key principles to observe in presenting translations are the facilitation of comparisons and the separation of distinct layers of information.

* * *

Part 4 has covered a wide range of topics, including various alternative music notational methods, tuning and timbre, texture and voice leading, schematic and procedural representations, pitch-class set tables, instrument ranges, and translation. Across the breadth of topics, several themes addressed earlier in the book have been reinforced, including the importance of showing comparisons (timbre, instrument ranges, translations, texture, temperament), of information richness (voice leading, instrument ranges, pitch-class set tables), and of user-focused design (alternative notational systems, voice leading).

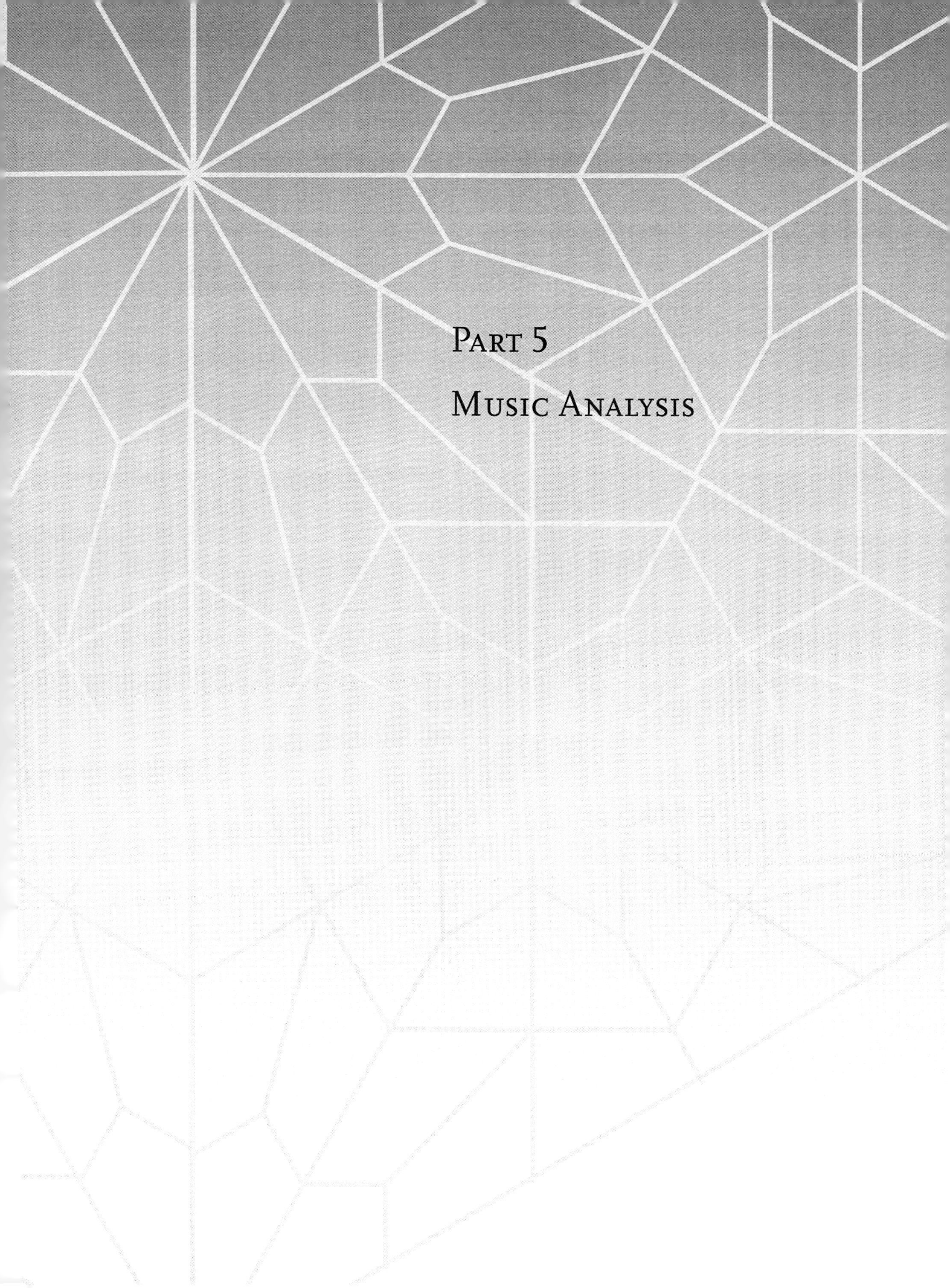

Part 5

Music Analysis

In part 5, we turn our attention to images and issues associated with music analysis, a topic of central concern to many of the book's readers. This is where art and craft meet. Although music analyses do not need to be graphical, they often benefit from illustrative images. Some of the chapters in this part of the book will revisit earlier topics, but now from an analytical perspective. We will start with a step-by-step redrawing of a detailed image (chap. 42), both to summarize many of the principles that have been explored to this point and also to foreshadow some of the issues that will arise in this part of the book.

Chapter 41 explores the general problem of annotating musical scores and reviews some of the basic concepts from part 1. In chapters 42–44, we look at the analysis of themes, melodic contour, and tonal plans. We continue with an examination of symmetry (chap. 45) and the analysis of rhythm and meter (chap. 46), before taking an extensive look at the analysis of musical form (chap. 47) and musical hierarchy (chap. 48). After a look at the analysis of serial music (chap. 49), we turn to the study of large bodies of music. Chapter 50 discusses corpus studies in music, while chapter 51 considers the depiction of musical styles. Finally, in chapter 52, we look at the uses of animation to visualize music.

While part 5 covers wide-ranging topics, the aims of the visualizations are the same: information-richness, clarity, efficiency, and (where possible) narrativity. Since so much Western music is goal-directed on some level, analyses that foreground that aspect of music—that is, those that find a way to direct the listening experience—are often especially compelling.

Chapter 40 Lutosławski's *Jeux Venitiens*

We begin part 5 with a redrawing case study. Figure 40.1 visualizes the first movement of Witold Lutosławski's *Venetian Games* (composed 1960–61), which employs limited aleatory and moment form. The image conveys the movement's pitch and formal organization.

The movement alternates between two types of music. The first involves wind-dominated aleatoric counterpoint, used in sections the composer labels in the score A, C, E, and G; I will refer to these as the "odd-lettered" sections. The second type features quiet, sustained string-dominated music, labeled B, D, F, and H, which I will refer to as the "even-lettered" sections. The beginning of each section is marked by a *fortissimo* percussion strike (not shown in the image), which also initiates the closing of the movement.

The score indicates the durations of the odd-lettered sections in seconds: 12, 18, 6, and 24. Section A features seven unmetered, rhythmically active, and texturally independent lines played by woodwinds (two flutes, oboe, three clarinets, and bassoon), whose pitch material collectively is drawn from a symmetrical twelve-note aggregate. The image, below A, shows the pitches of this chord and lists the number of semitones between adjacent pitches. It provides registral reference points (e.g., C2) along the image's left edge. This music repeats in each of the odd-lettered sections. The C section adds timpani, whose three pitches are not specified in the score

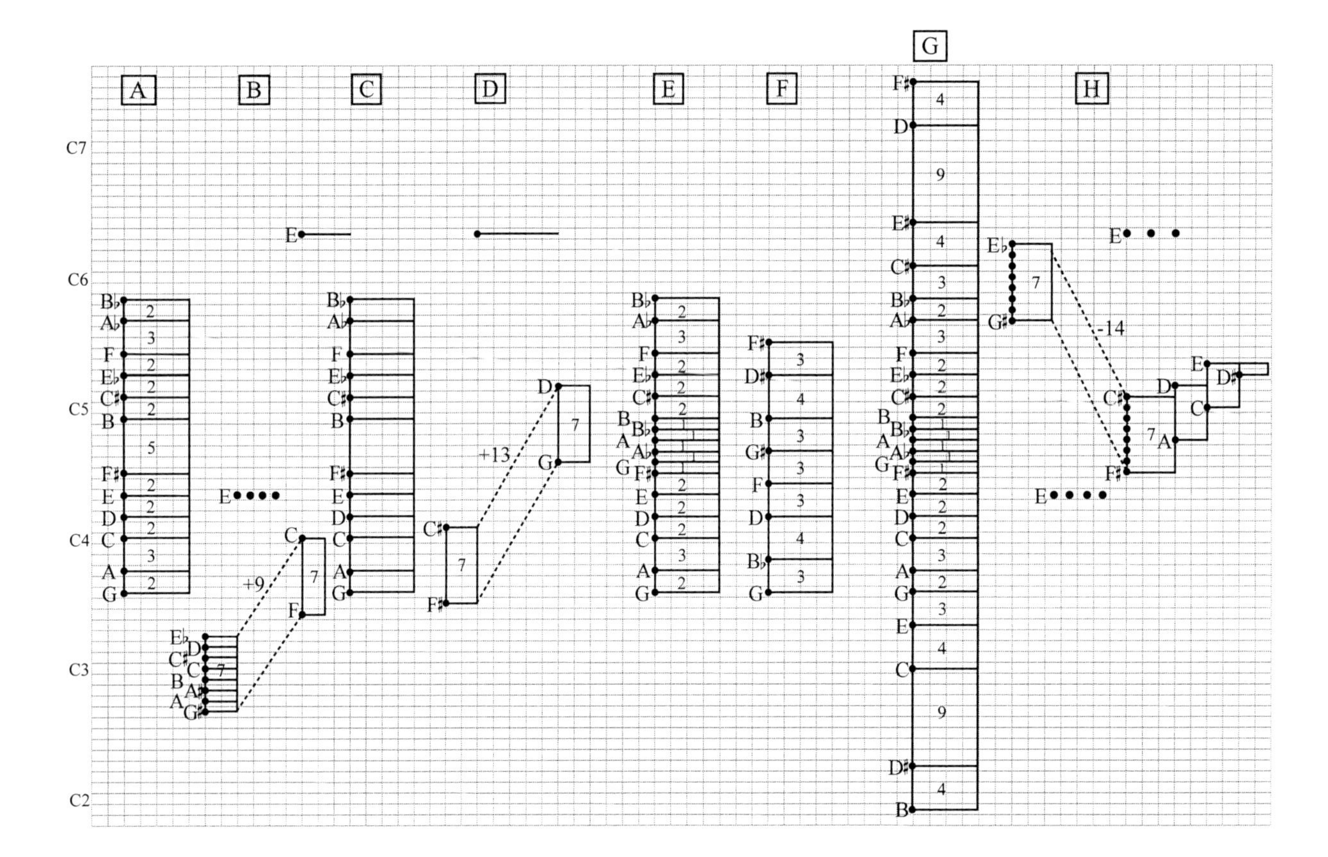

40.1 Miguel Roig-Francolí, *Understanding Post-tonal Music* (New York: McGraw-Hill, 2007), 290.

and therefore do not change the chord. The E section adds three brass instruments (trumpet, horn, trombone), which draw from the four pitches that chromatically fill in the twelve-note chord's central interval (5), preserving its vertical symmetry. The piano that enters in the G section adds eight additional pitches, four above and four below, which expand the sonority to twenty-four notes. Each pitch class sounds exactly twice, and the double aggregate, like the sonorities in the other odd-lettered sections, is registrally symmetrical.

Sections B, D, and H feature eight-note pitch clusters that span seven semitones. These are transposed note by note up nine semitones in B, up thirteen in D, and down fourteen in H. The transpositions sometimes involve an intervening note that falls outside the transposition interval. These do not appear in the image. During the last section, after the seven-note cluster has completed its transposition, it contracts to five, then four, then finally one semitone. Repeated or sustained Es are played in each of these three sections. The very brief F section (ca. two seconds, according to the score) also features the strings, but its symmetrical eight-note sonority comprises stacked minor (3) and major (4) thirds.

The conception that underlies the design is very attractive, and the image conveys many of the central aspects of the movement. The aim is so compelling, in fact, that it invites a closer look to see how the image might reach it more effectively. We could achieve this by removing some unnecessary elements and adding others that highlight certain features.

Let's start with the eraser. We know that gridlines are often unnecessary. In addition, the dots used to indicate each of the pitches serve the same purpose as the lines dividing the intervals and therefore can be removed. The vertical lines that connect the edges of the sonorities provide no information and can also be eliminated. The same is true of the boxes around the section labels. Figure 40.2 shows the result of these erasures. Clearly, we've gone too far. The section divisions, too stark before, are now too subtle. The pitches within each section do not cohere into sonorities. And the section headers now seem disengaged from the rest of the image.

We can address these issues, but before we do, let us consider ways of enrichening the image's information content. Can the image show the movement's

40.2 Fig. 40.1 with all noninformation ink removed.

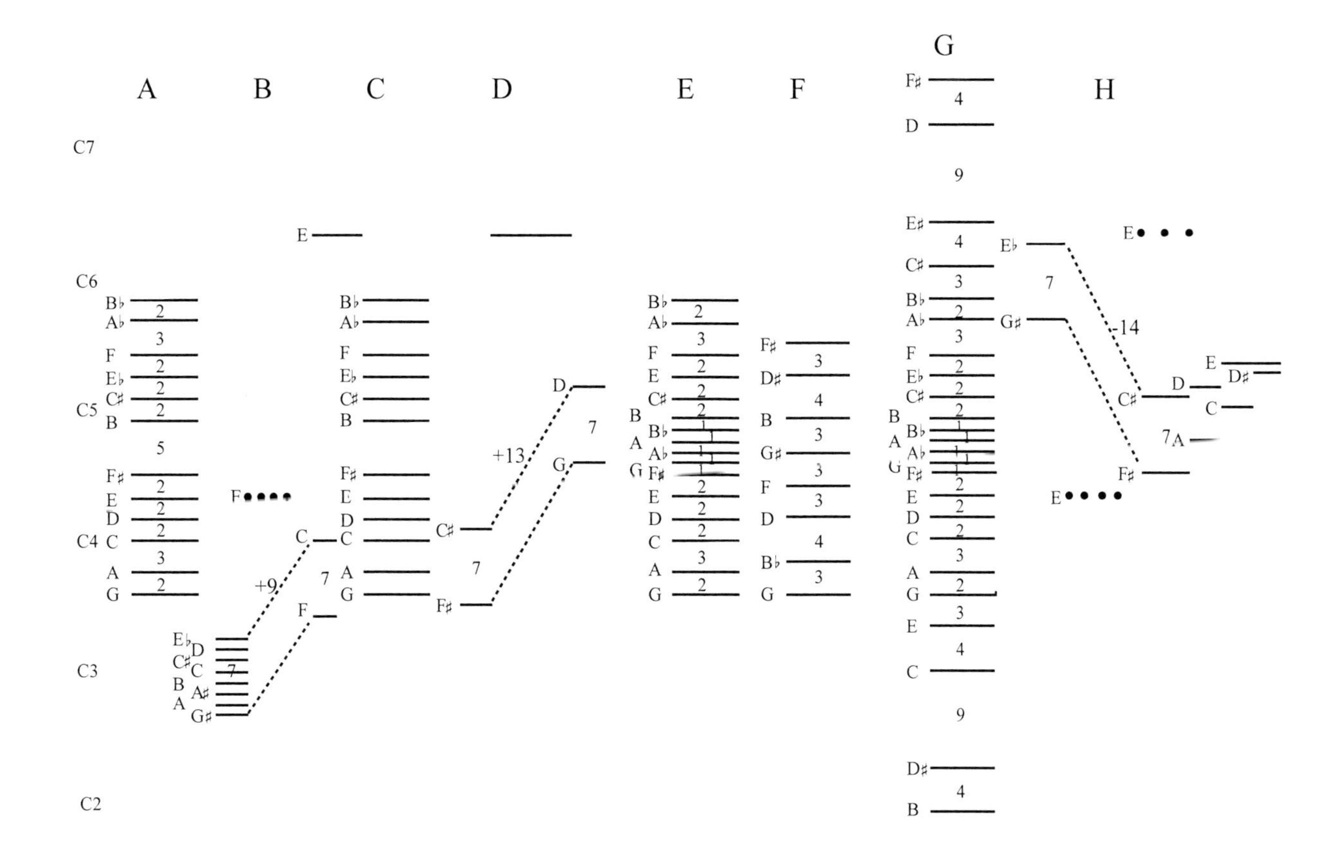

temporal proportions? Vertical intervallic symmetry plays a central role in the image's conception, so can that symmetry pop off the page more clearly? And since the symmetry is measured by intervals, can we emphasize these intervals over the pitches that define them? (The representation found in online fig. 40.3 does not improve the depiction of symmetry.) Finally, can we visually distinguish the odd- and even-lettered sections, which are musically so different?

Figure 40.4 (colorplate 10) attempts to do all this. The image contains all of the information from the original, avoids the features that detract from it, and makes the suggested enhancements. Contrasting muted background colors differentiate the odd- and even-lettered sections. The blue and yellow shading reintegrates the labels with the musical content of each section. Boxes around section labels (A through H) will not be missed. The section widths now indicate the proportions of the sections in the performance that is credited in the lower right. Timings from this performance have been added below each section header, and so has the instrumentation (or changes in instrumentation) for each section.

The image prioritizes the intervals between pitches, rather than the pitches themselves. The gray luminosity scale, in which larger intervals are darker, causes the intervals' symmetrical arrangement to pop off the page. The decision of whether to use a light-to-dark or dark-to-light scale is in a sense arbitrary. In fact, if one equates size with density, one could argue that smaller, "denser" intervals should be dark and larger ones light. However, because it is easier to make finer distinctions between light intensities than dark ones, and since the largest number of intervals are relatively small (1 to 3), the scale used here is easier to process visually. Note that, as in the original, the clusters in the even-lettered sections are treated as representing the intervals they span (7) rather than the individual minor-seconds that make them up, and they are shaded correspondingly. This makes the contraction of the cluster in section H more apparent.

The image eliminates not only the unnecessary dots representing each pitch but also the lines themselves, in favor of negative space between the interval blocks. It retains pitch names but presents them in a contrasting color of lower saturation. Because only the relative registers of the pitches are important to the visualization, a single reference pitch is sufficient to locate the structures relative to middle C (C4). The redrawing makes the excellent content of the original more apparent and, without substantial effort, adds valuable information.

The redrawing does not attempt to clarify one element omitted in the original: how the transpositions of the clusters in sections B, D, and H are actually carried out. Each of the seven notes moves at a different time, and some of the voices move twice before reaching their destination. This is addressed in online figure 40.5.

The original image is an excellent representation of the work it explains visually. As happens so often when we visualize musical phenomena, it seems not to have asked, "What additional information *could* I convey?" The redrawing honors the original but incorporates a number of answers to that question. (It is important to note, as I do from time to time in the book, that I have cheated by introducing color in an image where it was not possible in the first place.) And yet I have my own blind spots, and others might identify ways to further enhance the redrawing.

Chapter 41 Annotating Musical Scores

In books and articles, music scholars often illustrate points with excerpts from a musical score. Too often, those examples are unadorned, with the assumption that the point of the author's prose, which sometimes spans paragraphs or even pages, will be self-evident to anyone with the ability to read music. In nearly all such situations, however, authors could enhance their readers' understanding if their scores drew attention to and reinforced the authors' points.

Annotating a score requires careful design. Annotations represent a different layer of information and should therefore be visibly distinct from the score. That is, it should be clear which marks appear in the original and which have been added. The annotation layer needs to have appropriate weight, neither overwhelming the notation nor being overwhelmed by it.

We will explore these ideas in the context of pitch-class set segmentation in post-tonal music. These scores are often more intensively annotated than most others since the analytical objects in them are not understood by default as they tend to be in tonal music. Many analyses of atonal works provide a combination of segmentation and explanatory analysis. This makes it a useful topic with which to illustrate good principles to follow in annotating musical scores of any style.

The segmentation in figure 41.1, typical of this genre, will serve as the starting point of our discussion. It sets Anton Webern's *Five Pieces for String Quartet*, op. 5,

no. 4, on three staves, segments it in the manner of classic pitch-class set analysis, and adds some additional information. The image identifies and labels pitch-class sets to which the author wants to draw attention. Segments are marked in a variety of ways—brackets, solid and dashed closed shapes with straight lines, and one dashed circular line—though the methods used have no particular meaning. Text adjoining each segment provides the Forte number for the set.

41.1 David Beach, "Pitch Structure and the Analytic Process in Atonal Music: An Interpretation of the Theory of Sets," *Music Theory Spectrum* 1 (1979): 20.

The level of detail is impressive—more analyses would benefit from such ambition. The image has some visual challenges, however. Most significant, the segmentation often blends in with the score, making the image seem flat rather than multidimensional. Many of the segment boundaries are rectangular, their horizontal and vertical lines camouflaged among the staff lines, stems, and bar lines. Segments indicated by brackets do not have this problem, but those cases require careful visual inspection to see which notes are included in the segment. In the second and third measures from the end, eleven pitch events (discounting octave duplicates) are allocated to seven different sets that nest three deep. Rendering those coherently is a serious challenge.

Among segments that enclose all notes of a set, those drawn with dashed lines are somewhat easier to distinguish than those that use solid lines, since the difference in visual texture produces a pop-out effect. Even so, since the two dashed boxes in measures 4–6 overlap and run parallel to musical lines, especially the staff lines at the tops of the boxes, one cannot immediately ascertain that the point of overlap is not itself a segment.

Two segmentation styles used in the image stand out best. First are the diagonal brackets used in the callout below measures 4–6. The callout simply replicates the pitches of the lowest staff, but it provides a different segmentation. The diagonal brackets are easier to see than the horizontally oriented ones because they differ in orientation, a feature that produces a pop-out effect. The clearest segment is the lone set circled with a dashed line (m. 4). It stands out because it differs from its background in both texture (with a dashed rather than solid line) and shape (rounded rather than horizontal or vertical). This combination of differences provides the strongest pop-out effect among the segments in the example. That said, dashed lines in general are visually noisy, and it is often better to find an alternative because analyses full of them would create visual cacophony. We will return to this analysis at the end of this chapter. In the intervening images, we will consider other approaches to the segmentation problem, though I will use the opportunity to comment further on other strengths and challenges in visual design.

Figure 41.2 shows segmentation on a textual representation of the music rather than on the music itself. It shows fragments of serial row forms, using note names aligned by contrapuntal voice in four rows. The image clearly distinguishes row labels from pitches by placing them above and below the pitch notation, rather than within, and by setting them in a contrasting typeface. Inclusion of a score excerpt allows one to relate the segments to the music, a task made somewhat more difficult by the lack of vertical alignment. Better integration of the different parts would improve the image. In addition, the square-cornered segmentations do not pop out as well as rounded enclosures would. Compare the effectiveness of rounded shapes with the square shapes of figure 41.3 and the brackets of online figure 41.4.

The textural reduction in figure 41.3 makes it clear that the verticalities in the first two staves are chords, and the labels between those staves clearly refer to those chords. In the remainder of the image, however, the use of brackets is sometimes ambiguous. Below the middle staff of the upper system, 6-Z19 refers to the pitches in the two staves above it, 6-15, with legs extending both above and below the horizontal line referring to the pitches in all three staves, while 5-24 below the lowest staff refers only to pitches on that staff. Meanwhile, notations in the form ↑ 7-27 indicate a vertical slice through the

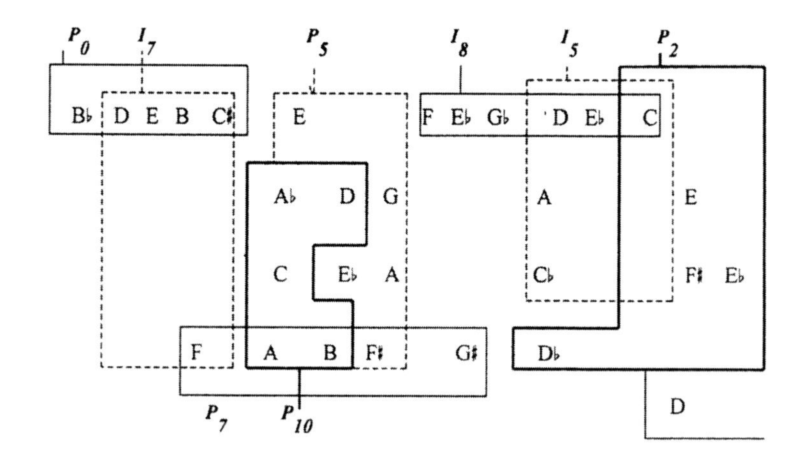

41.2 Bryan R. Simms, *The Atonal Music of Arnold Schoenberg, 1908–1923* (New York: Oxford University Press, 2000), 197.

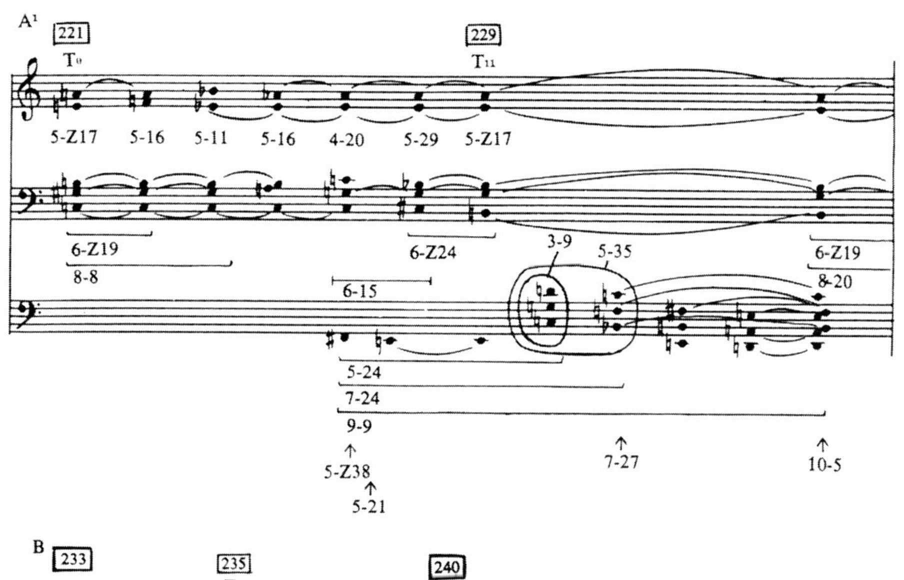

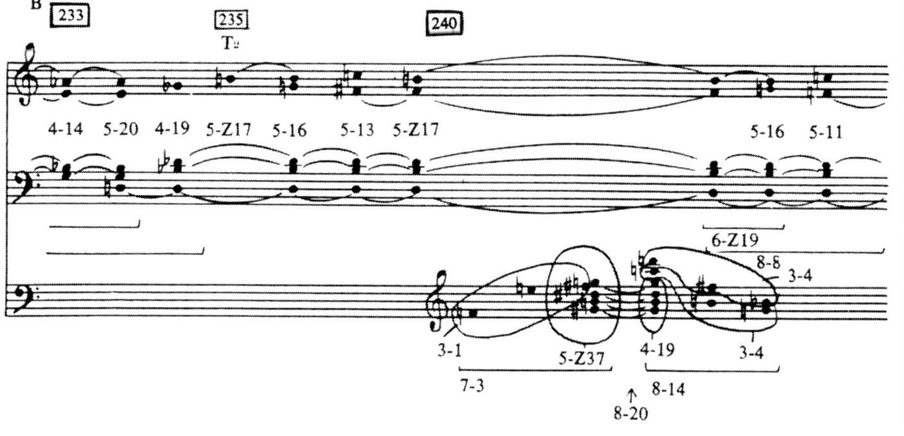

41.3 Allen Forte, *The Structure of Atonal Music* (New Haven, CT: Yale University Press, 1973), 167.

Set 1 [A, Bb, C, D, E, F#]
Set 2 [C, D, E, Gb, Ab, Bb]
Set 3 [C#, D, E, F#, G#, A#]
Set 4 [G, Ab, Bb, C, D, E]
Set 5 [Eb, F, G, A, B, (C#)]
Set 1 [A, Bb, C, D, E, F#]
Set 6 [C, Db, Eb, F, G, A]
accel.

entire texture. Each of these notations requires some degree of cognitive effort to interpret. Segments marked by circles, however, pop off the page readily because they fully enclose the pitches they refer to and because they appear in a contrasting shape. Online figure 41.4 illustrates a different type of problematic bracket.

Enclosures are generally better than brackets unless the segments are unambiguous. The segments in figure 41.5 are clear, in part because they do not overlap. The straight lines of the boxes blend with those of the music notation, however. It would be better if the different information layers were represented by different visual layers. Again, rounded enclosures would serve this purpose, as their curves would pop out from the notation's straight lines.

Facing, **41.5** Inessa Bazayev, "Scriabin's Atonal Problem," *Music Theory Online* 24, no. 1 (2018): ex. 4b1, https://mtosmt.org/issues/mto.18.24.1/mto.18.24.1.bazayev.html.

Even rounded enclosures must be used with care, however. In figure 41.6, the circles visually overwhelm the pitches they enclose; their size and thickness inadvertently call too much attention to them at the expense of the segments of music they mark off. A later edition of the same image (fig. 41.7) makes two excellent adjustments. The first collapses the notation from eight staves to two, which puts the pitches closer together. The second, a byproduct of the first, makes the enclosures less overpowering, though it would be better if they didn't touch at all. In both instances, while the prose accompanying the images makes clear that all the pitch sets are (014) types, the inclusion of that information on the image itself would benefit viewers. And the boxes around the row labels are not information-bearing and could be omitted.

Like figure 41.3, figure 41.8 uses a textural reduction, preserving only the pitches and their ordering. It employs letters keyed to a legend of set class names at

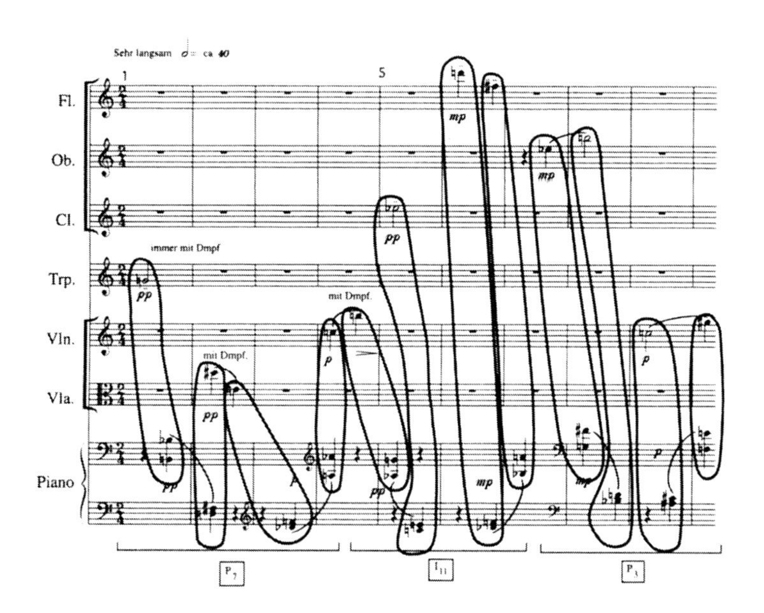

41.6 Joseph Straus, *Introduction to Post-tonal Theory*, 3rd ed. (Upper Saddle River, NJ: Prentice-Hall, 2005), 218.

41.7 Joseph Straus, *Introduction to Post-tonal Theory*, 4th ed. (Upper Saddle River, NJ: Prentice-Hall, 2016), 319.

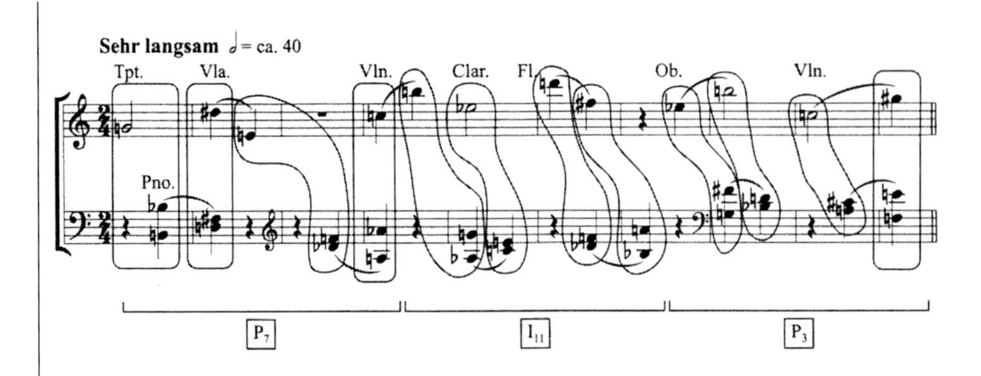

the bottom (as a reminder, it is always desirable to also include prime forms); each instance of a set class receives a unique subscript. In some instances when a segment represents a complete vertical cross section, it is rendered using an arrow ($\overset{F_2}{\downarrow}$); otherwise, brackets, circles, or boxes are used. The image employs a clever device to show several supersets of a sustained G augmented triad that result from melodic activity in one of the two upper staves (see segments A_5 through A_9). The added note is circled and a line drawn to the augmented triad, with the label attached to the line. While the extra lines contribute to a sense of busyness, the design also provides musical insight. Some of the pitches in measures 6 and 7 are duplicated on a separate system to show some larger segments. The method reduces clutter but makes it hard to see how the two sets of segments interact.

41.8 Christopher Lewis, "Tonal Focus in Atonal Music: Berg's op. 5/3," *Music Theory Spectrum* 3 (1981): 92.

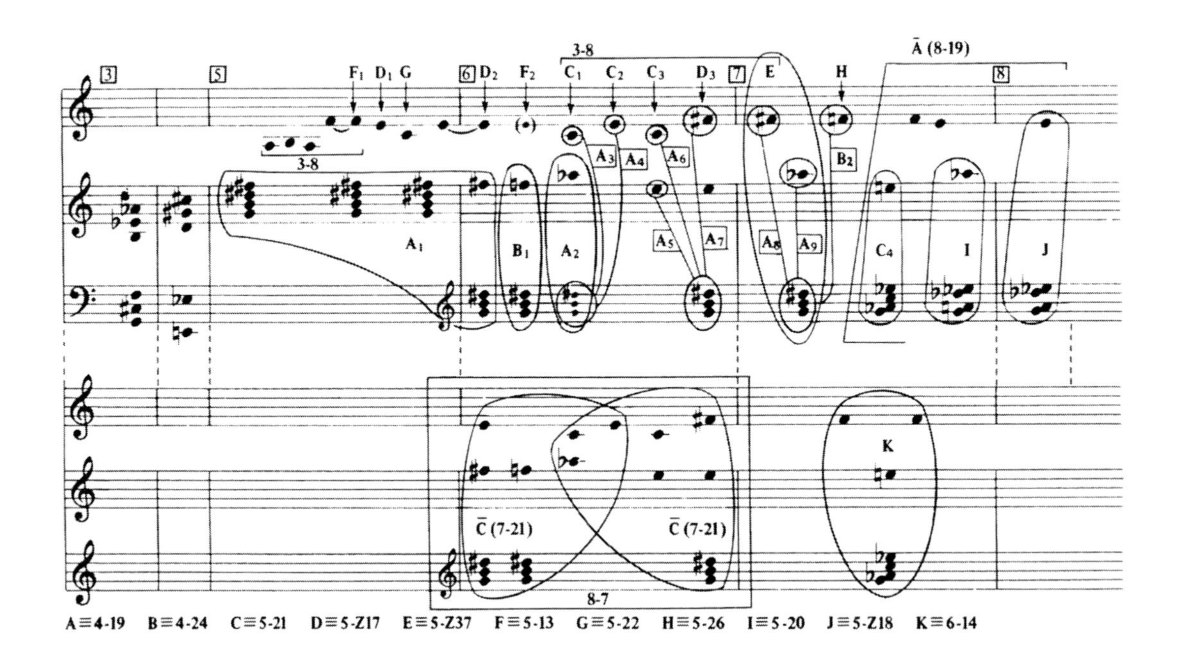

Color jumps off an otherwise black-and-white image into its own information layer, as the two images in online figure 41.9 illustrate. Meanwhile, online figure 41.10 employs rounded enclosures to group elements in an analysis, breaking free of the strictures imposed by a tabular structure.

While rounded enclosures are frequently the cleanest way to show segments, the novel approach in figure 41.11 sometimes works. The music appears above and a segmentation below. As in many other reductions, the figure eliminates temporal information aside from note order. Notes belonging to a segment are beamed together, with an appropriate label adjacent (the Forte name and an indication of which of the three octatonic collections the pitches come from). This approach does have a limitation: since a note can support only two stems, no pitch can belong to more than two segments. The segment labeled 4–20 in the second system is thus represented by a bracket, and although 4–19 could be rendered with a beam group, the author has chosen to use a parallel bracket.

The approach taken in figure 41.12 also works well. It highlights recurrences of three set types by pulling apart the music (a piano piece by Robert Morris) and putting the pitches forming each set type in the same row. It preserves pitch ordering

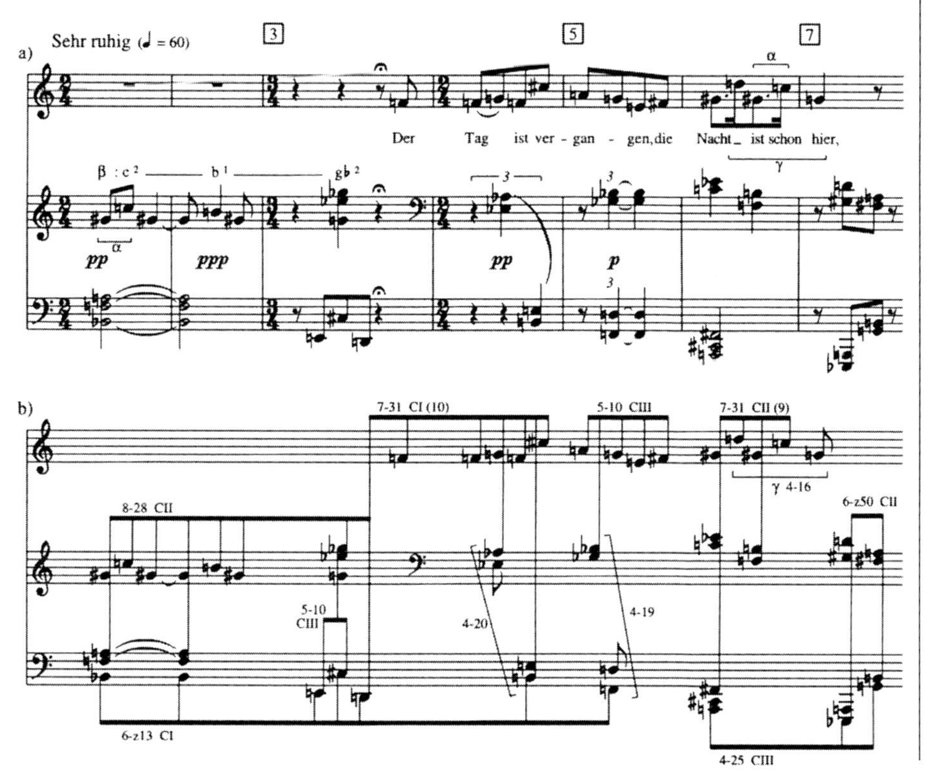

41.11 Allen Forte, *The Atonal Music of Anton Webern* (New Haven, CT: Yale University Press, 1998), 255.

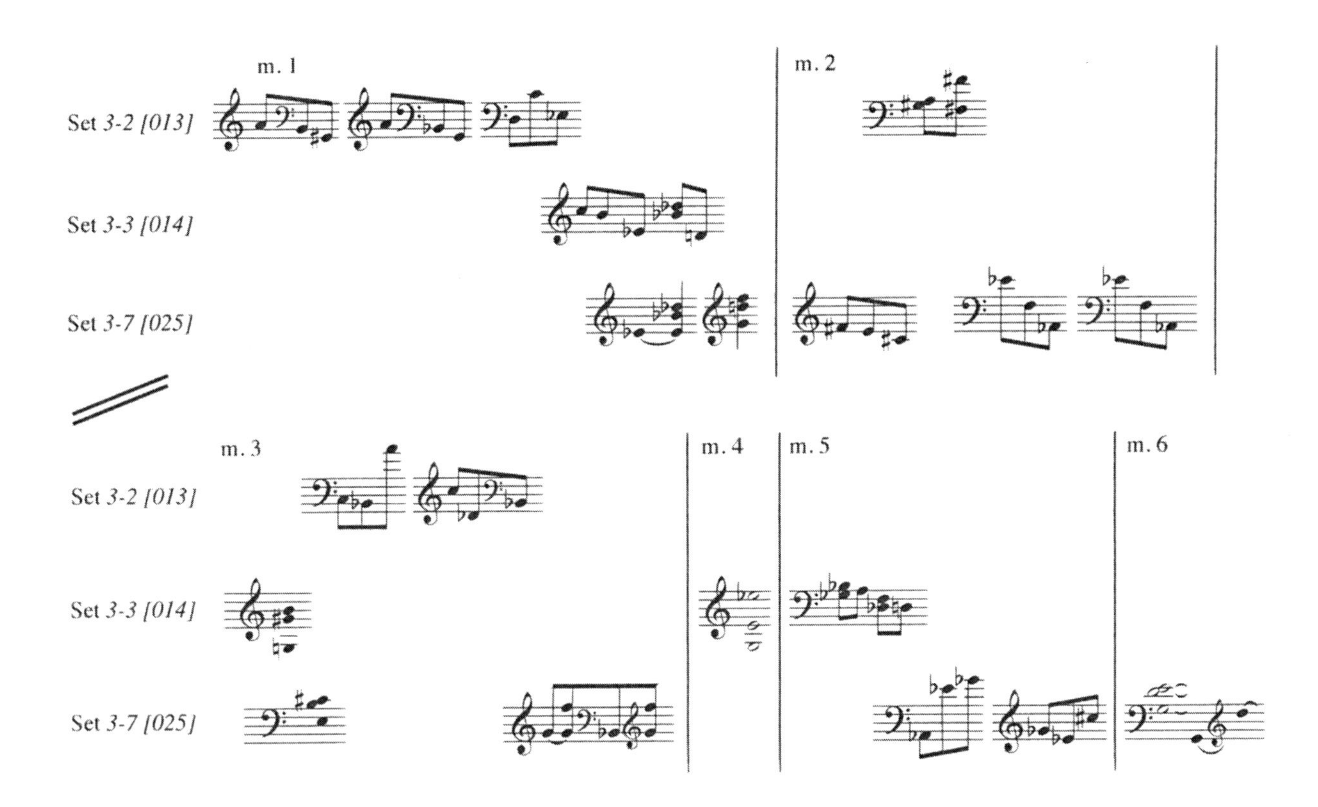

41.12 Dora A. Hanninen, *A Theory of Music Analysis: On Segmentation and Associative Organization* (Rochester, NY: University of Rochester Press, 2012), 380.

and register but omits rhythmic details aside from (useful) bar lines. Note also how easy it is to include the prime form along with the Forte label.

Now that we have seen a number of approaches to pitch-class set segmentation, we return to the chapter's first example (fig. 41.1) and provide a redrawing that employs some of the most effective practices we've examined (fig. 41.13, colorplate 11). In the redrawing, the use of rounded shapes rather than rectangles enhances contrast with the music's grid. Gray borders contrast with the black notation and provide important definition for each segment. The edges of segment shapes do not overlap, and subset/superset musical relationships are nearly always visually apparent. With very few exceptions, no staff or bar lines from the music have been occluded, and no set label appears on the staff. The shapes are filled with light-hued, semitransparent colors. The transparency allows the boundaries of occluded segments to come through. Solid shapes do not need to be heavily saturated to be visually prominent. In general, the larger a shape is, the lighter the shading should be. Sets of size 3, 4, and 5 therefore appear in slightly darker hues, cycling through a palette of red, green, and blue. The palette repeats using lighter hues for sets of size 6, 7, and 8. Set labels are larger than in the original, set in a sans serif font for better legibility, and they appear in the same place (bottom center) for each segment whenever possible. For segments that cross from the first to the second system, the set label repeats, but in parentheses to make clear that it does not demarcate a new set. The approach eliminates the need for the callout below measures 4 and 5. The repeated triplet figure in measures 7 and 8 is written out rather than using the ⁒ symbol. Measure numbers, essential for any musical example, were omitted in figure 41.1 but are given a distinctive setting in the color version here. Likewise, the revision gives formal labels, which are set in boxes in figure 41.1, a contrasting appearance and makes them more prominent. Ordinarily, it would be preferable to put prime forms directly in the segments, but it would be unwieldy to include them here, so one can look up prime forms from the Forte labels.

Most analyses do not include nearly fifty segments that stack up to five deep, so treatment this extensive will rarely be called for. The annotations very much emphasize the segments. Focusing on any one of them makes it easy to see the pitches that belong to it. Collectively, however, they push the music into the background. Whether this is desirable depends on the aim of an image.

The principles and practices outlined in this chapter, although focusing on segmentation on atonal scores, generalize to any situation that calls for score annotation: eliminate ambiguity regarding which musical events are included, favor rounded enclosures, employ color contrast when possible, and don't be afraid of novelty, so long as the viewers' understanding gets top consideration.

CHAPTER 42 Thematic Analysis

By *thematic analysis*, I mean the study of the recurrence of melodic material within a piece at the level of the motive or phrase. Because we understand things better in relation to other things, the best images compare two or more themes in the same work, though it is also possible to compare themes in more than one work or even against some idealized model. Ideally, such representations will show not just *that* two figures are related but *how*. The images on these pages achieve this in a variety of ways.

Thematic analyses of organ music by J. S. Bach are featured in figure 42.1 and online figure 42.2. In figure 42.1, square brackets mark chromatic motion from $\hat{5}$ to $\hat{1}$, while circled groups of notes indicate motion from $\hat{1}$ to $\hat{5}$. (I will not discuss the function of dashed shapes or individually circled pitches.) While the previous chapter advocated the use of rounded enclosures for marking off score segments, in tonal contexts like this, in which it is assumed that readers will recognize, say, tonic and dominant scale functions and in which the musical lines are easily traced, less ink-heavy brackets may suffice. Nevertheless, as discussed in chapter 41, putting the melodic labels directly on the music, rather than in a caption or in the prose, makes a better image. Online figure 42.2 does this.

Figure 42.3 features sets of type (014) (labeled as Forte number 3–3). It shows the opening figure from Arnold Schoenberg's *Piano Piece*, op. 11, no. 1, to be the "Zygotic" source of much of the rest of the piece. The semicircular arrangement projects a sense of the music figuratively spewing forth from the opening motive,

42.1 Marianne Kielian-Gilbert, "Inventing a Melody with Harmony: Tonal Potential and Bach's 'Das alte Jahr vergangen ist,'" *Journal of Music Theory* 50, no. 1 (2006): 93.

as if from a lawn sprinkler. The arrows and oversized *Z*s draw attention away from the main attraction, however.

One can show correspondences between themes or thematic statements by aligning them vertically and "connecting the dots," as in figure 42.4 and online figure 42.5. Both images connect corresponding pitches with dashed lines, which contrast effectively with the solid lines of the notation. Fig. 42.4 appropriately sacrifices the normal spacing of the first theme to make the alignment with the closing theme clearer, generally a good strategy.

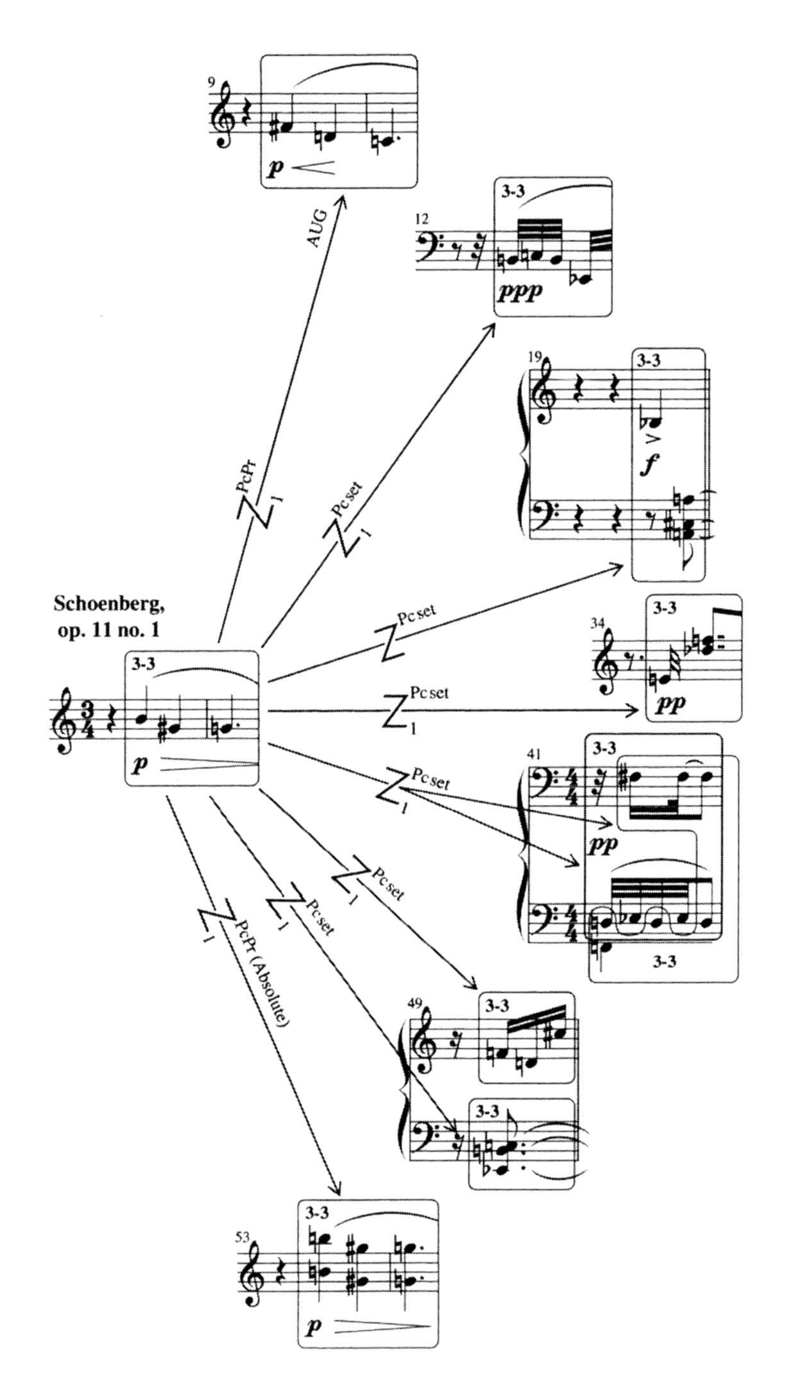

42.3 Adam Ockelford, *Repetition in Music: Theoretical and Metatheoretical Perspectives* (Burlington, VT: Ashgate, 2005), 112.

In an expanded and effective version of this technique, figure 42.6 stretches the notation out as needed to align thematic variants with one another and with background structures to demonstrate the many thematic interconnections. See also online figures 42.7 and 42.8.

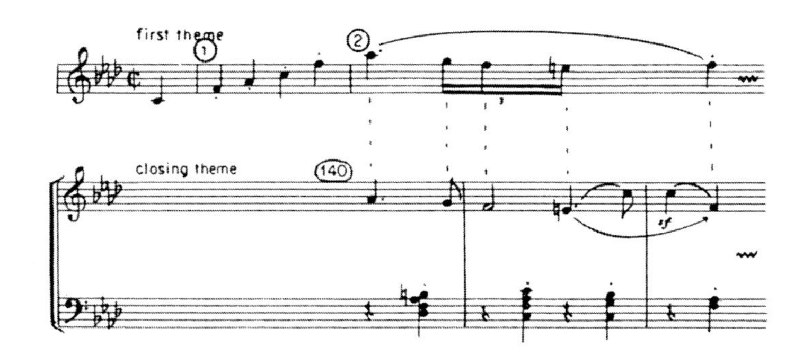

42.4 Charles Burkhart, "Schenker's 'Motivic Parallelisms,'" *Journal of Music Theory* 22, no. 2 (1978): 156.

42.6 Robert P. Morgan, "Coda as Culmination: The First Movement of the *Eroica* Symphony," in *Music Theory and the Exploration of the Past*, ed. Christopher Hatch and David W. Bernstein, 357–76 (Chicago: University of Chicago Press, 1993), 370.

Finally, figure 42.9 explores thematic recurrence at the movement level (the piece is Franz Joseph Haydn's *Sonata in A-flat*, Hob. XVI:43). It lists four distinct thematic ideas, and variations of them. Labels for these themes then appear in a timeline, showing their succession in the work. The three rows of the timeline represent the three main sections of the sonata form movement, while vertical lines mark off major divisions within those sections. The image would improve by overlaying standard formal labels, putting measure numbers at musically significant places rather than at arbitrary ten-bar intervals (note the use of multiples-of-10 plus 1 in the outer sections and multiples-of-10 plus 0 for the development!), and accounting more clearly for places where none of the thematic ideas are present.

42.9 Diether de la Motte, "Theory—Lehre, Wagnis Analyse," in *Musiktheorie*, ed. Helga de la Motte-Haber and Oliver Schwab-Felisch, 489–98 (Laaber: Laaber, 2005), 496.

Thematic analysis is almost invariably comparative. Ensuring equitable treatment of similarities and differences will help ensure that an image communicates effectively.

Chapter 43 Contour Analysis

The word *contour* conjures an imprecise sense of shape, in which points are measured in terms of one another rather than against an external scale. We judge the contour of a mountain range or metropolitan skyline by the relative positions of the highest points we see as we scan the visual field, not their absolute height measured against the horizon or sea level. In assessing contour, we are also mindful only that points are higher or lower than other points, not by how much. Contour works as a metaphor in music because of how we understand pitches to have the property of *height* (see chap. 2). The notes of a melody, the aspect of music most often associated with contour, are arrayed left to right and move up and down on the page, just like a mountain range or skyline.

One can create an informal sense of melodic contour simply by removing the measuring stick—that is, the staff. The line drawings in the lower portion of figure 43.1 show that the two melodies have the same gross shape: a short peak, a trough, a higher peak, and another trough, before a return to the starting pitch.

Music theorists have developed formalized notions of contour space that do not consider information about distance between pitches, only their relative

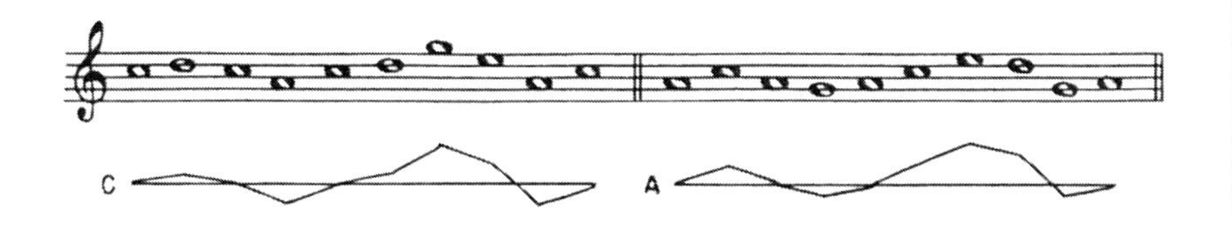

43.1 Howard Hanson, *Harmonic Materials of Modern Music: Resources of the Tempered Scale* (New York: Appleton-Century-Crofts, 1960), 57.

vertical position. As with pitch notation, different ways of representing contour have been suggested, each of which has benefits and drawbacks. One way uses the contour adjacency series (CAS; see online fig. 43.2). Limitations of the CAS are remedied by what is called (in online fig. 43.2) a contour class (CC) or (in fig. 43.3) a contour segment (CSeg). Figure 43.3 shows four CSegs, labeled A through D. The lowest pitch in a CSeg is labeled 0, the next lowest 1, and so on to n−1, which is the highest pitch. The contour numbers then appear in chronological order. The left-most contour can be realized in countless ways, so long as the first note is the lowest, the third note is next lowest, the note between them third lowest, and so on.

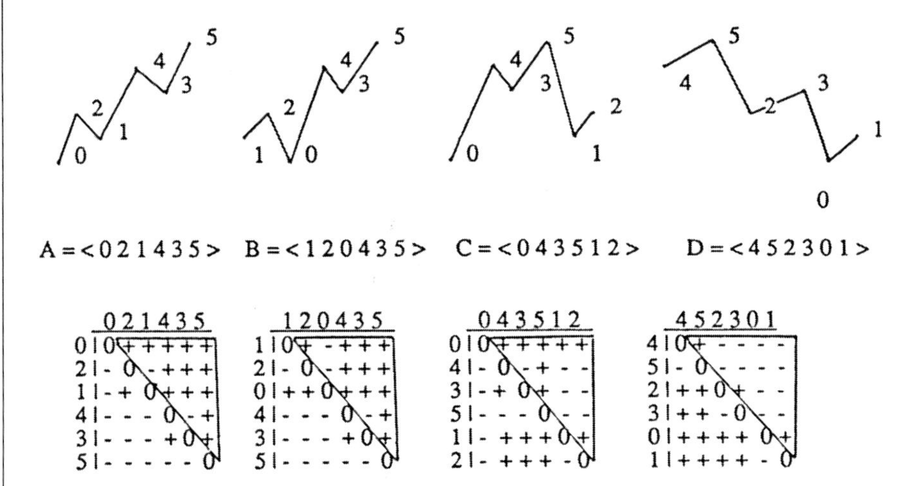

43.3 Elizabeth West Marvin and Paul A. Laprade, "Relating Musical Contours: Extensions of a Theory for Contour," *Journal of Music Theory* 31, no. 2 (1987): 233.

The graphical version of this representation at the top of figure 43.3 is clearer visually than a list of numbers. Inflection points (i.e., direction changes) in the line graph implicitly represent "contour pitches." Of course, if three pitches in the contour are in numerical sequence, there will be no inflection point for the middle one and the representation will fail. Online figure 43.4 shows an alternative that addresses this.

One can use the comparison (COM) matrix (bottom of fig. 43.3) to describe the total point-to-point contour of a melody, with + representing ascent and – representing descent. The more alike the entries in the upper right of the two COM matrices are, the more similar the contours will be considered. This is often calculated simply as a percentage of corresponding entries. Online figure 43.5 shows an effective way of visually representing the extent of that correspondence.

The contour theories cited above all collapse the notes into pitch order (numbered from 0 to n−1) and then space them vertically on the basis of their order number, regardless of their actual pitch height. The remaining images in this chapter retain the sense of pitch height but find ways to emphasize the overall shape of a melody as it occurs in pitch space. A common method is to eliminate any overt vertical grid while still plotting the pitches on that invisible grid, often then connecting the dots between pitches in some way. In figures 43.1 and 43.3, the angled lines do not represent what is sounding at that moment. There is no glissando between the pitches, after all. Rather, the lines connect the pitches suggested by the inflection points. What happens if we put the emphasis on the pitches and not their connectors? Figure 43.6 helps us explore this idea.

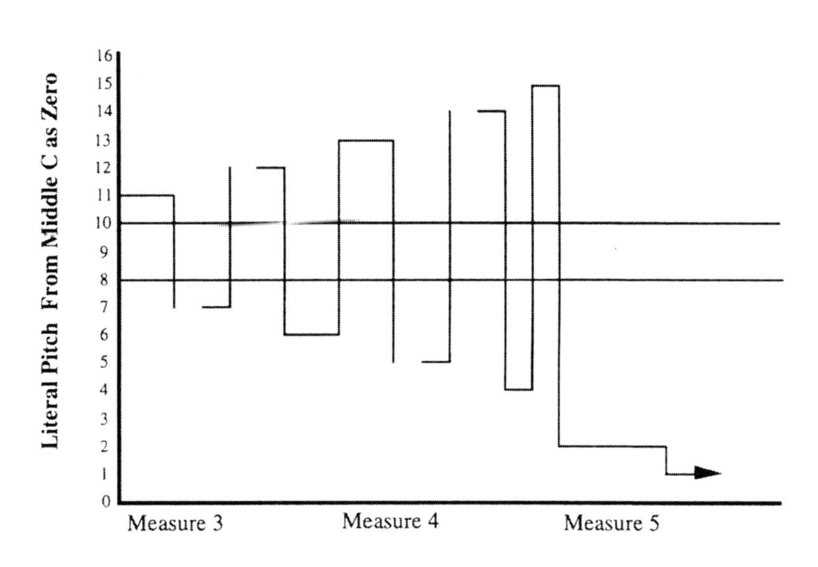

43.6 Jeffrey Johnson, *Graph Theoretical Models of Abstract Musical Transformation: An Introduction and Compendium for Composers and Theorists* (Westport, CT: Greenwood, 1997), 11.

In this piano roll representation, the long horizontal lines at 8 and 10 represent the repeated A♭–B♭ dyad in the upper staff, while the expanding squared line represents the lower-staff melody. Pitches are represented, not as inflection points, but by horizontal lines whose extent represents their duration. The vertical lines merely connect the notes and, by their length, measure the distances between them. The redrawing in figure 43.7 downplays the connecting lines and keeps the emphasis on the actual pitches. Does this accurately represent our conception of melodic contour, however?

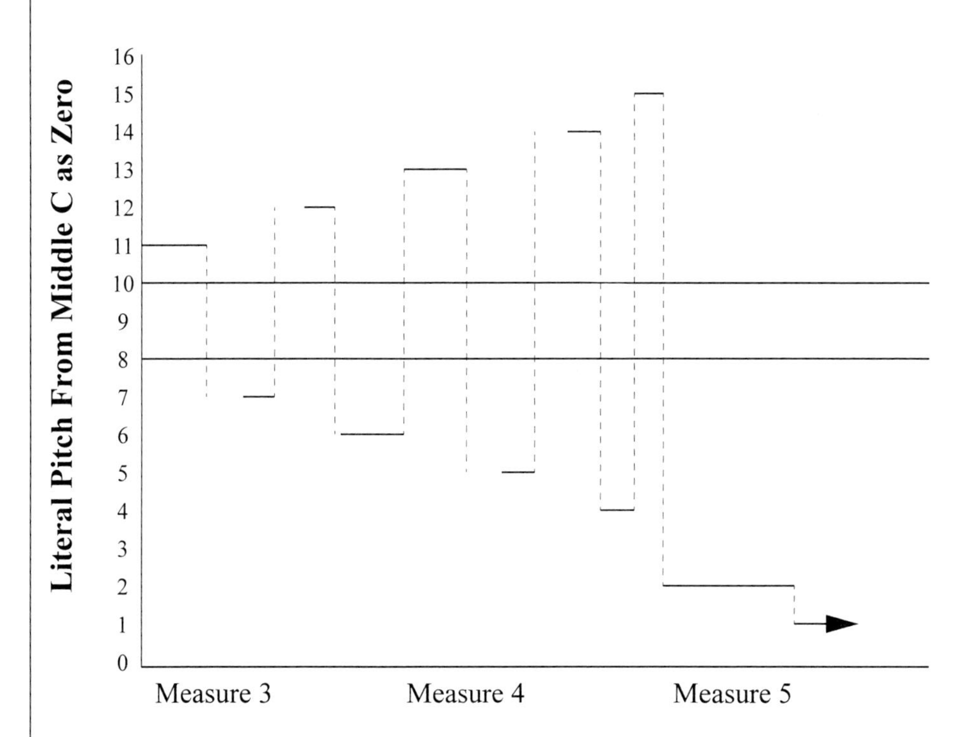

43.7 Redrawing of lower portion of fig. 43.6.

Figure 43.8 shows three representations of the opening melody of Claude Debussy's *Prelude*, "Bruyères," drawn by people who were asked to represent visually what they were hearing. The first is consistent with the listening effect expressed in figures 43.6 and 43.7. But our physical experience with music—whether as conductors, singers, instrumentalists, dancers, or listeners—is continuous, not discrete, as is reflected by the drawings of the second and third listeners. It therefore seems entirely appropriate that representations of musical contour connect the dots after all. See online figure 43.9 for an example.

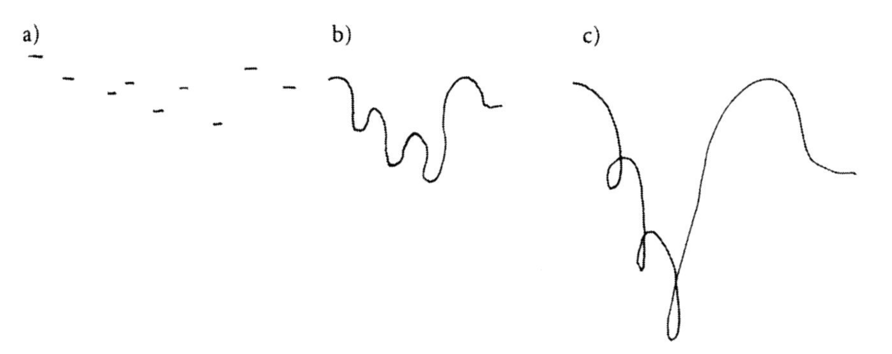

43.8 Alexandra Pierce, *Deepening Musical Performance: The Theory and Practice of Embodied Interpretation* (Bloomington: Indiana University Press, 2007), 51.

While several of the images depicting contour have benefited from the use of explicit markers to indicate pitches, this is not always needed. When a melody has consistently clear inflection points, the points representing notes are less critical and could in some cases be omitted without loss of information, as in figure 43.10, which depicts contour and other features in Arnold Schoenberg's *Six Little Piano Pieces*, op. 19, no. 4. Melody and accompaniment appear in contrasting line styles. The notation makes the similar openings of the five melodic fragments especially clear.

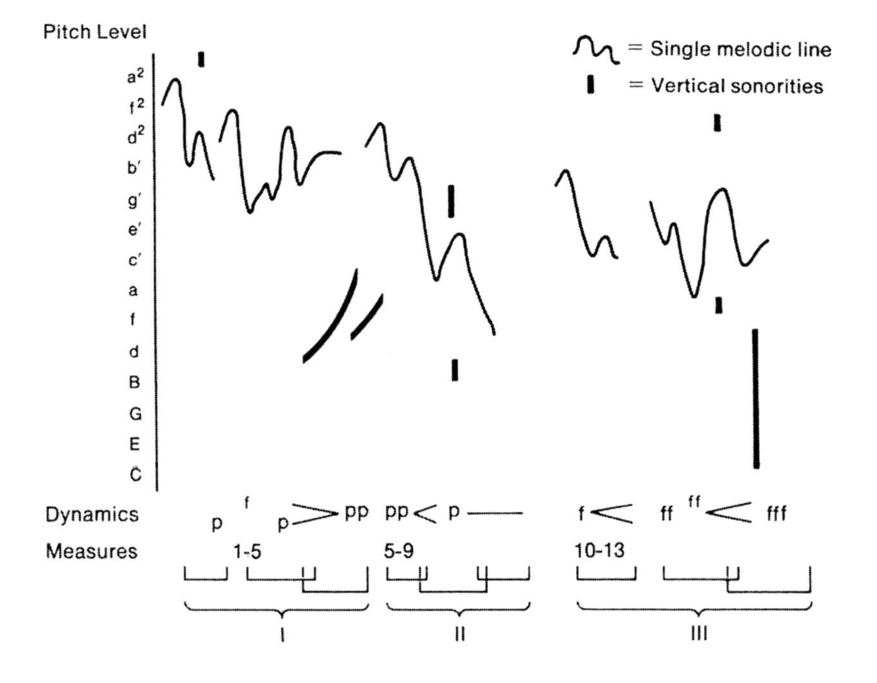

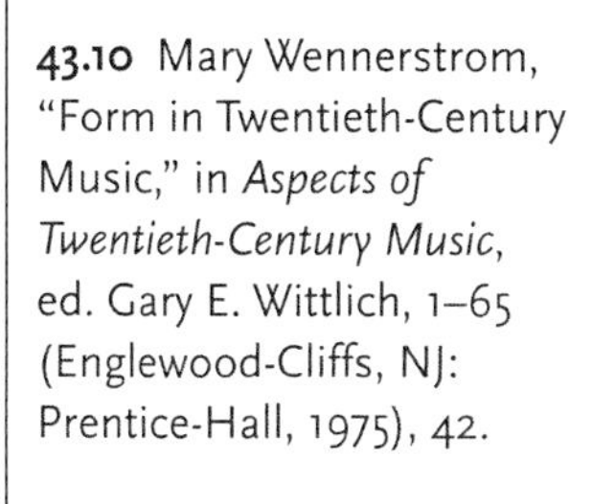

43.10 Mary Wennerstrom, "Form in Twentieth-Century Music," in *Aspects of Twentieth-Century Music*, ed. Gary E. Wittlich, 1–65 (Englewood-Cliffs, NJ: Prentice-Hall, 1975), 42.

The interweaving of two melodic lines is likewise enhanced by their mapping as contours in the excellent figure 43.11. Dotted and dashed lines trace the paths of soprano (S) and mezzo (M) lines as they crisscross, drift part, and intermingle dramatically before they end in parallel motion. At (c) the contour of attack density

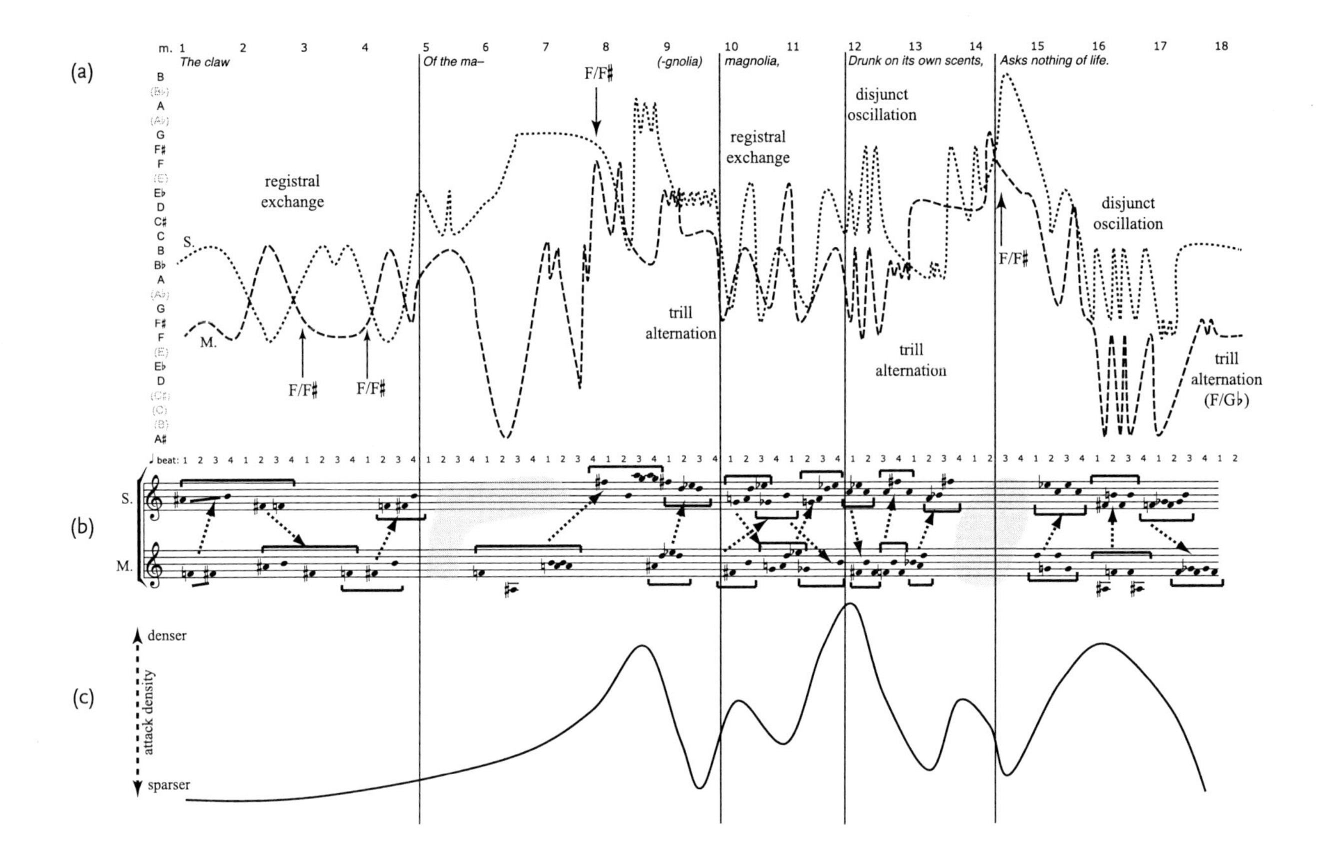

43.11 John Roeder, "Superposition in Kaija Saariaho's 'The claw of the magnolia . . . ,'" in *Analytical Essays on Music by Women Composers: Concert Music, 1960–2000*, ed. Laurel Parsons and Brenda Ravenscroft, 156–75 (New York: Oxford University Press, 2016), 162.

gives a clear picture of the passage's rhythmic shape. The tall lines that segment the music are just thick enough to do the job.

The examples so far have largely concerned the representation of melodic contour. We will now take a look at two contour analyses. Like a mountain landscape, musical contour can contain multiple layers, with local contours nested within larger contours, nested within an overall contour. Figure 43.12 illustrates one approach to representing this, using a contour reduction algorithm to determine the overall contour of six musical phrases (the same music as depicted in fig. 43.10, in fact). This process resembles that of a Schenkerian reduction in that it involves taking a series of snapshots of an iterative process, an often-successful approach. Stems and beams keep the upper and lower bounds of each contour distinct. The straight beams bulldoze the contours smooth, the way Ice Age glaciers leveled northern North America, however. A representation that kept the emphasis on the shapes of the successive reductions, perhaps along the lines of online figure 43.13, would work better.

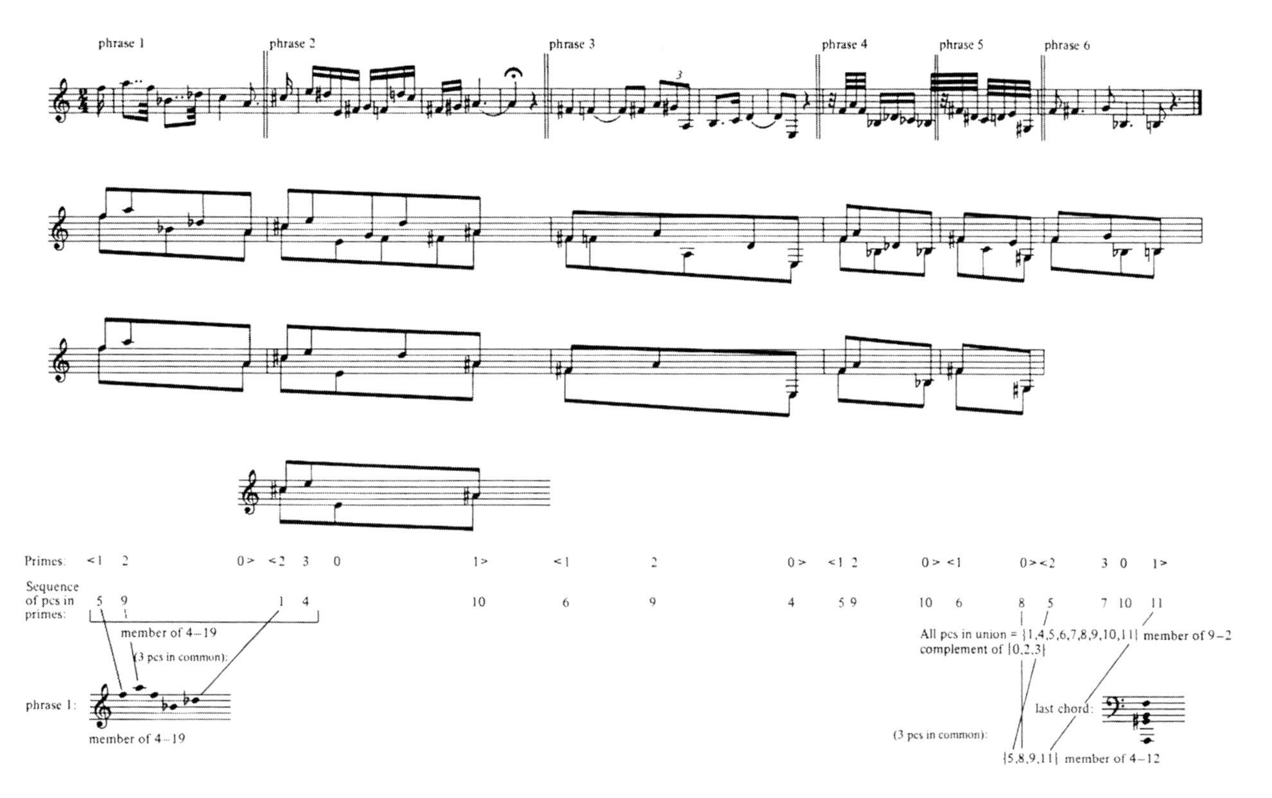

43.12 Robert D. Morris, "New Directions in the Theory and Analysis of Musical Contour," *Music Theory Spectrum* 15, no. 2 (1993): 214.

Contour similarity forms the basis of figure 43.14, which is a sophisticated meta-contour analysis of part of Steve Reich's *The Desert Music*. It analyzes the same melodies used in online figure 43.15. Each of the melodic lines there repeats some number of times before moving on to the next. Figure 43.14 represents a

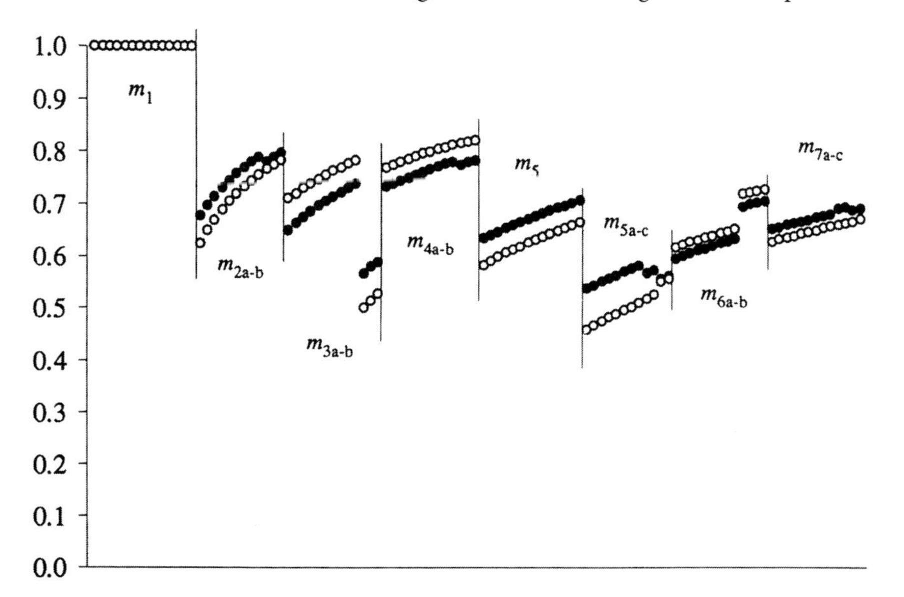

43.14 Ian Quinn, "Fuzzy Extensions to the Theory of Contour," *Music Theory Spectrum* 19, no. 2 (1997): 261.

"model of listening" to the passage (Quinn 1997, 261). As the music plays, the figure calculates a running "average" contour and plots the similarity of the current iteration to this running contour, measured on the 0.0 to 1.0 scale shown on the vertical axis. With each playing of a new melody, the accumulated average contour increasingly resembles the current melody, as is reflected in the generally ascending arch within each section. (Black markers indicate the melodies themselves, while white ones reflect a five-note reduction not shown here.) Nearly every mark in the image carries information. The horizontal and vertical axis lines provide a sense of comfort but could be omitted. The other vertical lines help to divide the sections from one another and are only as long as necessary. Design and execution are both excellent.

The study of musical contour is closely related to thematic analysis (chap. 42), but it foregrounds shape over specific pitches. Many successful images map pitches on an invisible or greatly muted grid and then visually emphasize the connectedness between pitches in some way. This keeps the representation closer to the musical surface. Showing similarity (or equality) between contours, however, sometimes requires spacing notes equally on the vertical dimension.

CHAPTER 44 Tonal Plans

The presentation of a synoptic view of a work's tonal plan is a common musical visualization. Such images typically show which tonal centers are established; where they are established, whether in relationship to one another, to a formal paradigm, or to a measure-grid; and how they relate to one another. It is not always possible (or desirable) to convey all of this, and one may want to do so in more or less detail and in conjunction with additional material. It is one thing to simply list a sequence of keys. It is another to tell a compelling story. Effective images do.

Figure 44.1 represents the second of three sections (measures 193–249) in the first movement's development in Anton Bruckner's Eighth Symphony. The

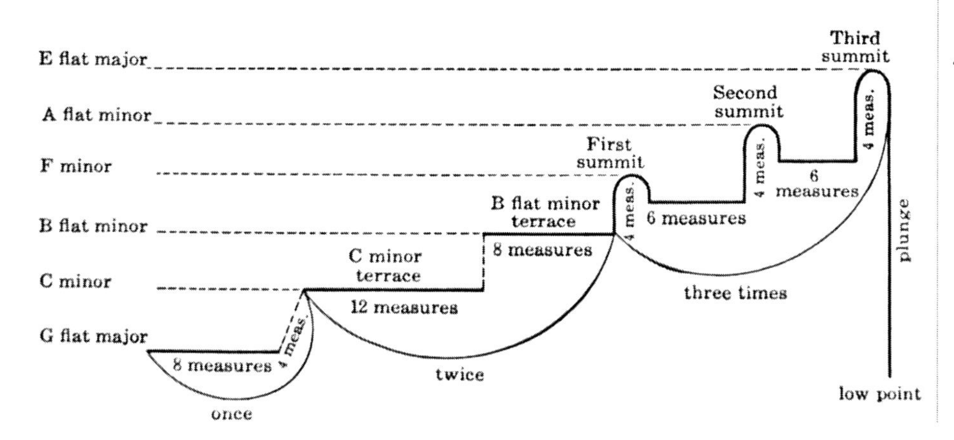

44.1 Hugo Leichtentritt, *Musical Form* (Cambridge, MA: Harvard University Press, 1951), 385.

topographical representation of a growth to a high point is vivid, enhanced by references in the prose to a "saddle-shaped hollow" and "a short valley [which] spreads over the E flat major region" (Leichtentritt 1951, 385). The figure lacks measure numbers linking the image to the music (they are present in the accompanying prose), as well as a sense of how the tonal centers relate to one another. Also, the undifferentiated list of tonal centers at the left downplays the climax-building analysis (one idea played once, a second played twice, and the final one three times).

Figure 44.2 features an undifferentiated list of key centers in Robert Schumann's song cycle *Dichterliebe*. The caption draws attention to the balance of keys related by third and fourth, but the image does not show these relations, which must be calculated one at a time. The distinction between major and minor keys and the locations of the nondiatonic relationships are not immediately clear. A redrawing in online figure 44.3 proposes another treatment.

44.2 David Neumeyer, "Organic Structure and the Song Cycle: Another Look at Schumann's *Dichterliebe*," *Music Theory Spectrum* 4 (1982): 95.

e). *Dichterliebe* roughly balances successions by third with those by fourth or fifth, almost all being diatonic. The only feature perhaps unusual is that there are no parallel key relationships or repetitions, except between the deleted 12b and 13:[16]

1. f# or A (?)
2. A
3. D
4. G
4a. E♭
4b. g
5. b
6. e
7. C
8. a
9. d
10. g
11. E♭
12. B♭
12a. g
12b. E♭
13. eb
14. B
15. E
16. c#/Db

(Nos. 4a, 4b, 12a, and 12b are songs contained in the original version but deleted before publication.)

Figure 44.4 depicts the tonal arrangement of a set of six sonatas by C. P. E. Bach. Each sonata is in three movements (thus forming a "triptych" in the author's formulation). Lines and brackets show three kinds of relationships:

1. Lines below the roman numerals that number the sonatas help group the three movements of each.
2. A long, dashed line shows the tonal return of the opening tonic, C (but minor).
3. Legged brackets above and below the diagram show the sonatas whose keys are related by thirds (the "tertian triptych plan") and fourths (the "quartal triptych plan").

Some minor adjustments eliminate the need for all of the lines and also the legend at bottom left (see online fig. 44.5).

44.4 Ellwood Derr, "The Two-Part Inventions: Bach's Composers' Vademecum," *Music Theory Spectrum* 3 (1981): 47.

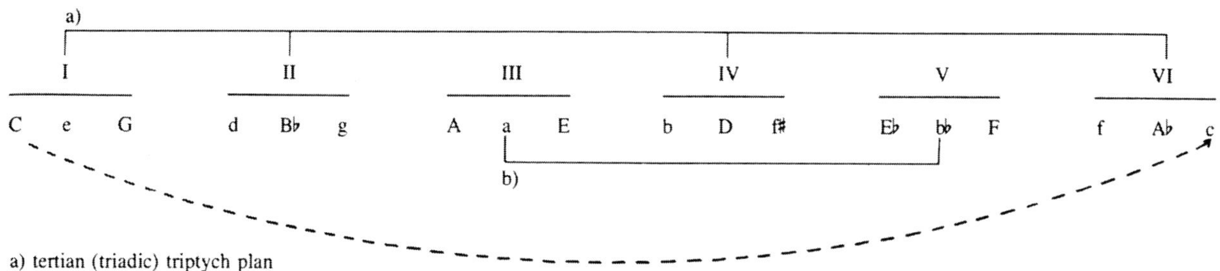

Online figure 44.3 put keys on a staff. Figure 44.6 takes the same approach, characterizing the tonal plan of the exposition and development from Wolfgang Mozart's Sonata, K. 332, movement 1. Duration symbols, as well as solid and dashed

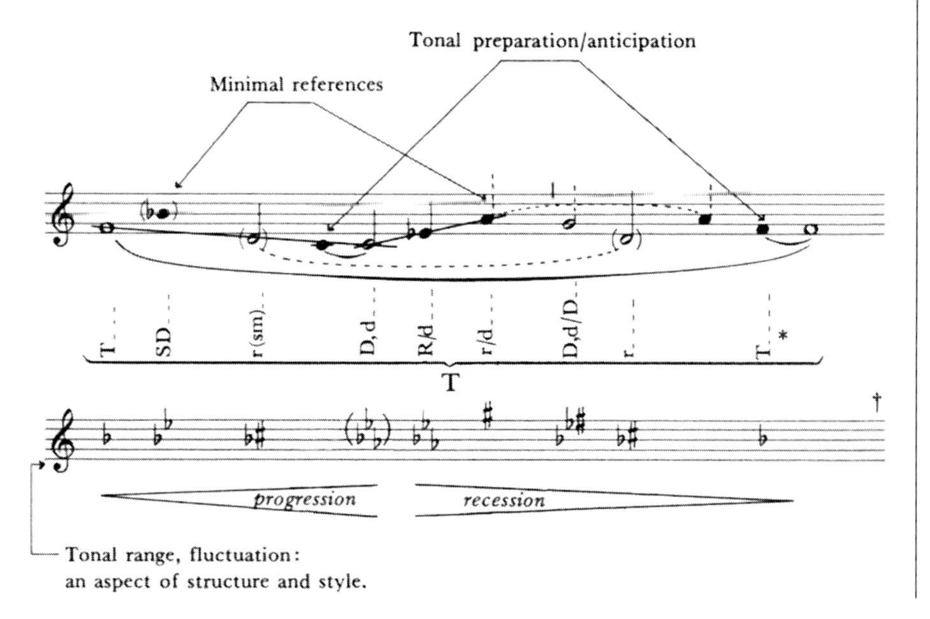

44.6 Wallace Berry, *Structural Functions in Music* (New York: Dover, 1987), 45.

stems, indicate proposed structural importance. The lower part of the image indicates "tonal range, fluctuation" through the use of key signatures, though what this means exactly is neither indicated nor apparent. The inclusion of the leading tone in the signatures for minor keys cleverly distinguishes them from major key signatures and merits wider adoption.

Both online figure 44.7 and figure 44.8 trace harmonic motion in a musical work through a musical space. Figure 44.8 features a nodes-and-arrows image that traces the tonal path of the first movement of Beethoven's *Appassionata* sonata through the circle of fifths. Numbered arrows walk viewers through the succession of tonal centers in the movement. The concept is excellent, but the design introduces challenges of information layering. The information present includes (1) the underlying circular pitch space, for which three concentric circles are two too many; (2) keys that are visited; (3) those that are not; (4) the arrows that convey the directed tonal motion, which find themselves camouflaged by the large circles and sometimes one another; and (5) the labels that count their steps, which are enclosed by yet more circles. Online figure 44.9 teases these layers apart and addresses other issues. (The lack of measure numbers prevents the viewer from taking the image to the score, by omitting information on where the keys are established and how long they last.) See also online figure 44.10.

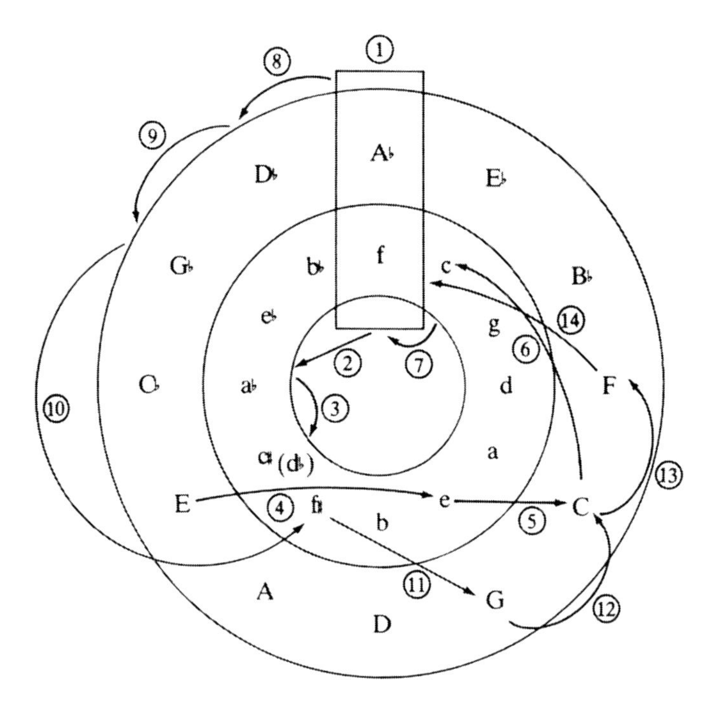

44.8 Patricia Carpenter, "Grundgestalt as Tonal Function," *Music Theory Spectrum* 5 (1983): 19.

Chapter 45 Symmetry in Music Analysis

In chapter 19, we explored the visualization of symmetrical pitch structures. Here we explore examples of symmetry in music analysis. As noted previously, when depicting symmetry visually, one should ensure that any axes of symmetry are apparent and that points that correspond to one another are clear. These do not need to be explicit if they are sufficiently implicit.

In figure 45.1, E♭s wedge inward in register to a central octave. The wedge makes the axial pitch clear. The compressed vertical scale (the entire image spans four octaves) is entirely appropriate. Contrasting line styles distinguish figure and ground, though the grid could be eliminated altogether and the dashed lines replaced with solid ones.

Larger-than-needed brackets in figure 45.2 highlight the symmetrical structuring of a hypothetical choral concert. Pairs of unaccompanied soprano-alto-baritone (SAB) or soprano-alto-tenor-bass (SATB) works (4–5 and 7–8) surround

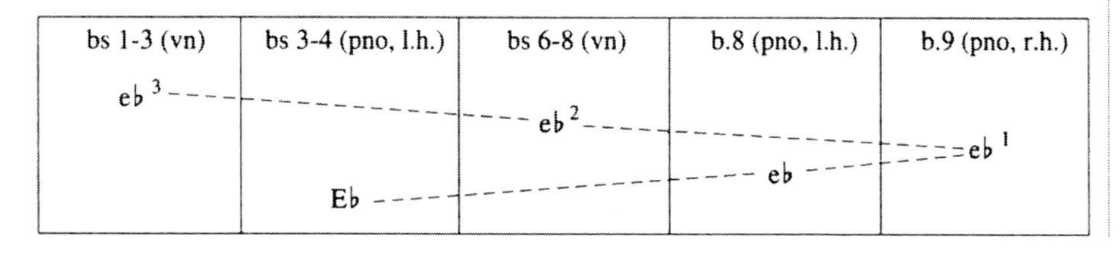

45.1 Felix Meyer and Anne Shreffler, "Performance and Revision: The Early History of Webern's Four Pieces for Violin and Piano, op. 7," in *Webern Studies*, ed. Kathryn Bailey, 135–69 (Cambridge: Cambridge University Press, 1996), 152.

an accompanied unison piece. Outside of those are the pieces with the densest texture (3 and 9–10) and then another pair of a cappella SATB pieces (1–2 and 11–12), with a trio of accompanied pieces serving as a finale. The image is inconsistent, using individual lines for the 1–12 and 2–11 pairings but a single one for the 4/5–7/8 pairing. The central axis is perfectly clear, but the stark brackets are visually out of balance with the information, and one must make two turns tracing the lines from one end to the other.

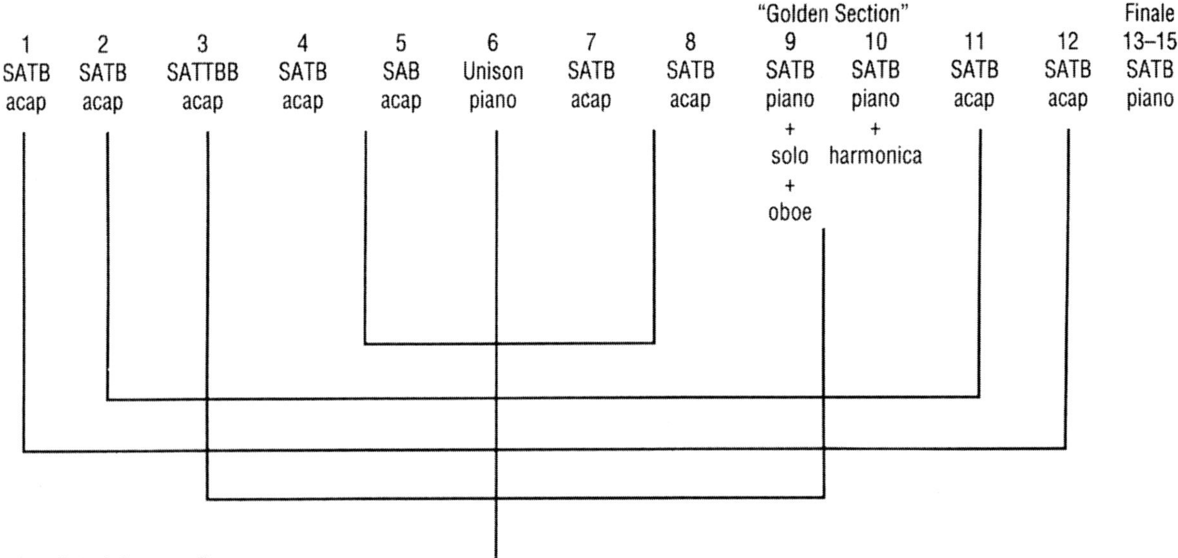

45.2 David L. Brunner, "Choral Program Design Structure and Symmetry," *Music Educators Journal* 80, no. 6 (1994): 49.

Figure 45.3 takes a softer approach when showing the contrapuntal pitch structure in the theme of Anton Webern's *Symphony*, op. 21, movement 2. The example is used to illustrate that symmetry results when ordered pitch intervals (i<x,y>) in the passage are reduced to unordered pitch intervals (i(x,y)) and when ordered pitch-class intervals (ip<x,y>) are reduced to unordered pitch-class intervals (ip(x,y)). The symmetries are shown with nested arcs, an approach that works well here. The central axes (flavors of 6) are clearly apparent. It is not clear, however, whether the second row of numbers in each pair is being treated as symmetrical. They are not in the sense that they are not palindromic, and this is how they are discussed in the text accompanying the image, but of course they are inversionally symmetrical. See also online figures 45.4–45.6.

Figure 45.7 uses circular arrows to show various self-symmetrical collections in Webern's *Five Pieces for Orchestra*, op. 10, nos. 3 and 4, which themselves are related symmetrically to another set, as with the two collections labeled with the

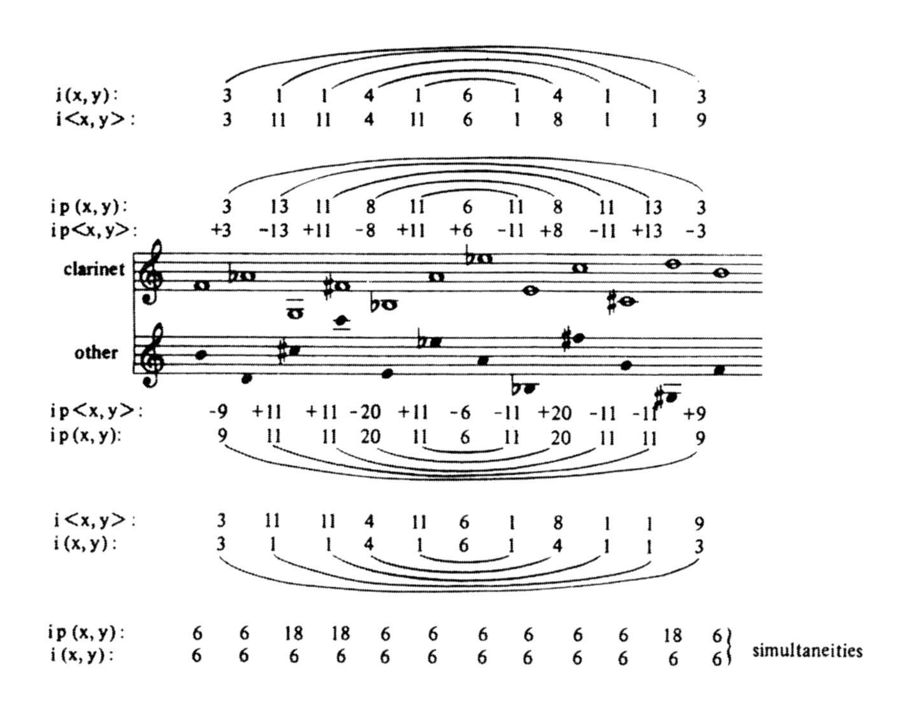

45.3 John Rahn, *Basic Atonal Theory* (New York: Longman, 1987), 30.

two-headed I arrow. Inversion operators (M, I) are defined contextually within a particular analysis, so one needs to read the accompanying prose (or do some independent calculation) to know that M represents inversion around A, while I represents inversion around B♭. The figure reasonably omits inversional axes and pitch-to-pitch correspondences in the interest of clarifying the story involving M and I operations, though the image would benefit from the inclusion of the axes of the M and I operators.

Symmetry can be expressed in numerous ways. Because we are surrounded by symmetry and recognize it readily, subtle clues can cue the viewer toward its presence. Be careful not to make them too subtle, however.

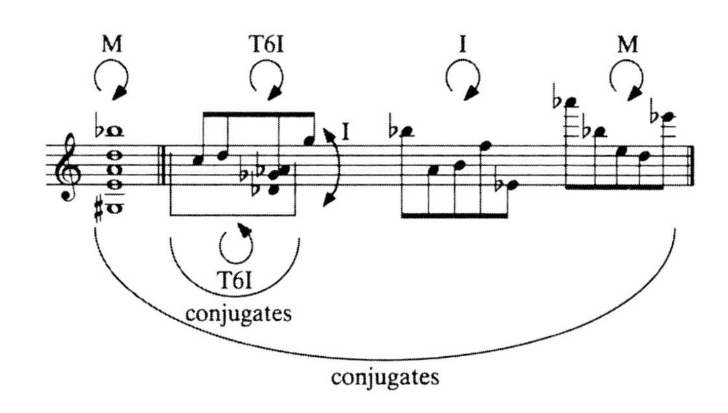

45.7 David Lewin, *Musical Form and Transformation: 4 Analytic Essays* (New Haven, CT: Yale University Press, 1993), 89.

Chapter 46 Rhythmic Analysis

In recent years, sophisticated theories of rhythm and meter have emerged, many of which involve visualization. The simple approach used in figure 46.1 works well. It presents a reading in which perceived downbeats do not appear every two beats as suggested by the meter signature. The proposed new groupings are indicated above the score, by the rebarring of the music with new meter signatures. This method is easily interpreted: it clearly shows the signatures, and they communicate directly in the language of the musical property being studied. See also online figure 46.2.

Figure 46.3 depicts polymeter in a way similar to online figure 46.2. The upper part of the image pairs a normative meter notated in $\frac{2}{4}$ on the second staff with changing, irregular metrical groupings above. It indicates individual durations by numbers of sixteenth notes, the lowest common denominator, while groupings are indicated by beam groups and measured with circled numbers. Two line styles differentiate where the *violão* patterns do and do not align with the basic metric

46.1 Joel Lester, "Articulation of Tonal Structures as a Criterion for Analytic Choices," *Music Theory Spectrum* 1 (1979): 78.

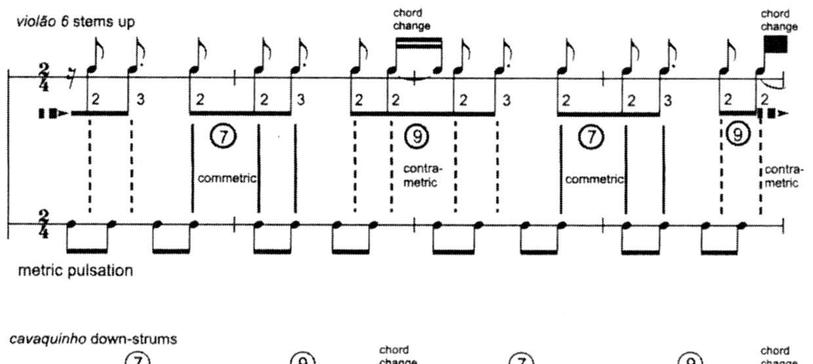

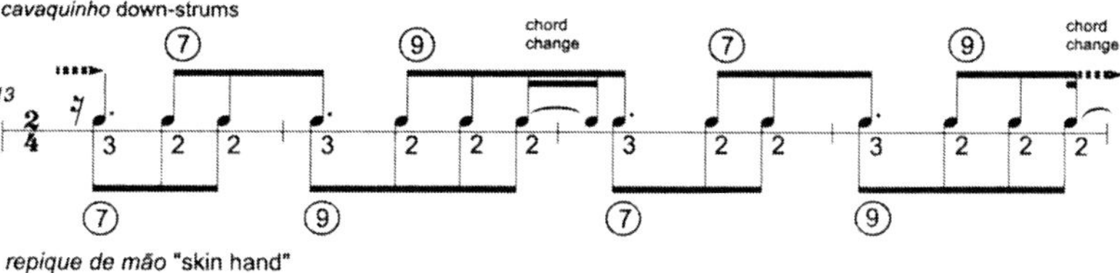

46.3 Jason Stanyek and Fabio Oliveira, "Nuances of Continual Variation in the Brazilian Pagode Song 'Sorriso Aberto,'" in *Analytical and Cross-Cultural Studies in World Music*, ed. Michael Tenzer and John Roeder, 98–146 (New York: Oxford University Press, 2011), 129.

pulsation. The placement of circled numbers quasi-centered below the beam groups is not as clear as their placement at the start of each quasi-metrical beam group in the similar lower part of the image. See also online figure 46.4, in which rebarring replaces notated meter entirely.

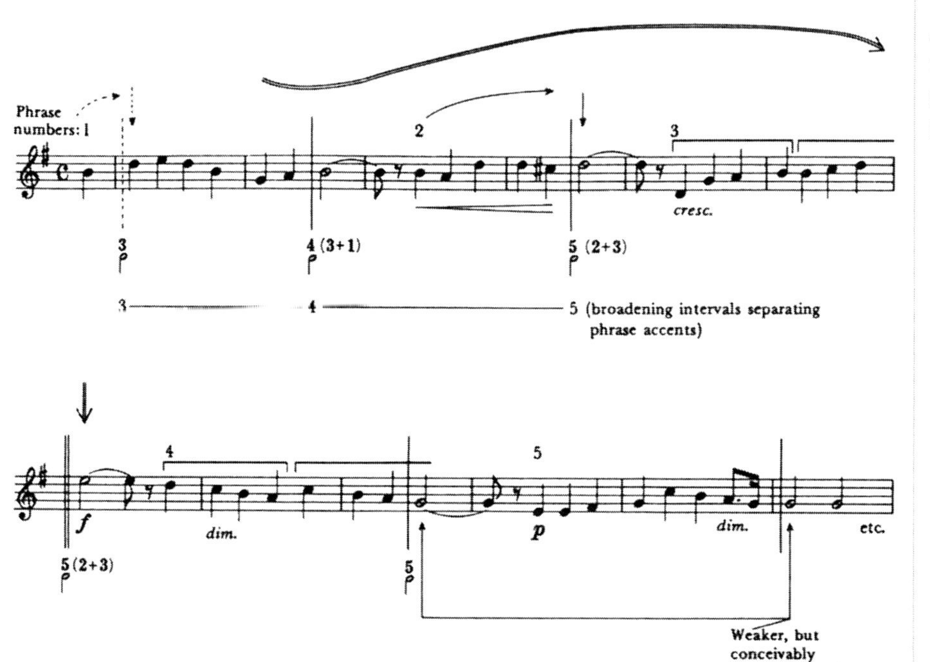

46.5 Wallace Berry, *Structural Functions in Music* (New York: Dover, 1987), 334.

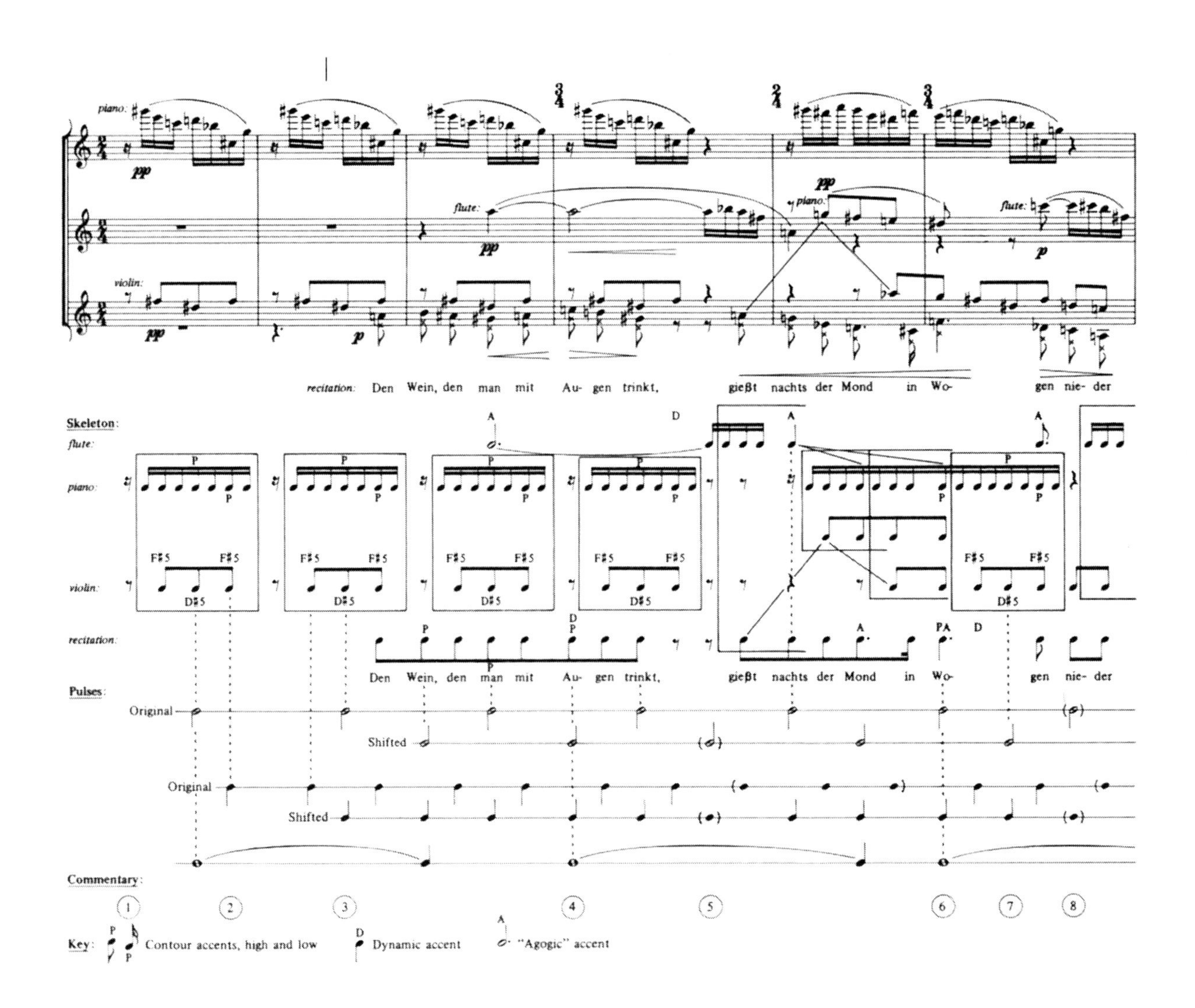

46.6 John Roeder, "Interacting Pulse Streams in Schoenberg's Atonal Polyphony," *Music Theory Spectrum* 16, no. 2 (1994): 236.

The previous images display metrical structure in a primarily quantitative way. Figure 46.5 adds qualitative indications. Arrows—first dashed, then solid, then double—show how the flow of successive "downbeats" increases in intensity as time between them increases from 3 to 4 to 5. Prior images invited us to map our hearing against those in the image. This image more actively guides our listening.

Figure 46.6 rewards careful study. It reduces a six-staff score to three staves, indicating meter changes only above the bar line for the score rather than within each part. Below that ("Skeleton"), the rhythmic values repeat, aligned precisely below the score for easy reference. Three types of accent are indicated via simple mnemonics, from which several proposed pulse streams are derived. The words *original* and *shifted* suffice to convey the image's meaning. Dashed lines gently guide the eye from the pulses back to the skeleton; from there, the music itself is readily available.

Finally, circled numbers at the image's bottom provide access points for extended commentary in the accompanying prose. The image is simultaneously efficient and richly detailed.

In much Western music, we perceive temporal periodicity at levels both below that of the division and above that of the measure (see online figs. 46.7 and 46.8). The latter are often referred to as hypermeter. Figure 46.9 shows an analysis from the early twentieth century that relates hypermetrical structure to a prototypical eight-bar phrase. Numbers in parentheses below the music (from Ludwig van Beethoven's Sonata no. 12, op. 10, movement 3) indicate which measure in the prototype the bar corresponds to. The notation allows for elisions (4=5) and extensions (8b . . . 8c . . . 8d). This image and online figure 46.10 both show deviations from a normative structure, a powerful basis for hearing the push and pull of music against a learned paradigm. The notation in figure 46.9 works well for music that evokes the eight-bar default and viewers who understand it. It is elegant and simple. It invites viewers to rehearse the music themselves and partake in the interpretation.

Figure 46.11 depicts multiple hypermetrical layers at the level of an entire movement. The image represents metric hierarchy in Ludwig van Beethoven's *Sonata*, op. 13, movement 2. Whereas the lowest level in online figure 46.8 is the eighth note, with the measure representing the fourth level, figure 46.11 starts with

46.9 Hugo Riemann, *L. Van Beethoven sämtliche Klavier-solosonaten: Ästhetische und formal-technische Analyse mit historischen Notizen* (Berlin: M. Hesse, 1919), 329.

the measure and at the highest level (g) marks off a single point, m. 51. The image is more ambitious in scope than online figure 46.8. It is also harder to tease information from. The lack of music notation is one obvious reason. While it would be impossible to provide that here, the omission makes it harder to read the notation and experience it personally, discouraging verification. Additionally, the vertical dashes are harder to count quickly than the dots of online figure 46.8 as they tend to blur perceptually into lines that must be actively separated into their constituent segments. Finally, it can be difficult to see which measures exist within the same hierarchical level. The redrawing in online figure 46.12 offers an improved approach.

Given that hypermeter is conceptually just meter at a higher level, and given that the Western notation is optimized for the display of meter (but not hypermeter), online figure 46.13 offers a novel and clever approach.

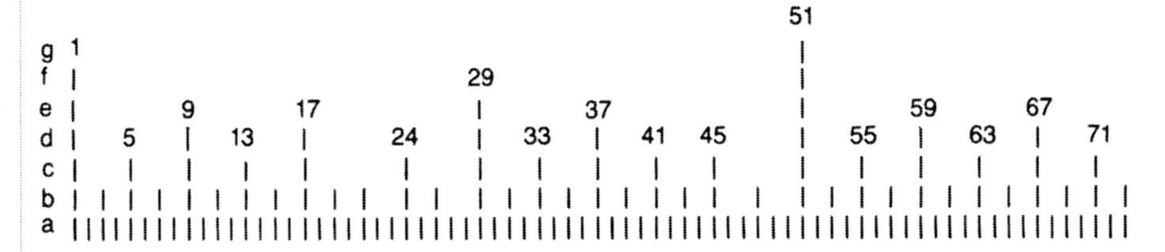

46.11 Jonathan D. Kramer, *The Time of Music: New Meanings, New Temporalities, New Listening Strategies* (New York: Schirmer Books, 1988), 119.

Perhaps more than any other aspect of music, we think about its temporality on a physical level, to feel the ebb and flow, the succession of strong and weak pulses, and so on. Successful images depicting rhythm and meter not just facilitate but invite and even encourage that kind of engagement.

CHAPTER 47 Formal Analysis

We looked at schematic representations of musical form in chapter 36. Here, we examine depictions of form in actual pieces. When creating a visualization of musical form, one must decide how detailed to be in the division of the piece—that is, what the smallest units should be and whether to group these into hierarchical levels. An image might present various properties of the sections, properties that are often consistent within one hierarchical level but that might vary across different levels. Formal analysis emphasizes forest over trees, and clarity of the large structure should be the foremost concern. At the same time, it is important to provide enough detail to clarify the features that are unique to a piece. Images generally show sections to be related or unrelated on the basis of things like melodic material, tonal centers, or formal function. In this chapter, we will see different tactics used to make sections clear and to show how they are related to one another. We will also see different levels of detail. We will look first at depictions of form in which time flows vertically from top to bottom, then some in which it flows horizontally from left to right, and finally some in which the horizontal and vertical dimensions combine in informationally meaningful ways. Sometimes the decision of which orientation to use is purely pragmatic, and sometimes one is clearly superior to the others.

In the first several images, time is expressed top to bottom. This orientation is useful when an image is text-heavy or when there are too many sections to fit effectively into the more space-constrained horizontal orientation. We will see that vertically oriented images are more likely to focus on the succession of time than the measurement of time and, therefore, are less likely to convey a sense of music's proportions.

In figure 47.1, a depiction of the form of Ives's *The Unanswered Question* separates three musical strands: the opening and closing string sections, the seven "questions" posed by the trumpet, and the answers given by the woodwinds. The overall top-to-bottom layout allows one to follow the passage of time, while the columns distinguish instrumentation and function. The figure is elegant in its simplicity. The colons down the first column imply the continued presence of the strings; a hazy line would better express the strings' affect in the piece. The time indices clearly imply that the image is based on a particular performance. A proportional arrangement of the sections would be easy to produce and would enhance the image. A similar image appears in online figure 47.2.

47.1 Erik Christensen, *The Musical Timespace: A Theory of Music Listening* (Aalborg: Aalborg University Press, 1996), 1:52.

Strings:	Trumpet Questions:	Woodwind Answers:
0'00-1'35 Strings alone		
:	*1'35* Question	
:		*2'04* placid, gentle answer
:	*2'27* Question	
:		*2'44* calm, slightly dissonant answer
:	*3'13* Question	
:		*3'30* hesitating statement
:	*3'52* Question	
:		*4'07* firm statement
:	*4'23* Question	
:		*4'33* lively polyphonic discussion
:		*4'38* soft sustained cluster.....*4'52*
:	*4'47* Question	
:		*4'53* hectic activity
:	*5'37* Question	
5'47-6'03 Strings alone, Fading		

The next five images use a tabular format to present information about the sections of the works they are studying. Figure 47.3 depicts the structure of a poem set by Anton Webern in *Drei Lieder nach Gedichten von Ferdinand Avenarius*. The columns include the verse, text, rhyme scheme, line numbers of the poem, bar numbers from the setting, and rhythm of the accents in the setting. The rhyme scheme is the only column that conveys structural similarities and differences among the sections. As commonly found in tables, the gridlines complete for attention with the information and could be minimized or, particularly in the case of the vertical lines, removed altogether.

verse	text	rhyme	lines	bars	rhythm of accents
1	Ertrage du's,		1a	1–2	2
	lass schneiden dir den Schmerz	a	1b	3–6	4 5 6
	scharf durchs Gehirn und wülen Hart durchs Herz	a	→2		
			–	7–8	piano interlude
	Das ist der Pflug, nach dem der Sämann sät,	b	3	9–12	10 11 12
	dass aus der Erde Wunden Korn ersteht.	b	4	13–15	14 15
			–	15–17	piano interlude
2	Korn, das der armen Seele Hunger stillt -	c	5→6	18–23	19 20 21 22 23
	Mitt Korn, o Vater, segne mein Gefild:	c			
	doch wirf auch ein in seine Furchen Saat!	d	7	23–5	24 25
	reiss deinem Pflug erbarmungslos den Pfad,	d	8	26–8	27 28

47.3 Susanne Rode-Breymann, "'. . . Gathering the Divine from the Earthly . . .': Ferdinand Avenariou and His Significance for Anton Webern's Early Settings of Lyric Poetry," in *Webern Studies*, ed. Kathryn Bailey, 1–31 (Cambridge: Cambridge University Press, 1996), 24.

Figure 47.4 is an information-poor representation of the structure "surface organization" in Igor Stravinsky's *Piano-Rag-Music*. The table format presents only "large divisions" (which in some cases are not particularly large) and smaller divisions. Unfortunately, for sections III, IV, and VII the second column wraps to a third rather than simply continuing on new lines. The table provides no information about how the sections relate to one another, so we do not know whether there are recurrences of any kind. In the second column, brackets link sections that are related through a close "surface relationship, such as a literal repetition" (Joseph 1982, 79). Because brackets conventionally group sets of contiguous items rather than noncontiguous items, curved lines with arrowheads would better indicate these discontinuous connections. On its own, the picture doesn't tell much of a story. Is this because the music conveys no story? Or is that that most of the interesting information has been relegated to the surrounding prose? When we are constructing a detailed narrative of a piece, and here the whole twenty-five-page article is devoted to a discussion of this music, we must decide whether to show a detailed, synoptic view, one that creates a rich picture inviting detailed and lengthy scrutiny, or to introduce the attentional elements in a more piecemeal way. While there isn't a single right answer to that question, in general, fewer, more detailed images are preferable, as that makes it easier to relate overview and detail, forest and trees. Unfortunately, figure 47.4 is not very informative. Online figures 47.5 and 47.6 provide different, excellent representations of form in movements by Wolfgang Mozart. See also online figure 47.7.

47.4 Charles M. Joseph, "Structural Coherence in Stravinsky's *Piano-Rag-Music*," *Music Theory Spectrum* 4 (1982): 79.

Large Divisions	Internal Organization*	
I. Mm. 1 - 9	Mm. 1-2	
	3	
	4	
	5-7	
	8-9	
II. Mm. 10 - 14	Mm. 10-13	
	14	
III. Mm. 15 - 23	Mm. 15-16	22-23
	17	24
	18-19	
	20	
	21	
IV. Mm. 25 - 54	Mm. 25-28	50
	29-32	51-54
	33-36	
	37-41	
	42-49	
V. Mm. 55 - 82	Mm. 55-66	
	67-72	
	73-76	
	77-80	
	81-82	
VI. Measure 83 (unbarred)		
VII. Mm. 84 - 113	Mm. 84-87	96-98
	88	99
	89-92	100-101
	93	102
	94	103-112
	95	113

Figure 47.8 shows the structure of the Gloria from Heinrich Schütz's *Musicalische Exequien*, an early funeral mass setting (written ca. 1636). The Gloria includes sections for one or more solo voices and sections for chorus. The image depicts the structure in four columns: section numbers, name, scoring, and mode. In columns 2, 3, and 4, the image groups sections according to various criteria and draws arrows to highlight correspondences of one kind or another. Drawing attention to connections is an entirely worthy activity. In this image, however, the abundance of brackets and arrows, their visual weight relative to the text, the fact that they nearly collide between the columns, and the way they overlap in the rightmost column make the patterns harder to see than necessary. Putting more space between the columns, removing the arrowheads, and using different styles for lines that mean different things would all improve the image's effectiveness.

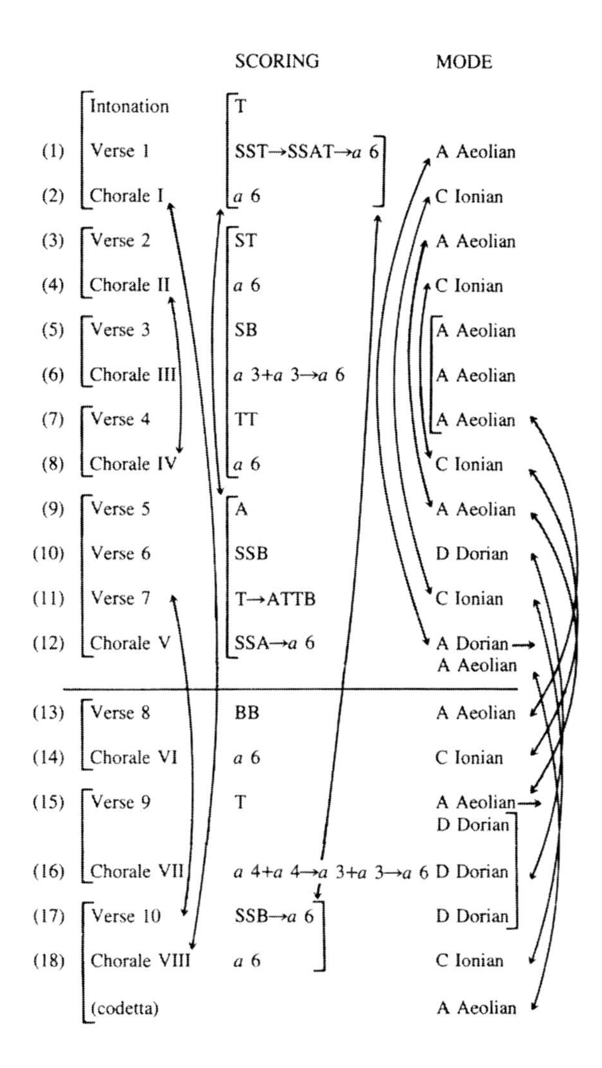

47.8 Thomas Bernick, "Modal Digressions in the *Musicalische Exequien* of Heinrich Schütz," *Music Theory Spectrum* 4 (1982): 61.

The next images arrange time on the horizontal axis. This orientation usually results in more compact images that can be better taken in holistically. For this reason, it should be considered the default choice, unless circumstances warrant otherwise. It more effectively conveys temporal proportion (see chap. 26) and expresses formal hierarchies (see online fig. 47.7). For reasons of space, such images tend to make more use of symbols than of text, which consists at most of short phrases.

Figure 47.9 was produced at a time when shading was harder to print, so publishers (and software designers) would sometimes resort to hash marked fill textures of this sort. It produces a disconcerting effect that both distracts and creates figure-ground issues. Better today would be a very light gray fill. Other improvements are also possible. Measure numbers use a time signature typeface when a standard face would be better, and it would be better if the numbers used were meaningful to the music, rather than to an arbitrary five-bar grid. The image is taller than necessary. Shrinking this dimension would bring the content closer together. A redrawing might employ a single horizontal line, perhaps thicker for text lines, thinner for piano passages, with very short tick marks to indicate these and other internal divisions. Below the line, text sections could be set in bold ("line X") and piano passages in regular type ("piano interlude"). The expressive text could appear in italics above the line ("voll Leidenshaft"). See also online figures 47.10 and 47.11.

47.9 Susanne Rode-Breymann, "'. . . Gathering the Divine from the Earthly . . .': Ferdinand Avenariou and His Significance for Anton Webern's Early Settings of Lyric Poetry," in *Webern Studies*, ed. Kathryn Bailey, 1–31 (Cambridge: Cambridge University Press, 1996), 26.

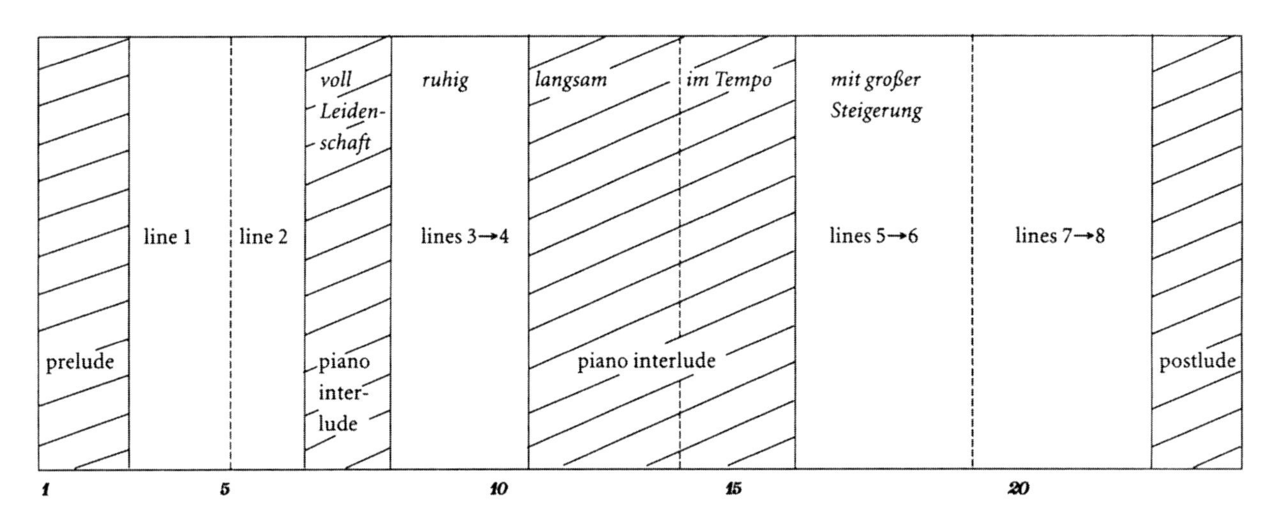

Figure 47.12 depicts the palindromic design of Bach's Two-Part Invention in D minor. The simple design conveys only basic information. Five sections, marked off by horizontal lines, span segments of various lengths that are indicated by ranges of measure numbers hiding within horizontal brackets. The sections form a

palindrome, as evident by the letter names above and the larger brackets below. In the Y section, the (7) over the dashed line indicates a passage that does not have a corresponding passage in Y′.

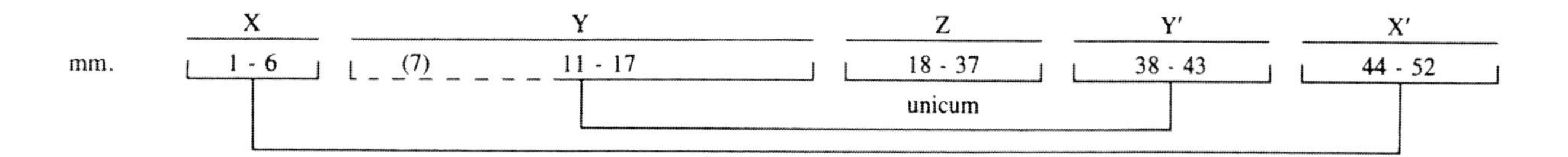

47.12 Ellwood Derr, "The Two-Part Inventions: Bach's Composers' Vademecum," *Music Theory Spectrum* 3 (1981): 31.

Even in a simple image like this, one can improve clarity. The measure number ranges are of course needed to locate the sections in the music. The letters expressing formal recurrence are likewise essential. The lines below the letters currently only organize the information that lies below *them*, however. Their unequal length implies something about proportion that isn't true. The eleven-measure Y section appears to be two and a half times the length of the twenty-measure Z section. It would be better if the sections either were equal in length or, better, reflected the proportions of the music. The horizontal brackets beneath the measure ranges can be eliminated by a simple change in the drawing of the measures for the first Y section: "(7–10) 11–17." *Unicum* is simply a fancy word for the center of the palindrome and is not necessary, as the context makes the center point clear. Since the evidence of the palindrome appears in the letters and not the measure ranges, those letters should be linked by lines and not the (now erased) brackets. Rounded lines would be visually clearer than squared lines.

Figure 47.13 represents musical recurrence in a rather different way. Designed by a digital artist, the diagram uses arches to connect repetitions of musical material in a Chopin Mazurka. Annotations appear below the image. Whereas form diagrams typically show thematic repetitions only at the sectional level, this one connects repeated events of any length. The more extensive the material that is repeated, the thicker the ribbon. For example, the two broad shaded regions show that A repeats at B and that C repeats at D. On the other hand, the series of tall, thin arches spanning most of the length of the figure indicate material in E, all quite brief, that was first heard in B. Although the diagram is visually appealing, it falls short as a depiction of musical design. Most significantly, it is selective in the recurrences it shows. For example, if B repeats A, why do the materials at E refer to the repetition B and not the original A (and similarly throughout the image)? This leads to a strangely nonhierarchical view of the piece that is surely at odds with its structure. Also, because the height of an arch is related only to the distance between the events it connects, the diagram gives a sense of importance to repetitions that are far apart in the music that may or may not be justified.

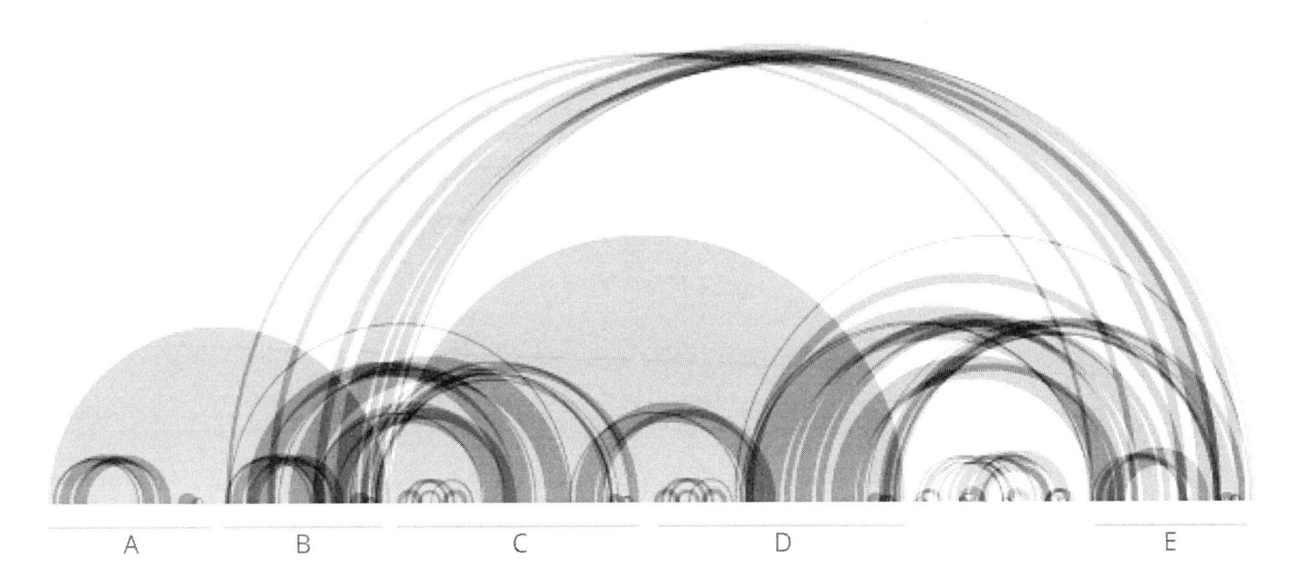

47.13 Martin Wattenberg, "The Shape of Song," Turbulence, accessed October 12, 2009, http://turbulence.org/Works/song/gallery/gallery.html. Annotations added.

Figure 47.14 (with annotations added) shows information about a blues tune and a performance of it. The tune itself, a typical twelve-bar blues, is at *a*. The structure of the tune is given in *b*. Vertical ticks mark off each measure. Brackets above show the tune's A–A–B structure, reflecting the schematic design of the blues, while brackets below show the division of A into parts 1 and 2. Brackets below a summary of the harmonic plan (*c*) imply a temporally symmetrical arrangement, with A(1) and B on tonic, the two A(2) sections on I^7, and the central A(1a) on the minor tonic, except the scheme at *c* does not follow the proportions established in the formal plan (*b*).

The image then represents the structure of a particular performance of the tune at *d*. In the first system, it marks off the introduction and two statements of the theme in four-bar units (4 T = *Takte*, followed by two groupings of three four-bar units: A A B = 12 *Takte*). After this come ten improvisations. Ticks now mark off twelve-bar units, not four-bar units, with flute and piano alternating improvisations, as indicated by brackets below the line. These improvisations shorten (*Abnehmende Formgrößen*) from 5 to 4 (except 3 for the flute's second improvisation) to 3 to 2. Oddly, as the number of choruses shortens, the space they occupy horizontally increases. Improvisations 7 and 8, totaling four choruses, occupy the same space as the six choruses of improvisations 5 and 6. The last improvisations (9 and 10) are each just one chorus long, but they cover more than three times the width of improvisations 5 and 6 directly above, though they match the width of the twelve-bar theme statements that end the performance, as well as those that opened it. The drawing thus gives exactly the opposite impression of the shortening process. It would also be better if the entire performance were displayed in a single row, which, together with a consistent horizontal scale, would make the gradual

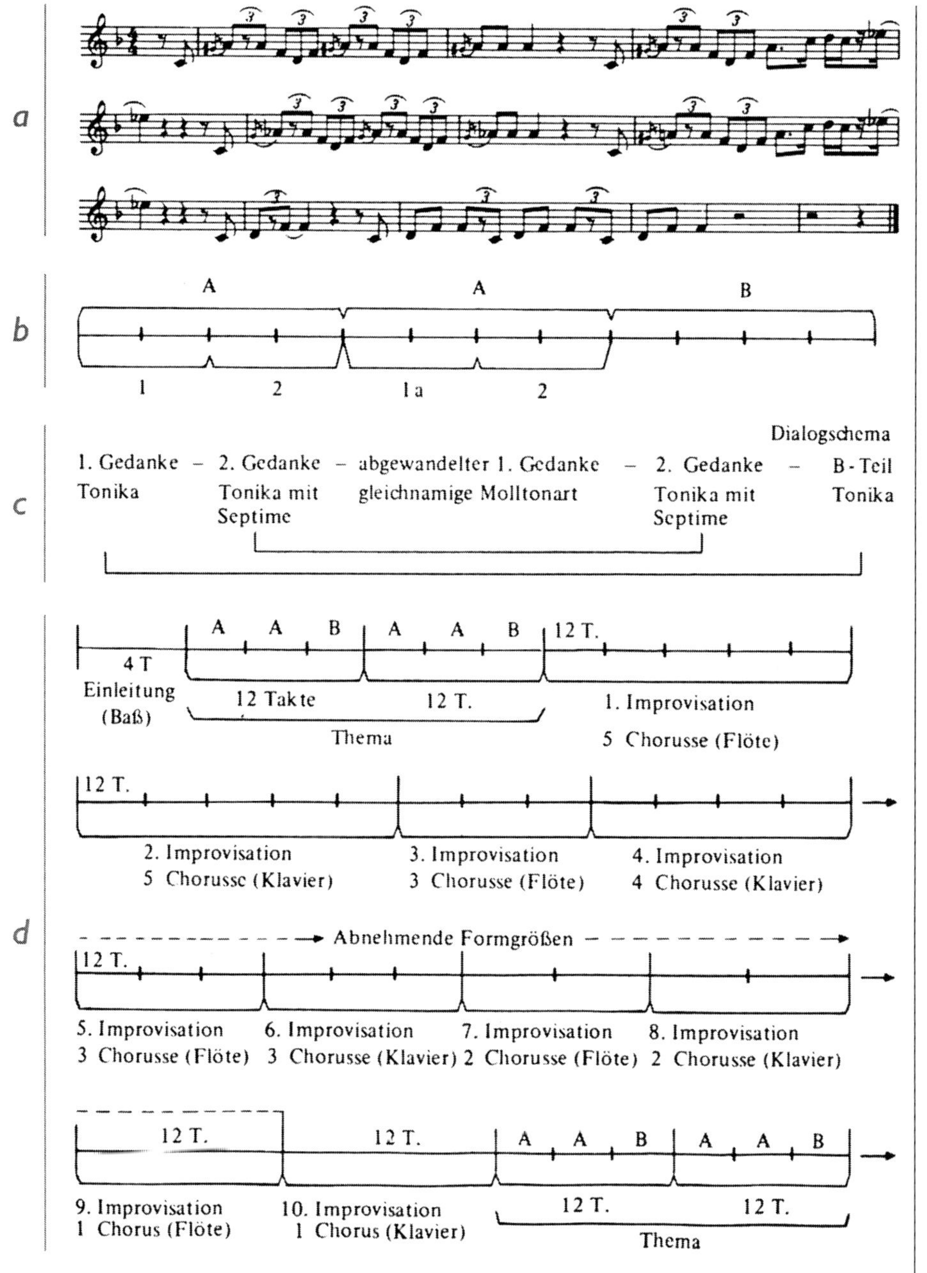

47.14 Carl Gregor zu Mecklenburg and Waldemar Scheck, *Die Theorie des Blues im modernen Jazz* (Strasbourg: Heitz, 1963), cited in Werner Breckoff, *Musik Aktuell: Informationen, Dokumente, Aufgaben* (Kassel: Bärenreiter, 1971), 240. Annotations at left edge added.

shrinking of the improvisations easier to see. This would eliminate the need to find the misplaced arrows in *d*; the arrows at the right end should each be moved up one line.

The visually appealing outline of George Crumb's *Black Angels* in figure 47.15 says as much about formal *process* as it does form itself. It conveys a complex narrative with clarity. Among the image's strengths is its careful layering of information. The image's core lies at the vertical center, where the most prominent lines and text in boldface caps anchor the rest of the image. The work's thirteen pieces form three groups with differing programmatic significance. Roman numerals invite a sequential reading that guides the viewer from departure to absence to return. The numbers of the pieces within each grouping are given unobtrusively immediately below.

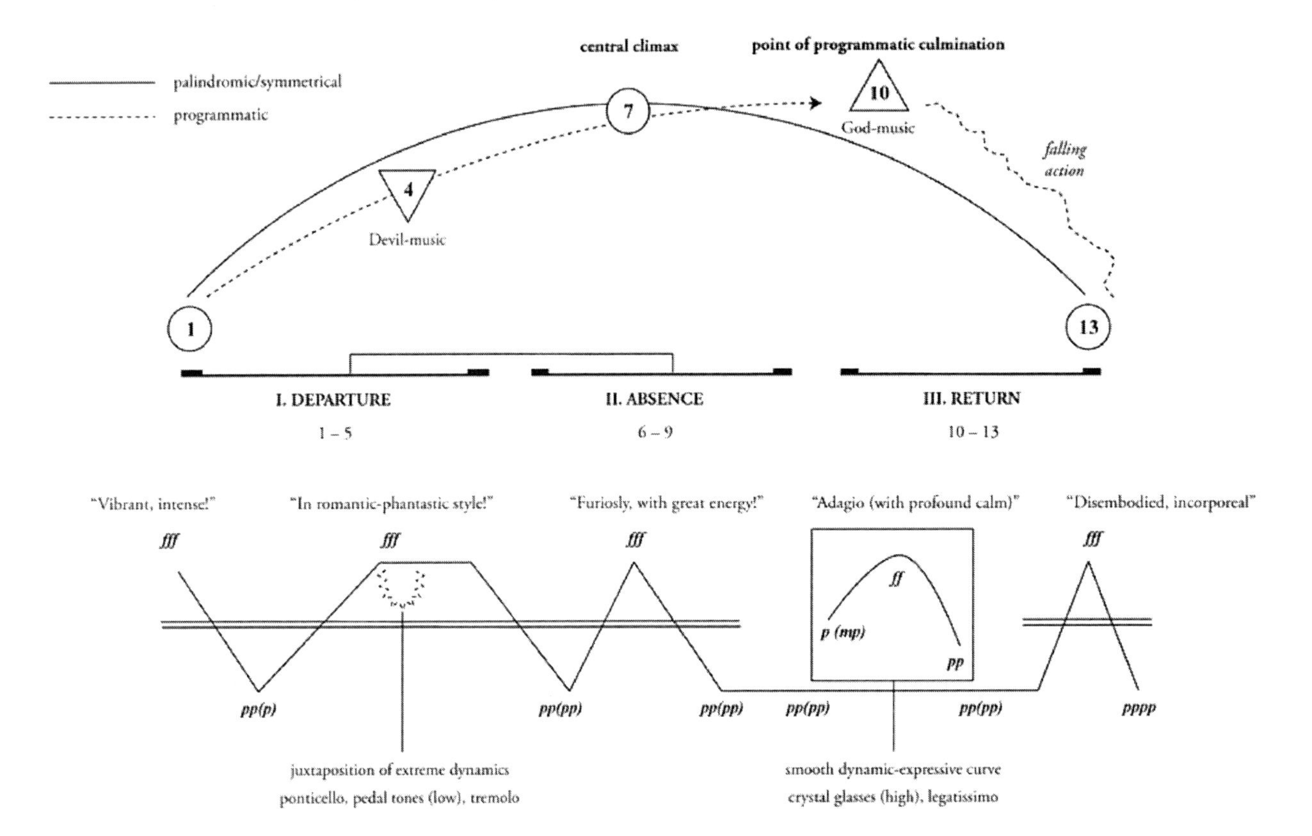

47.15 Blair Johnston, "Between Romanticism and Modernism and Postmodernism: George Crumb's *Black Angels*," *Music Theory Online* 18, no. 2 (June 2012): ex. 2, http://www.mtosmt.org/issues/mto.12.18.2/mto.12.18.2.johnston.html.

Below this, the image shows the dynamic contour of the entire work, not just with traditional dynamic markings but also in graphic form. A double line that differs in style from other lines in the image provides a visual fulcrum around which the dynamics flip. While a thin gray line would have been less jarring, this choice highlights the work's extreme dynamic contrast. Five movements that feature dynamics of ff or stronger, but which contrast in character (as indicated by brief quotations from the score) are then the subject of the narrative(s!) sketched at the top of the image. Here, visually contrasting line styles and enclosures depict two kinds of structure in the work: a solid line connects pieces that form the central and end points of a palindrome, while a dotted line traces a trajectory through the Devil-music movement to the programmatically climactic God-music (note the use of the inverted triangle for the former, subtly implying moral inversion of the latter). A squiggly dashed line falls irregularly from this climax, in a visual denouement. There is little wasted ink in this drawing. It efficiently presents multiple layers of meaning in a simple, easily understood design.

The remaining images employ both horizontal and vertical dimensions for deliberately communicative purposes. Figure 47.16 (colorplate 12) is one of a series of graphical listening guides intended for use by audiences at public performances, such as professional symphony concerts. Multilayered and appropriately detailed, the image is delightful. On the largest scale, it is organized from top to bottom. The large sections of sonata form consist of up to three lines of graphical information that, while it can be read as graphical prose, generally puts line breaks at structurally significant places and therefore scans more as poetry.

Principal keys and primary themes receive colorful prominence, while other keys and themes remain neutrally gray. Themes are rendered without staff lines in a way that recalls the heightened neumes of the Middle Ages, though here they are intended to trigger recognition upon listening rather than recall for performance. Simplified but not dumbed-down terms label the themes. Icons identify instruments to listen for, which are also listed below the themes they play. Dynamic markings add another layer of information. It's a lovely piece of "public" music theory. I only wish that the large-scale structure had been highlighted more prominently.

Figure 47.17 and online figures 47.18–47.20 all show correspondences between two halves of a composition in which the second half substantially repeats the first. In images of this kind, time is depicted essentially left to right but is broken deliberately at the point that allows corresponding sections to be displayed over each other to the extent possible. Such a design readily facilitates comparison. It is the strategies for showing parts that do *not* correspond that differentiate the four images.

Figure 47.17 shows structure in an unusual sonata form movement from Leopold Koželuch's *Keyboard Sonata*, op. 1, no. 1. The image maps similar sequences of events in the two halves of the movement, with sections in the second half shown below their corresponding sections in the first. The consistency of construction facilitates ready comparison between these parallel sections. A break in the second half allows an arrow to pass through that shows where a twenty-five-bar insertion has been made. Nevertheless, there is room for improvement. The grid needlessly imprisons the information; the horizontal lines in particular could be eliminated altogether. The width of the cells is determined by the width of the text they contain rather than the amount of music they correspond to. In the primary theme area, the twelve-bar A idea, the four-bar B idea, and the four-bar A′ idea appear to be the same length, and all appear longer than the thirteen-bar transition, which is longest of them all. When these four sections return at m. 63, each encompasses a different visual width than in the exposition, even though they are all the same musical length, except TR, which is one bar longer the second time. When using a proportional representation, one should include the length of each section in measures, since it is difficult to visually compare lengths that are not parallel to one another. Finally, highlighting the salient differences in the repeat would help distinguish aspects that are retained and aspects that are different (mainly related to tonality) in the two halves.

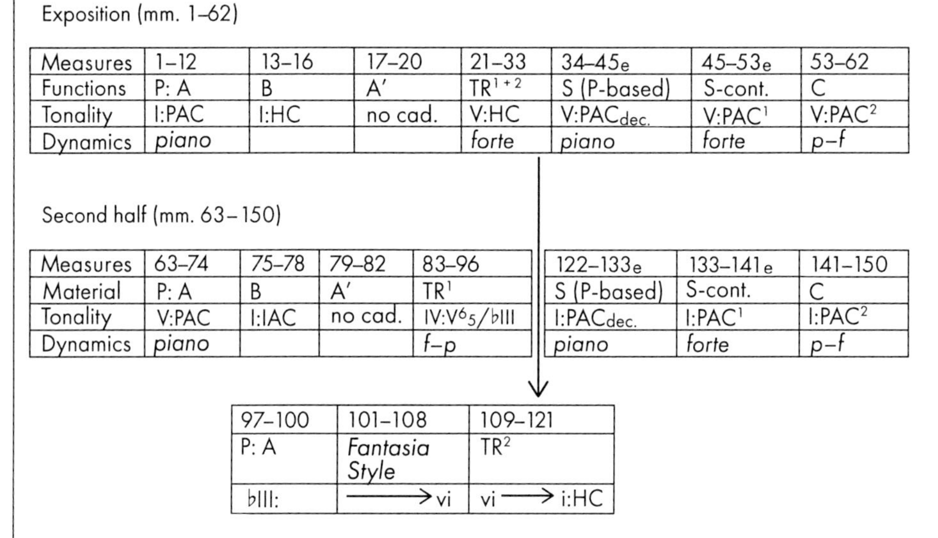

Exposition (mm. 1–62)

Measures	1–12	13–16	17–20	21–33	34–45e	45–53e	53–62
Functions	P: A	B	A′	TR^{1+2}	S (P-based)	S-cont.	C
Tonality	I:PAC	I:HC	no cad.	V:HC	V:PAC$_{dec.}$	V:PAC^{1}	V:PAC^{2}
Dynamics	*piano*			*forte*	*piano*	*forte*	*p–f*

Second half (mm. 63–150)

Measures	63–74	75–78	79–82	83–96	122–133e	133–141e	141–150
Material	P: A	B	A′	TR^{1}	S (P-based)	S-cont.	C
Tonality	V:PAC	I:IAC	no cad.	IV:V$^{6}_{5}$/♭III	I:PAC$_{dec.}$	I:PAC^{1}	I:PAC^{2}
Dynamics	*piano*			*f–p*	*piano*	*forte*	*p–f*

97–100	101–108	109–121
P: A	*Fantasia Style*	TR^{2}
♭III:	⟶ vi	vi ⟶ i:HC

47.17 Markus Neuwirth, "Surprise without a Cause?: 'False Recapitulations' in the Classical Repertoire and the Modern Paradigm of Sonata Form," *Zeitschrift der Gesellschaft für Musiktheorie* 10, no. 2 (2013): 280, http://www.gmth.de/zeitschrift/artikel/722.aspx.

A similar image (fig. 47.21) compares the structure of the three verses of a Brazilian song, "Sorriso Aberto." Structuring the information the same way for each verse, a structure explained in the box in the upper left, allows for easy comparison. The image would be improved, however by moving the verse number and measures from above to the sides so the information about the subsections appears closer together. As always, the gridlines are better omitted. See online figure 47.22.

Figure 47.23 depicts the large-scale form of "Farben," the third movement of Arnold Schoenberg's *Five Pieces for Orchestra*. The vertical layout suggests a three-part division of the piece. Time within each division proceeds prose-like, suggesting a narrative. The design has consequences for how we interpret the music: it downplays differences, emphasizing subtle changes in keeping with the music's nature and the gently rippling waves suggested by the movement's subtitle, "Summer Morning by the Lake." Unlike nested form diagrams such as online figure 47.6 and figure 11.3 (colorplate 4), the design pushes the hierarchical levels further apart,

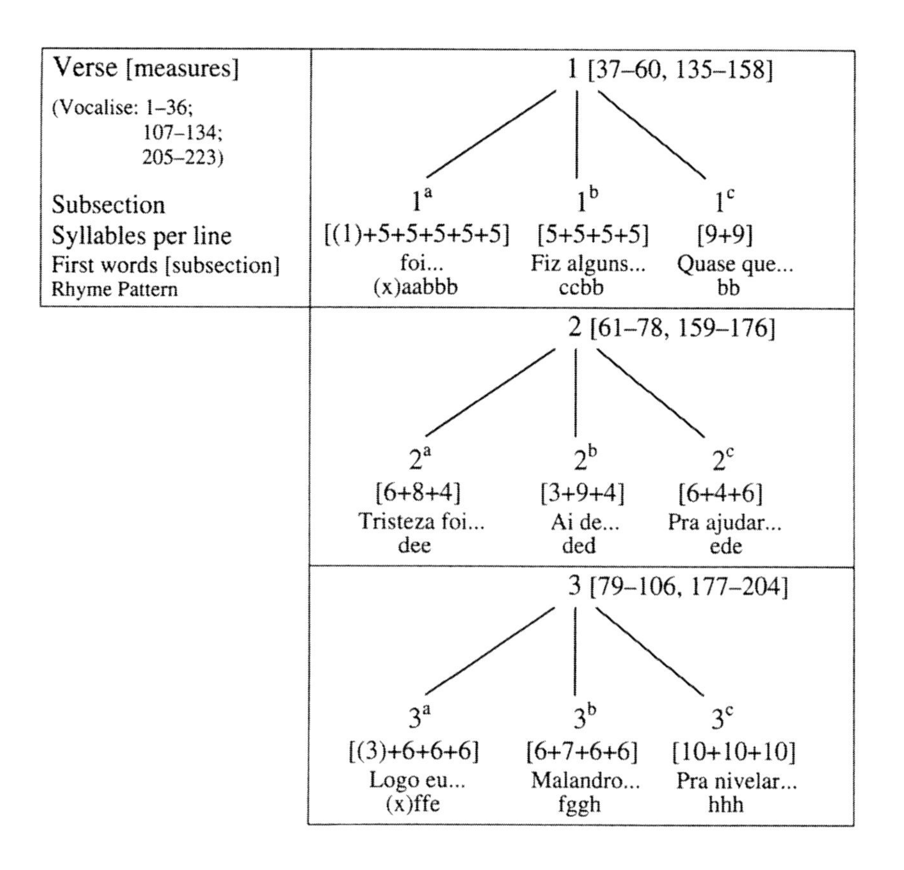

47.21 Jason Stanyek and Fabio Oliveira, "Nuances of Continual Variation in the Brazilian Pagode Song 'Sorriso Aberto,'" in *Analytical and Cross-Cultural Studies in World Music*, ed. Michael Tenzer and John Roeder, 98–146 (New York: Oxford University Press, 2011), 108.

0'00-1'04:

Gently waving surface, subtly changing in light and color. *0'32* and *0'43* slight disturbances; *0'52* movement ceases. *0'56* Completely static, rather dark and dense sound, *1'01* fading.

1'04-2'46:

Renewed movement. *1'11* Surface colors in gliding wave motion, 1'*27* sense of pulse briefly evoked by a few "spots of light". *1'48-1'53* Glittering splashes break the surface twice (like the tail of a fish); 2'*06* darkening .... 2'*11* brief splashes and a bright streak of light. *2'16* Gradually increased agitation... 2'*40* calming down in static sound.

2'47-4'07:

Clear, tinkling "tic-toc" sound appears. *2'51* Waving, gliding surface colors, *3'28* and *3'34* soft, deep disturbances; *3'38* bright reflection. *4'00* calming down, fading.

47.23 Erik Christensen, *The Musical Timespace: A Theory of Music Listening* (Aalborg: Aalborg University Press, 1996), 1:57.

by suggesting that the internal divisions within the sections are less significant, perhaps considerably so, than the larger structure.

Missing from the image is any sense of how the three sections themselves cohere as sections. It would, of course, be nice if the description were tied to the recording itself, though this can be done by the reader (the recording is of the City of Birmingham Symphony, Simon Rattle conducting). A list format (bullet points) would make it easier to track successive events. While that would make the time points easier to see, the affect when reading would be quite different.

Figure 47.24 shows the structure of Rued Langgaard's *Music of the Spheres*. Again, successive passes through similar music are aligned to allow comparison. The table should be read top to bottom, then left to right. Here, the columns provide the macrostructure, with each large section including two to five sections. The layout makes clear that sections 2 through 4 include subsets of the sequence found in the first: a "tremolo space," a slow melody section, a section featuring repetitive patterns, a climactic section, and finally a section conveying relief. Improvements would include (1) typographically separating the three kinds of information contained in each box, for instance by making the time indexes bold, putting the

47.24 Erik Christensen, *The Musical Timespace: A Theory of Music Listening* (Aalborg: Aalborg University Press, 1996), 1:64.

0'00 Tremolo space	*9'55* Tremolo space		
4'15 Slow motion melody		*16'16* Melodic idyll	
6'31 Repetitive patterns	*10'54* Repetitive garlands	*17'54* Endless repetitive patterns	
7'54 Climax Repetitive melody Crescendo	*12'32* Climax Repetitive patterns	*U3'09* Climax Repetitive patterns, Crescendo, accelerando	*29'12* Climax Repetitive fanfares, Crescendo, Great noise
8'38 Relief Distant music	*14'20* Relief Distant orchestra	*24'34* Relief Distant heavenly music with soprano	*31'59* Relief Heavenly music, Angelic chorus, Great light

functional labels "Climax" and "Relief" in italics to distinguish them from the descriptions of musical materials, and leaving the latter in regular type; (2) finding a way to show proportional relations: the section of "Endless repetitive patterns" in the third column is three and two times the length of the parallel sections in columns 1 and 2, and the "Relief" sections grow from 1:17 to 1:56 to 4:36, with the length of the final section unknown, since only the start time is shown; and (3) reducing the grid (of course).

This chapter has organized form diagrams into three categories: those in which time flows top to bottom, those with a horizontal orientation, and those that employ both dimensions for purposes of facilitating internal comparisons. The best choice depends on the content and the intended messaging. It is worth spending time at the start of the design process to weigh the pros and cons of each. Once that is decided, the fun begins.

CHAPTER 48 Hierarchy in Music

Most scholars view Western tonal music as hierarchical. Some chords or melodic pitches embellish, connect, or prolong other chords or melodic pitches that are more foundational in some way. Parallels with architecture are obvious: a house is built on a solid surface, which supports load-bearing walls, to which framing is attached, from which drywall is hung, which is then painted and, finally, adorned with art. Remove the art and the house still stands; remove the foundation and the house collapses.

It is often impractical to capture all aspects of a hierarchical music analysis in a single image. Therefore, visual depictions of musical hierarchy need to decide (1) which aspects of hierarchy to display, (2) how many layers to present, (3) how to distinguish the hierarchical layers from one another, (4) which elements of the musical surface (that is, the notation) to include and how, and (5) what explanatory annotations to include and how. Then a depiction needs to present all of this in ways that clearly differentiate the informational figure and ground as clearly as the musical figure and ground.

Among the simplest type of musical hierarchy is the melodic step progression in which figurations elaborate a scalar passage. Figure 48.1 illustrates one way to depict the structure of such a passage. Stems extend upward and downward beyond the notation and are linked with a long beam. The figure preserves the melodic surface in its entirety, with deeper-level stepwise motion highlighted unambiguously. (The circled pitches reference an aspect of the analysis not pertinent here.)

Hierarchical organization of melody is not restricted to tonal music. Figure 48.2 illustrates the composing out of the first four melodic pitches across the piece's first four phrases. As above, the image uses stemlike lines stretching upward out of the score's frame into its own information layer. Incomplete beam-like lines connect these notes. One could more easily see the mapping of the pitch sequence of these stemmed notes to the first four notes of the passage if the names of the latter had not burrowed themselves within the notation. Online figure 48.3 applies a similar technique to poetic scansion.

48.1 Ellwood Derr, "The Two-Part Inventions: Bach's Composers' Vademecum," *Music Theory Spectrum* 3 (1981): 35.

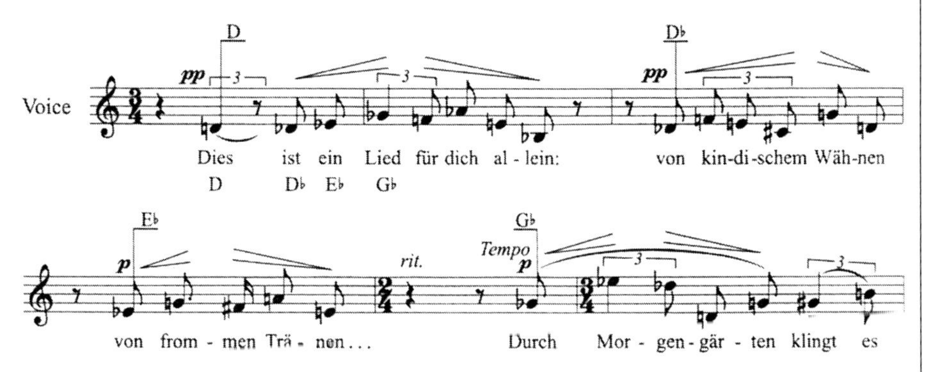

48.2 Joseph N. Straus, *Introduction to Post-tonal Theory*, 3rd ed. (Upper Saddle River, NJ: Prentice Hall, 2005), 103.

It is not immediately apparent how to read figure 48.4. It expresses a schematic hierarchy of tonalities found within the third movement (minuet and trio) of Franz Joseph Haydn's Symphony no. 104. The image is unrelated to the passage of time in the movement. Instead, beginning with the movement's tonic, D major, it provides the various tonics suggested in the movement. The keys organize their signatures around the circle of fifths, with separate staves for relative keys where both are present. Tonicized chords are presented left to right in ascending order (mostly; it seems that B-flat should precede B minor), generally preceded by their dominant.

For the most part, these tonal centers have boxes around them that connect the local tonics back to the D major tonality of the minuet, the B-flat tonality of the trio, or in one case both. Finally, the key areas are assigned a hierarchical ordering on the basis of their relative durations and their cadential functions, from the overall tonic D (1st) to the fleetingly referenced E-flat (9th). The image expresses hierarchical strength using ordinals in the rightmost column and durational symbols in the notation: whole notes for the first level, eighth notes for the last, with durations in parenthesis being inferior to the same duration without parentheses.

Unfortunately, the image's design is full of challenges. Although the decision to abstract tonal information from the music itself was intentional (Berry 1980, 26), a series of footnotes in the image tries, ineffectively, to relink the image to the music. Among other things, the meaning of "major division A/B" in notes 1 and 2 is unclear, when the terms "menuetto/trio" and "division a/b" are used in subsequent footnotes. Also, the hierarchical levels, the main point of the image, are not clearly conveyed. While the mapping of *duration symbol* to *structural significance* works well in online figure 48.3, it works less well here. There, the durations appear on the same staff, so the eye processes them in relation to one another. Here, the need to jump back and forth between staves negates the effect. Between this and the use of ordinals (1st, 2nd, and so on) rather than simple numbers, the visual cues needed to reinforce the sense of hierarchy the image proposes prove inadequate. One appealing aspect of the image is the use of the parenthetical leading-tone accidental in the key signatures of minor keys.

The visual representation of large-scale tonal hierarchy is most famously found in the work of Heinrich Schenker. Before continuing, it is important to acknowledge that Schenker was an avowed racist and that, as Philip Ewell (2020) has persuasively argued, Schenkerian analysis is a central player in a white racial frame that Ewell argues permeates the music theory discipline. I accept Ewell's claims and embrace any effort to address them. I am also mindful that in valorizing an analytic method that was birthed and nourished by racism, I am helping perpetuate the racist frame it has helped support. Nevertheless, because Schenkerian notation is one of the most widespread, and effective, of all methods developed for visualizing music, I must discuss it here.

Schenker's analytic apparatus can convey nuanced interpretive differences and is versatile in its application, accessible, and unambiguous (when deployed correctly). While mastering the subtleties of Schenkerian analysis requires extensive practice, one can learn to read Schenkerian analyses at a functional level with relatively minimal training because the system is based on symbols drawn almost entirely from common Western notation and employs them in ways that are metaphorically approachable and fairly intuitive. Figure 48.5 provides an example.

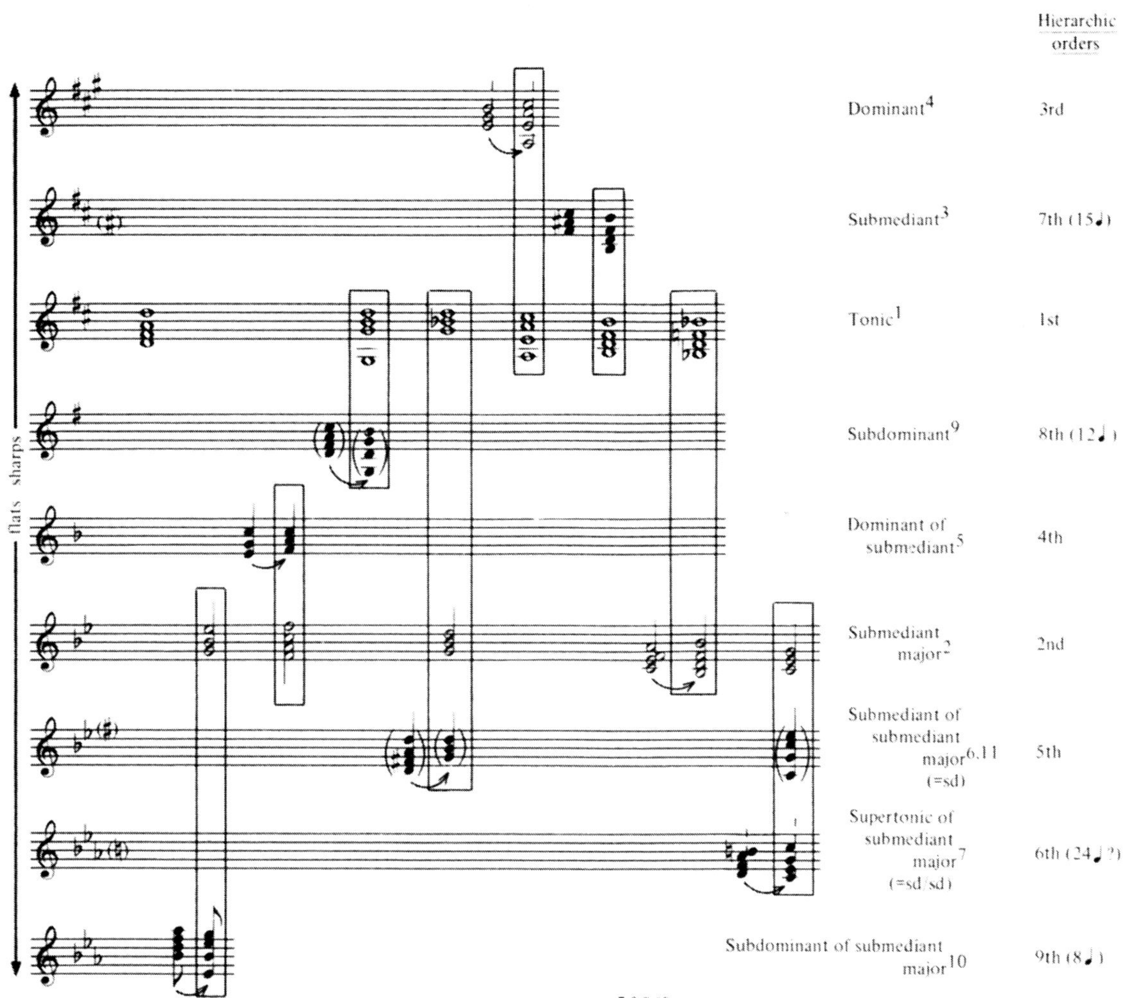

[7,8,9,10] Lesser internal tonics; durations given as to quarter-note values. (Note that G and E-flat, subdominants of the primary and secondary tonics, respectively, are tonicized at markedly deviant internal stops within the menuetto and trio, respectively.)

[11] I.e., same as subdominant minor.

[1] Final cadence of major division A.
[2] Cadence of major division B.
[3] Cadence of division a of menuetto.
[4] Cadence of division b of menuetto.

[5] Cadence of a of trio.
[6] Cadence of b of trio.

Noteheads and stems, sometimes combined with beams and slurs, express a pitch's hierarchical significance. The use of open noteheads to represent pitches that remain operative at deeper levels of structure (that is, over longer spans) and closed noteheads to represent notes operative only closer to the surface (that is, over shorter spans) is analogous to rhythmic notation (notes with open noteheads sound longer than notes with closed noteheads). Notes with stems operate at deeper levels than those without. Sometimes longer stems represent notes at even deeper levels. The exception is the flagged note. Rather than serving as structurally

48.4 Wallace Berry, "On Structural Levels in Music," *Music Theory Spectrum* 2 (1980): 27.

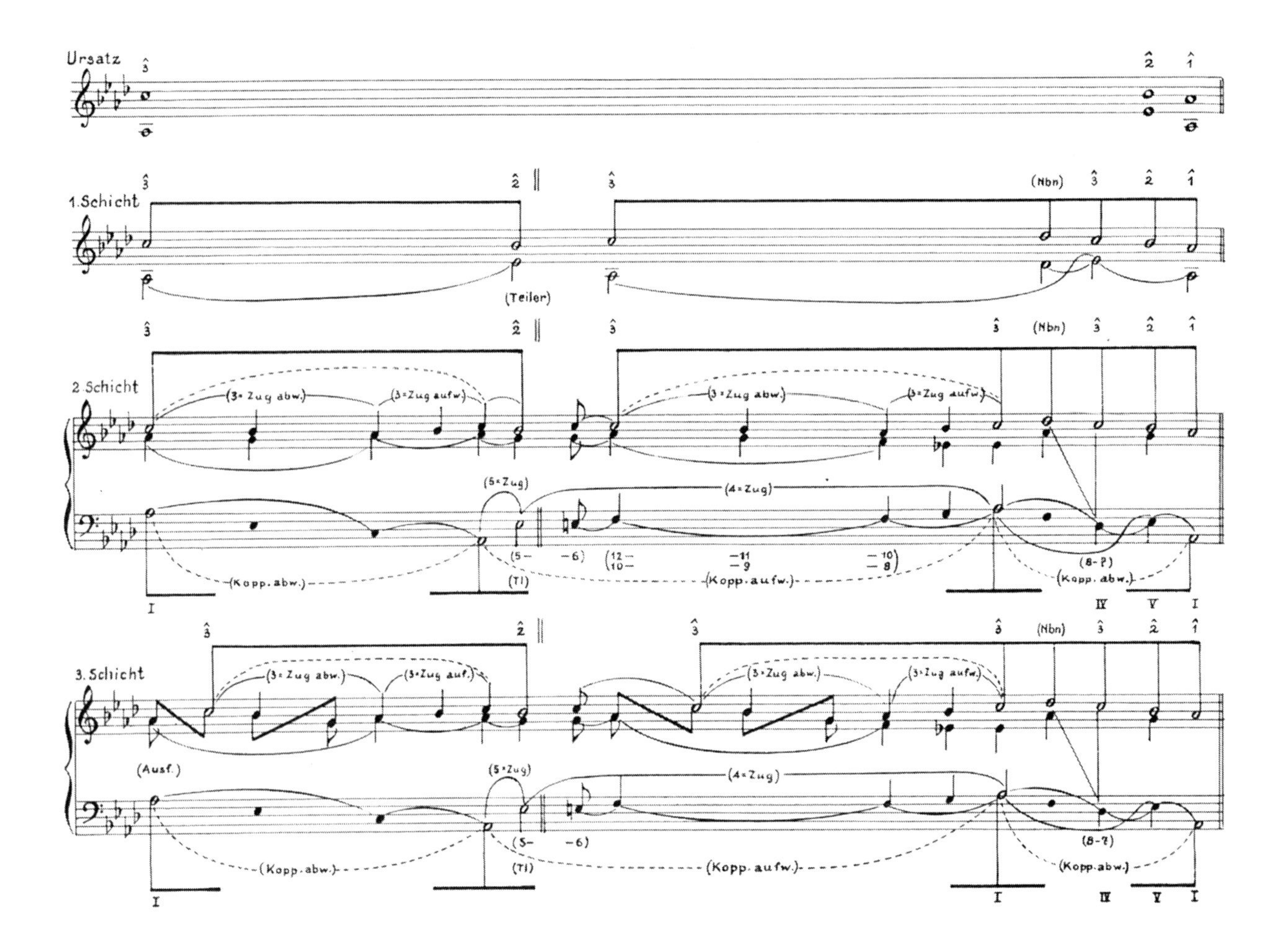

48.5 Heinrich Schenker, *Five Graphic Music Analyses (Fünf Urlinie-Tafeln)* (New York: Dover, 1969), 32.

inferior to the stemmed solid notehead ("quarter note"), it serves more like an exclamation point in chess notation (or a raised-eyebrows emoji) than an indicator of location in a hierarchy. Beams and slurs connecting two or more noteheads imply dependencies that represent prolongation.

While the system enables the presentation of several hierarchical levels on a single staff, multiple staves are sometimes employed to represent different levels, as in figure 48.5. Rarely is it beneficial to render these levels in a way that invokes depth perception, as in online figure 48.6.

The Schenkerian analytic system is exceptional at what it does. It supports high information density. The layering of visual information is as clear as the layering of the musical content it conveys. Virtually every mark in the typical sketch is meaningful and intuitively grasped (though I confess that the unfolding symbol has always stymied me a little). Because a structural analysis is represented using music notation, it permits relatively easy connection to the musical score, though this is not always as easy as one would hope.

The annotated reduction in figure 48.7 is unusually effective. Adding to the nicely set reductive apparatus, the image includes large-scale formal information ("[RIT[1]: MT]") and more localized descriptors of formal process ("lead in," "antec."), which enhance understanding. It also unobtrusively acknowledges another analysis ("Bribitzer-Stull").

Certain practices can help make Schenkerian notation even clearer. Figure 48.8 renders less structurally important notes with smaller noteheads, so the main counterpoint is more apparent. In addition, it places the notation in a metric context, providing both bar lines and selected measure numbers, to facilitate the mapping of analysis to score. In this case, doing so makes possible a Riemannian metrical analysis (see fig. 46.9), which is rendered unobtrusively between the staves.

The excellent figure 48.9 adds the actual music to a reduction, which, like figure 48.8, displays a metric grid and employs different-sized noteheads. Doing so eliminates one of the primary challenges in following a typical linear analysis—linking the analysis to the score. The image spreads the analysis across multiple systems, at the cost of some of the visual continuity normally associated with such images, but the benefits are substantial. As always with musical notation, gray staff lines would be better than black.

In general, Schenkerian diagrams would benefit from treatments that help separate the information layers even further. Online figure 48.10 provides some thoughts in this direction.

Lerdahl and Jackendoff (1983) propose another major hierarchical analytical system. Drawing on the field of linguistics, the theory borrows ideas pertaining to generative grammars to explain the perception of tonal music by experienced listeners. Four kinds of analytic structures are derived by application of an extensive set of well-formedness rules and preference rules. Figure 48.11 illustrates three of these four components.

Metrical structure is represented using a dot notation. Events with more dots are more strongly weighted metrically. The method is simple and effective. One can instantly read the number of dots at a musical moment, which makes it easy (for instance) to find other moments with the same weight. At the same time, because they are top-aligned, we can conceive of groups of dots as lines and therefore recognize by their lengths the relative metrical strength of adjacent events without the cognitive mental effort that would be required if the metric weights were represented using, say, numbers instead.

The grouping structure, represented by nested (and sometimes overlapping) brackets below the music, assembles short musical units into larger units hierarchically, including the possibility of alternate (competing) hearings and elisions. Because our visual system groups parts into wholes, the eye readily understands the notation. However, the curves marking endpoints do not pop out very well as

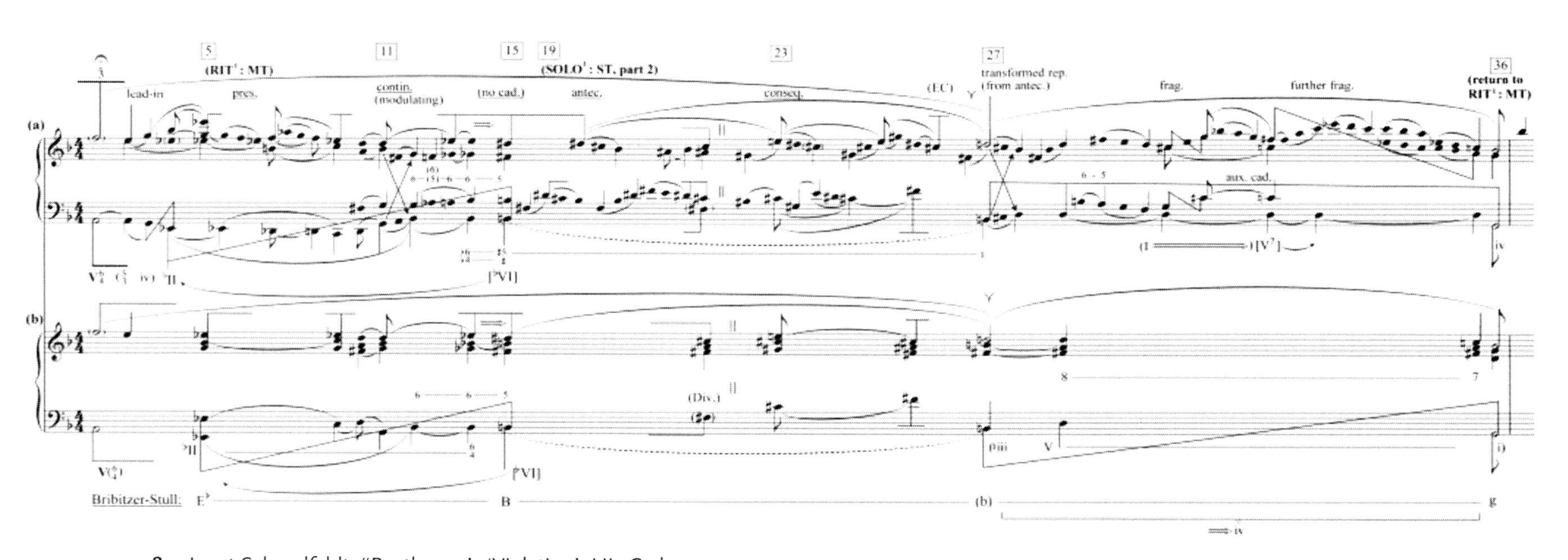

48.7 Janet Schmalfeldt, "Beethoven's 'Violation': His Cadenza for the First Movement of Mozart's Concerto in D Minor, K. 466," *Music Theory Spectrum* 39 (2017): 10.

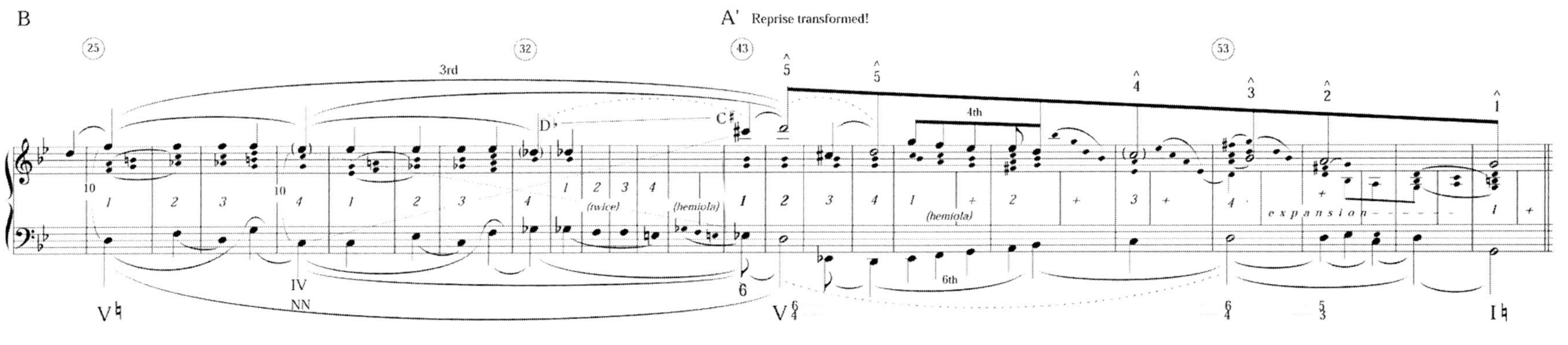

48.8 Frank Samarotto, "Sublimating Sharp $\hat{4}$: An Exercise in Schenkerian Energetics," *Music Theory Online* 10, no. 3 (2004): ex. 7b, https://mtosmt.org/issues/mto.04.10.3/mto.04.10.3.samarotto.html.

48.9 John S. Reef, "Subjects and Phrase Boundaries in Two Keyboard Fugues by J. S. Bach," *Music Theory Spectrum* 41, no. 1 (2019): 51.

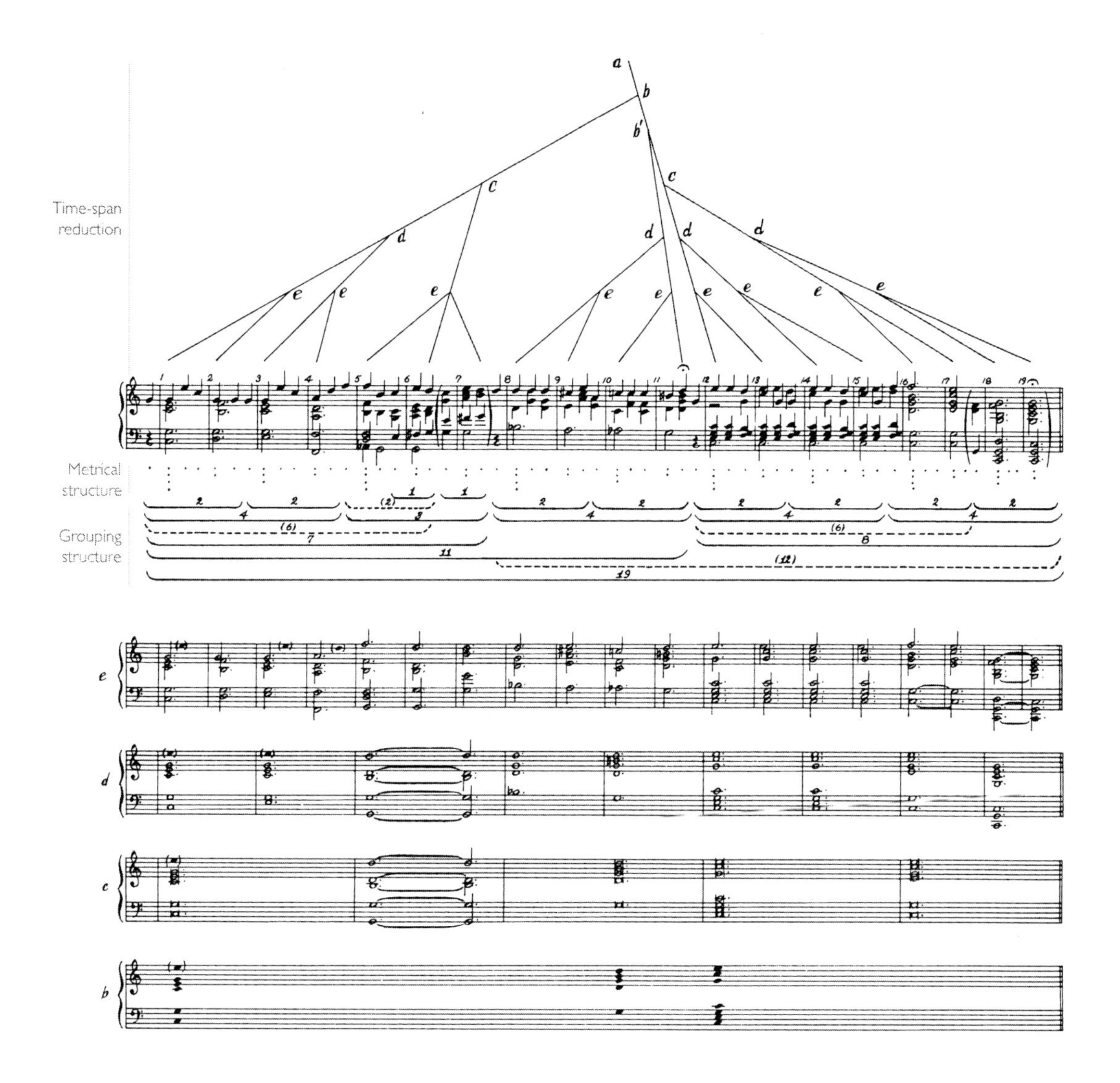

represented here, particularly in the diagram's interior, and added space between levels would help. A clean typeface would enhance legibility of the numbers representing group lengths.

The time-span reduction is a tree structure. Derived from the metrical and grouping structure, the time-span reduction chooses one of the (usually two) parts of a group as the head, with that component propagating up to the next level in the tree structure. In the end, a single event serves as the head node for the entire passage. The layers of the reductive tree structure can be represented in notation as at the bottom of the image. Here, the various reductive layers are cross-referenced to the tree diagram with letters. The tree notation is largely clear, though

48.11 Fred Lerdahl and Ray Jackendoff, *A Generative Theory of Tonal Music* (Cambridge, MA: MIT Press, 1983), 265. Annotations added.

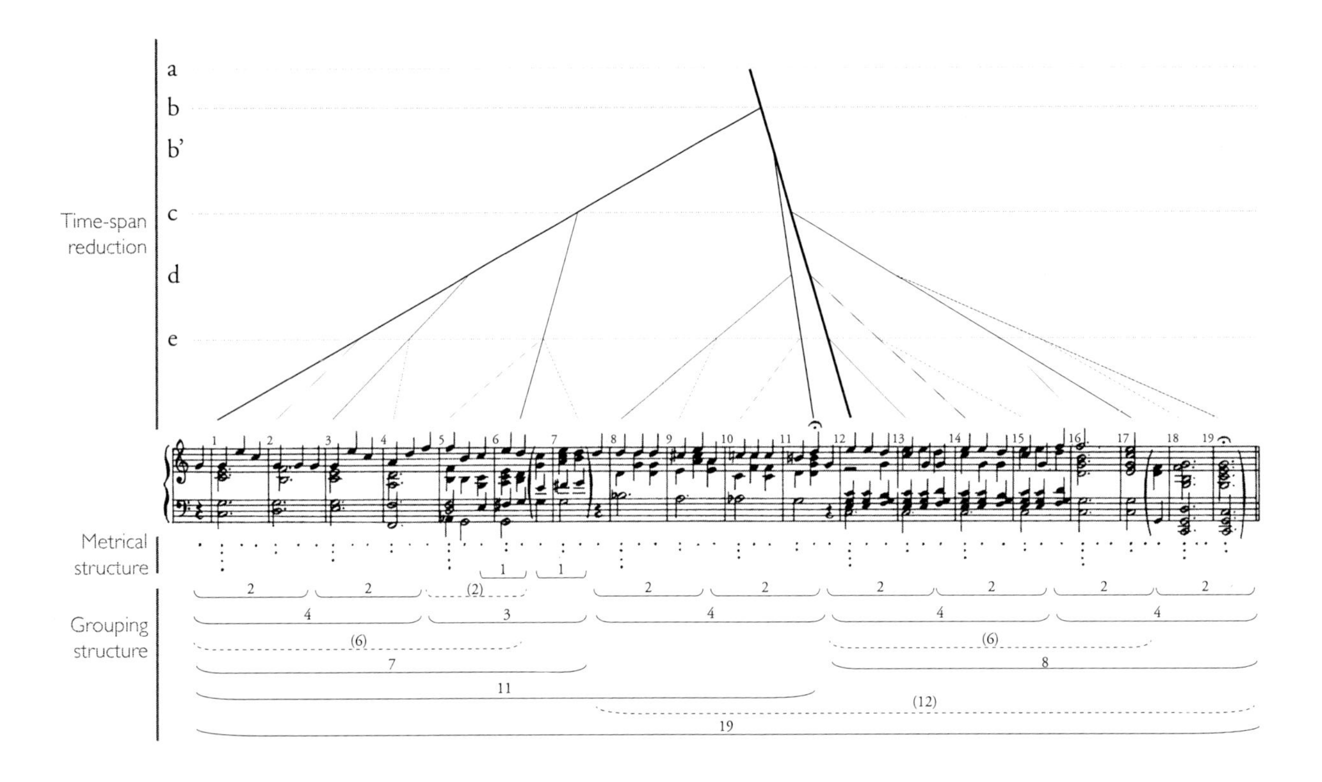

48.12
Redrawing of fig. 48.11.

Facing top, **48.13**
Redrawing of time-span reduction in fig. 48.12.

the uniformity of the various lines makes it hard to see the relative significance of the branches. Also, the repetition of the level labels at each node is redundant.

The redrawing in figure 48.12 (which omits the secondary notation layers) addresses these issues. In the time-span reduction, it renders structurally inferior lines less boldly than the line representing the head of the span. The redrawing replaces repeated level labels with lines that connect to a single letter that identifies that level. The letters are retained to allow linkage to the secondary notation omitted here. The grouping structure employs thinner lines that are spaced further apart and also a tidier font (the same font used to reset the measure numbers).

Even with this treatment, other features of time-span reductions might be addressed. For instance, the amount of vertical height associated with them results in a lot of extra white space. Also, the branches themselves do not carry information; only the nodes where they join do. Finally, the branching structure creates a horizontal distorting effect. The node at *b* appears directly above m. 10, while the events referenced by that node occur in mm. 1 and 12, and the musical event in m. 1 is the head of nodes that are drawn above measures 3, 5, 7, and finally 11.

Figure 48.13 suggests an alternate notation. This proposed reconception of the Lerdahl and Jackendoff time-span reduction replaces the tree structure with a set of horizontal bars, one for each level *a* through *e*. The node that will become the head is darker. The musical event that serves as the head of a particular span is

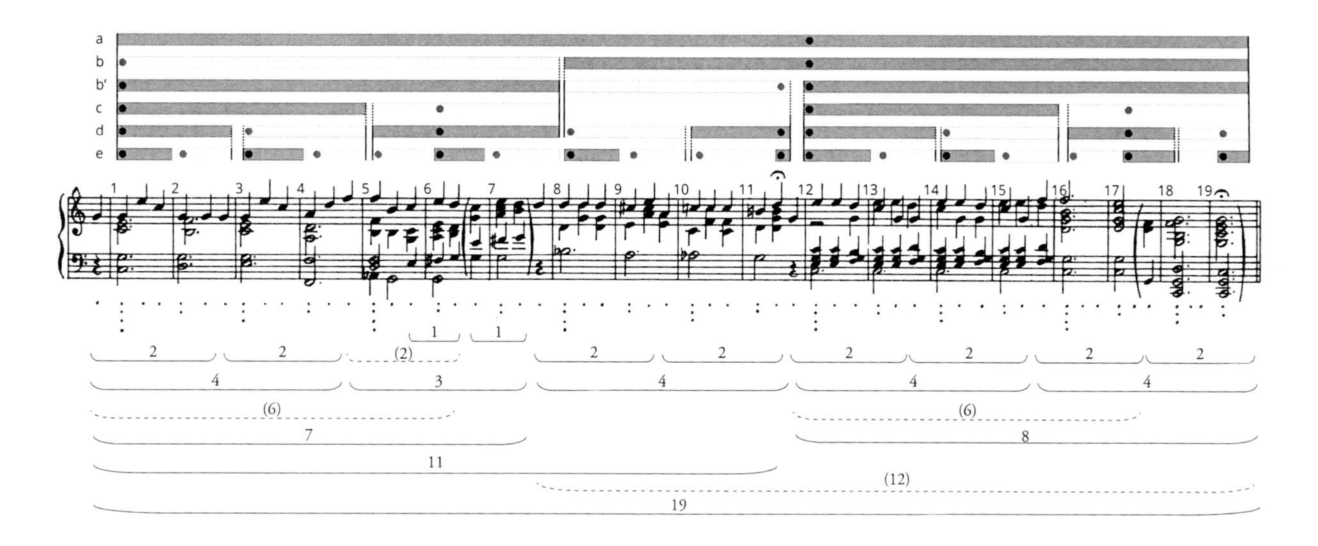

shown with a dot that travels into higher levels until another event subsumes that span. The event that serves as head of the entire passage thus appears above m. 12 at every level.

I started this chapter with an example of "composing out" that represents a kind of melodic hierarchy in atonal music. I end with another atonal example, this one focusing on harmonies in an Alexander Scriabin Prelude (fig. 48.14). Row 1 of the analysis traces inversions and transposition of the first chord through the

***Below*, 48.14** Joseph N Straus, "Voice Leading in Atonal Music," in *Music Theory in Concept and Practice*, ed. James M. Baker, David Beach, and Jonathan W. Bernard, 237–74 (Rochester: University of Rochester Press, 1997), 261.

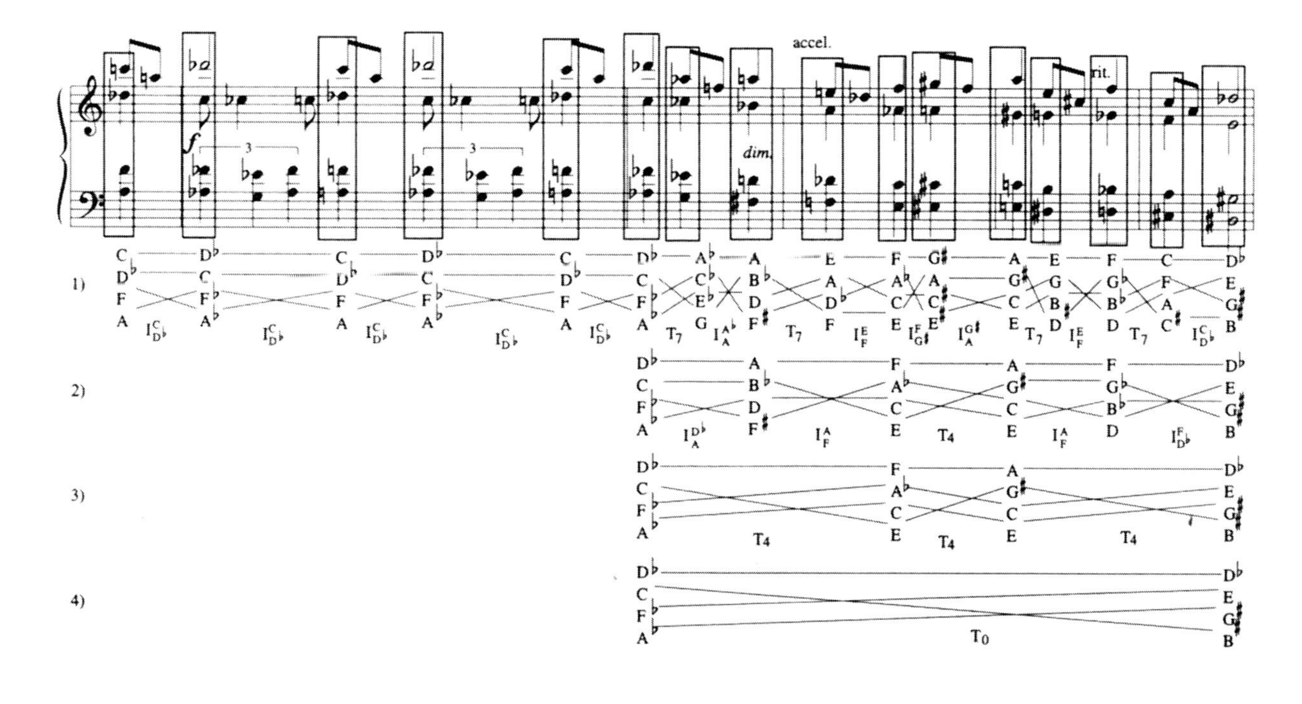

remainder of the passage. Row 2, focusing on the second half, reduces out every second chord, revealing an underlying symmetry in the relationships that connect these chords. Row 3 eliminates the second and fifth of these, yielding a cycle of T_4 transformations. Reducing out the two T_4 transformations (row 4) reveals that all of that activity simply connects the first chord to a revoicing of itself. The layering of different analytical stages and gradual elimination of all but a small set of elements parallels other images in this chapter. The treatment of voice leading is novel, however. By focusing on how particular pitches participate in the various transformational links, this figure overlooks the semitonal motion of the musical voices. On the other hand, it also makes clear that, while the various transformations shuffle pitches from one part to another, the soprano voice participates fully in the transformational story of the passage. (Rounding off the corners of the rectangles and thinning the lines slightly would help declutter the notation.)

Hierarchical views of music in the Western classical tradition are among the most common, and certainly among the most complex, musical visualizations one can find. Deciding how to reflect the musical layers in the visual layers is the central design question for such images.

Chapter 49 Serialism

Images pertaining to twelve-tone and other serial music typically involve one of several issues: properties of the tone row itself, the twelve-tone matrix or analogues to it, exploration of the relationships among versions of the row (often those used in a particular piece), and analysis of the music itself.

Images of row properties should of course highlight those properties. Figure 49.1 focuses on the intervals between adjacent pitches. (Today, we are well past the point where pitch-class numbers need to be listed explicitly when the notation is provided.) This approach is eminently clear, though the image could show explicitly the symmetrical arrangement of the intervals.

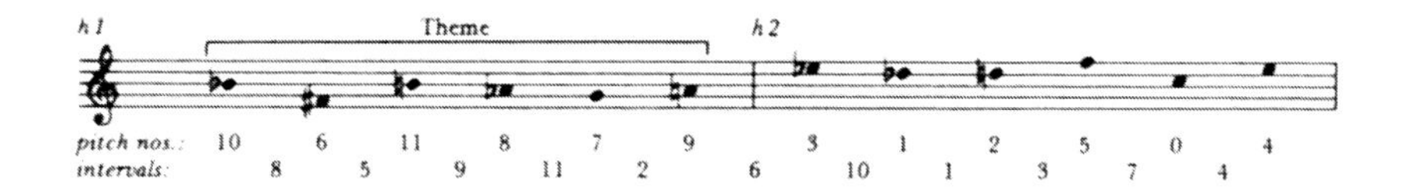

49.1 Christopher Wintle, "Milton Babbitt's *Semi-simple Variations*," *Perspectives of New Music* 14/15 (1976): 119.

Figures 49.2 and 49.3 provide two ways of showing selected subsets from a row, both employing brackets and set types. Figure 49.2 arranges the subsets in a way that aligns recurrences, making them discoverable. (Hyphens between the pitches are unnecessary.) Figure 49.3 pulls each subset class into its own staff. The dashed lines that help the eye track upward are distracting, however. The redrawing in figure 49.4 spaces the notes out horizontally, brings the staves closer together, and reconceives the ends of the brackets in a way that makes them and the noteheads they are attached to self-aligning, eliminating the need for the dashed lines.

49.2 Christopher Wintle, "Milton Babbitt's *Semi-simple Variations*," *Perspectives of New Music* 14/15 (1976): 124.

(015) (012) (012) (015)

Basic Set: 10 - 6 - 11 - 8 - 7 - 9 3 - 1 - 2 - 5 - 0 - 4

(025) (025)

(014) (014)

49.3 Joseph Straus, *Introduction to Post-tonal Theory*, 3rd ed. (Upper Saddle River, NJ: Prentice Hall, 2005), 208.

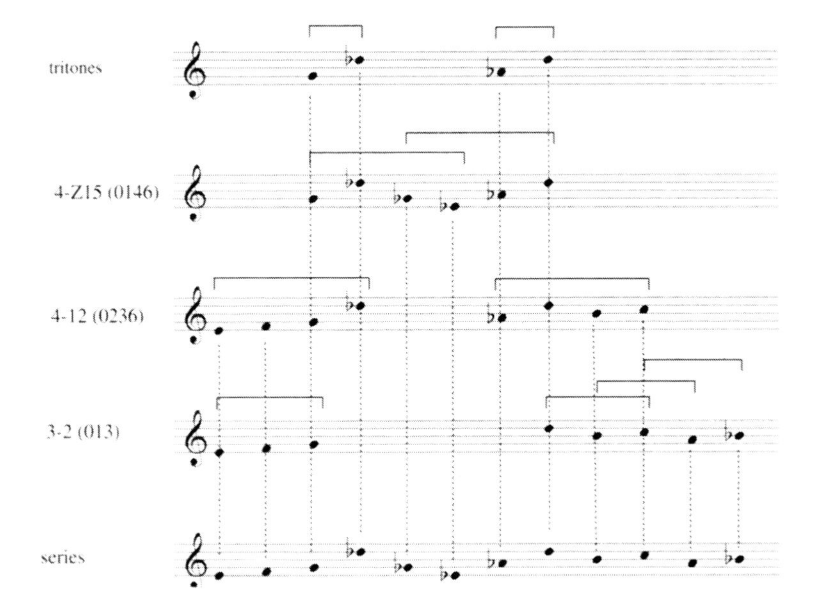

49.4 Redrawing of fig. 49.3.

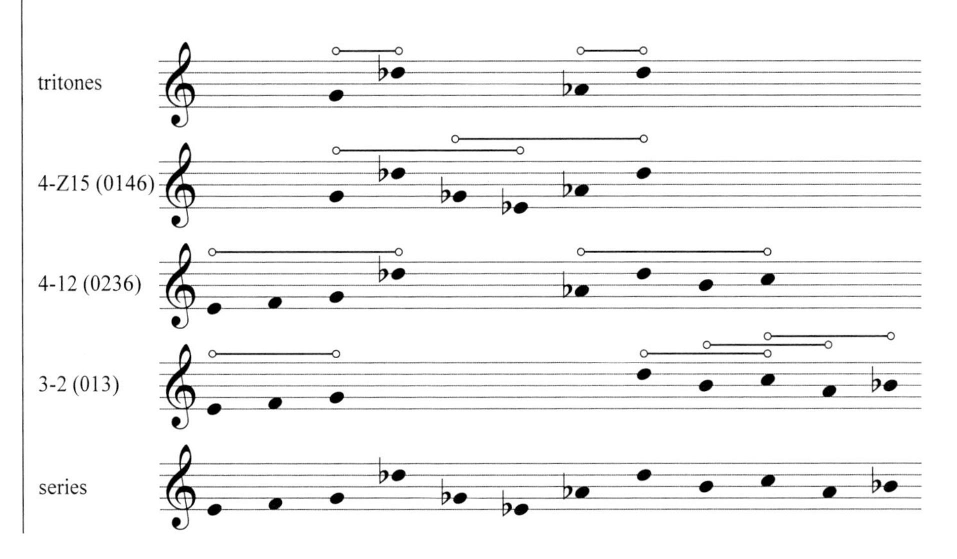

The standard twelve-tone matrix efficiently gathers the forty-eight row forms into a single grid. The primary function of a matrix is to facilitate score analysis. Therefore, it is better to list the pitches by pitch names rather than pitch-class numbers. The standard twelve-tone matrix in figure 49.5 conveys all the essential information in a tidy package. It renders gridlines in gray and omits them along the top and left edges. It separates row form labels from content by the use of boldface. The use of a serifed font distinguishes the letter *I* from the numeral 1. Use of actual sharp and flat symbols rather than the too-often-used # and b is always superior. The choice of which enharmonic spelling to employ is largely arbitrary. If the music is reasonably consistent, employing the accidentals found in the score makes it slightly easier to identify row forms in the matrix. (The Stravinskian rotational array is more complicated because Stravinsky's treatment of the row is more complicated. We will not take it up here, however.)

	$\mathbf{I_0}$	$\mathbf{I_1}$	$\mathbf{I_{11}}$	$\mathbf{I_9}$	$\mathbf{I_3}$	$\mathbf{I_8}$	$\mathbf{I_6}$	$\mathbf{I_7}$	$\mathbf{I_2}$	$\mathbf{I_5}$	$\mathbf{I_4}$	$\mathbf{I_{10}}$
$\mathbf{P_0}$	A♭	A	G	F	B	E	D	E♭	B♭	C♯	C	F♯
$\mathbf{P_{11}}$	G	A♭	F♯	E	B♭	E♭	C♯	D	A	C	B	F
$\mathbf{P_1}$	A	B♭	A♭	F♯	C	F	E♭	E	B	D	C♯	G
$\mathbf{P_3}$	B	C	B♭	A♭	D	G	F	F♯	C♯	E	E♭	A
$\mathbf{P_9}$	F	F♯	E	D	A♭	C♯	B	C	G	B♭	A	E♭
$\mathbf{P_4}$	C	C♯	B	A	E♭	A♭	F♯	G	D	F	E	B♭
$\mathbf{P_6}$	D	E♭	C♯	B	F	B♭	A♭	A	E	G	F♯	C
$\mathbf{P_5}$	C♯	D	C	B♭	E	A	G	A♭	E♭	F♯	F	B
$\mathbf{P_{10}}$	F♯	G	F	E♭	A	D	C	C♯	A♭	B	B♭	E
$\mathbf{P_7}$	E♭	E	D	C	F♯	B	A	B♭	F	A♭	G	C♯
$\mathbf{P_8}$	E	F	E♭	C♯	G	C	B♭	B	F♯	A	A♭	D
$\mathbf{P_2}$	B♭	B	A	G	C♯	F♯	E	F	C	E♭	D	A♭

49.5 Twelve-Tone Matrix of Luigi Dallopiccola, *Goethe-Lieder*, no. 2.

The choice of which row form to assign the privileged label P_0 is not standardized. A "fixed-do" approach would be neutral, though one must still decide which of the two inversionally related forms beginning on C should have the privileged P_0 label. Sometimes, however, choosing P_0 on the basis of musical prominence, or on evidence from a composer's sketches, reveals relationships among row forms used in a piece and suggests an intentionality that is best made apparent. I therefore prefer a "movable-do" approach when choosing the form to call P_0. Building matrices using order numbers rather than pitches, as in online figure 49.6, is neither standard nor particularly informative.

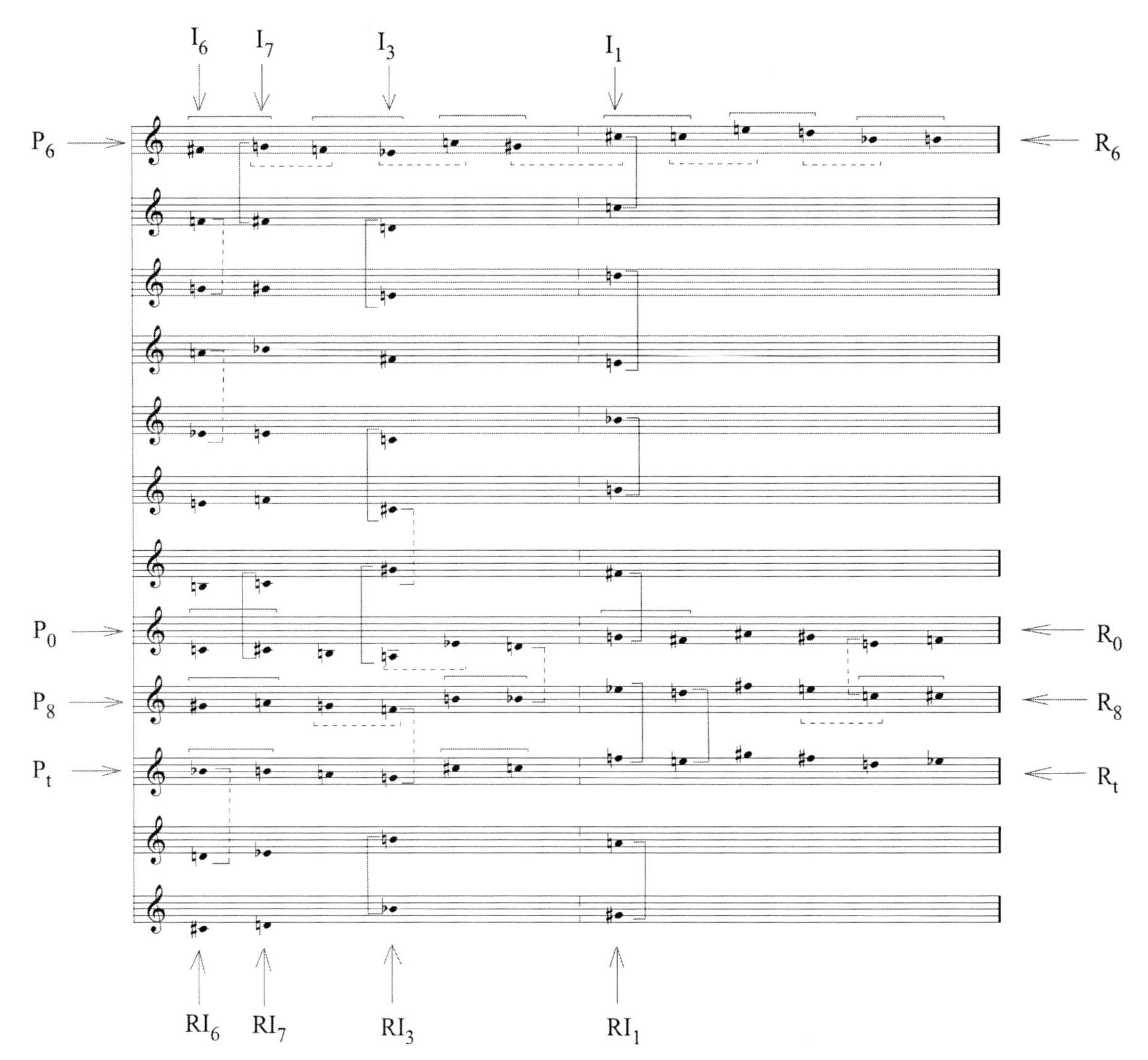

49.7 Christoph Neidhöfer, "Improvvisazioni Concertanti no. 1 by Norma Beecroft: Serialism, Improvisatory Discourse, and the Musical Avant-Garde," in *Analytical Essays on Music by Women Composers: Concert Music, 1960–2000*, ed. Laurel Parsons and Brenda Ravenscroft, 33–66 (New York: Oxford University Press, 2016), 50.

Figure 49.7 provides an appealing alternative representation of a twelve-tone matrix. It represents the row forms in notation rather than text, and it shows only the forms used in the music. The example highlights dyads that occur in multiple row forms, a topic addressed below.

When studying serial music, one must determine which row forms the music employs. This pre-analytical activity is sometime called *twelve counting* (though I often use the term *row chasing* when teaching the practice). Showing a twelve count simply requires making the row form clear and then numbering the pitches using order numbers 1 through 12 (some like to count from 12 down to 1 for retrograde and retrograde-inversion forms). An approach like the one taken in figure 49.8 is simple and often adequate, although some changes to typography would help. The row forms used appear above the piano score, facilitating reader verification. Numbering the pitches here or adding a subtle divider every four or six pitches would make it easier to find, for example, the eighth note.

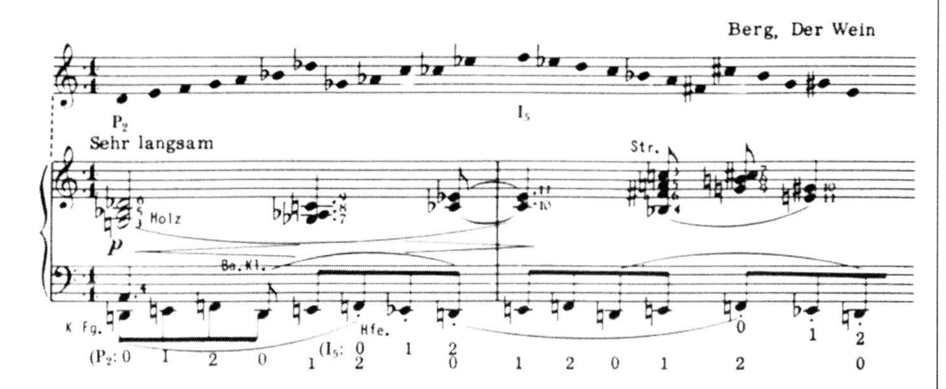

49.8 George Perle, *Serial Composition and Atonality: An Introduction to the Music of Schoenberg, Berg, and Webern* (Berkeley: University of California Press, 1991), 89.

In figure 49.9, twelve counting needs to be distinguished from duration labels (measured in sixteenth notes), which are another layer of score annotations. The figure does so successfully by putting the twelve-count figures directly next to the noteheads and making them large, bold, and in italics. Durations are smaller, in parentheses, and placed consistently with respect to the staves.

Online figure 49.10 separates the twelve count from the score, while online figure 49.11 does away with the score altogether, to its detriment. Figure 49.12 shows, however, that this can be done. Arranging the pitches in register, with instrumentation indicated, makes the analysis clear, and the layout makes it easy to refer back to a score if desired. Online figure 49.13 illustrates the unsuccessful use of shape notes to trace four row forms.

Often, it makes sense to stipulate (and thus omit) the twelve count and show just the row forms used, to focus attention on higher-level observations. This raises the question of how best to show which row forms are used (and perhaps why). In texted music, connecting the row forms used to the text as in figure 49.14 often suffices, since it links the serial analysis directly to the score, in this case covering eighty-five bars of music in a single image.

The similar image in figure 49.15 relates the row usage in Milton Babbitt's *Three Compositions for Piano*, movement 1. It lists six sections, with inclusive measure numbers for each, along with the row forms found in each section, in order and arranged according to their position in the texture. A person wanting to verify the analysis would need to go back to the score and trace the row forms. The picture

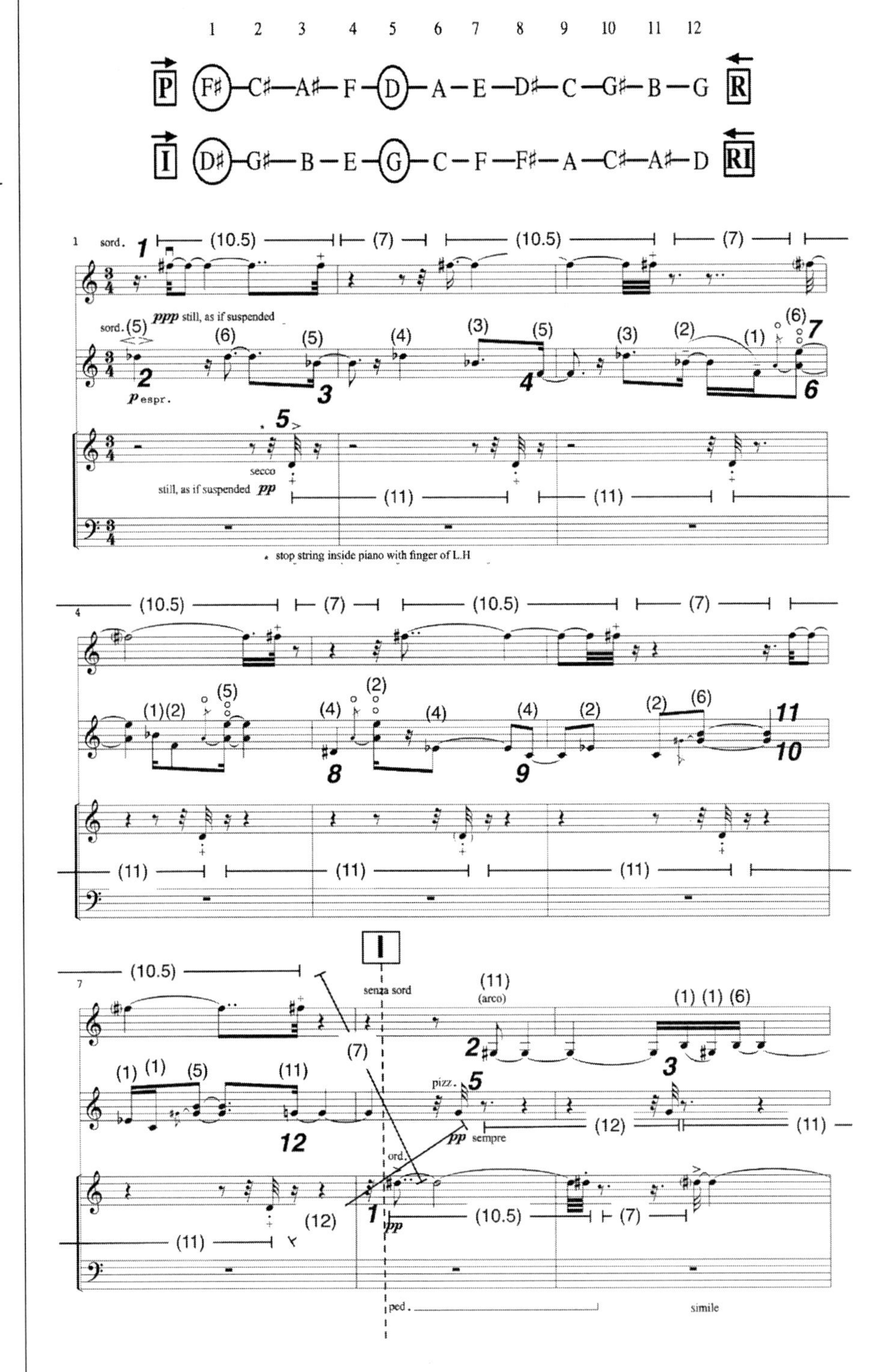

49.9 Joseph N. Straus, "'Twelve Tone in My Own Way': An Analytical Study of Ursula Mamlok's Panta Rhei, Third Movement, with Some Reflections on Twelve-Tone Music in America," in *Analytical Essays on Music by Women Composers: Concert Music, 1960–2000*, ed. Laurel Parsons and Brenda Ravenscroft, 18–31 (New York: Oxford University Press, 2016), 26, excerpt.

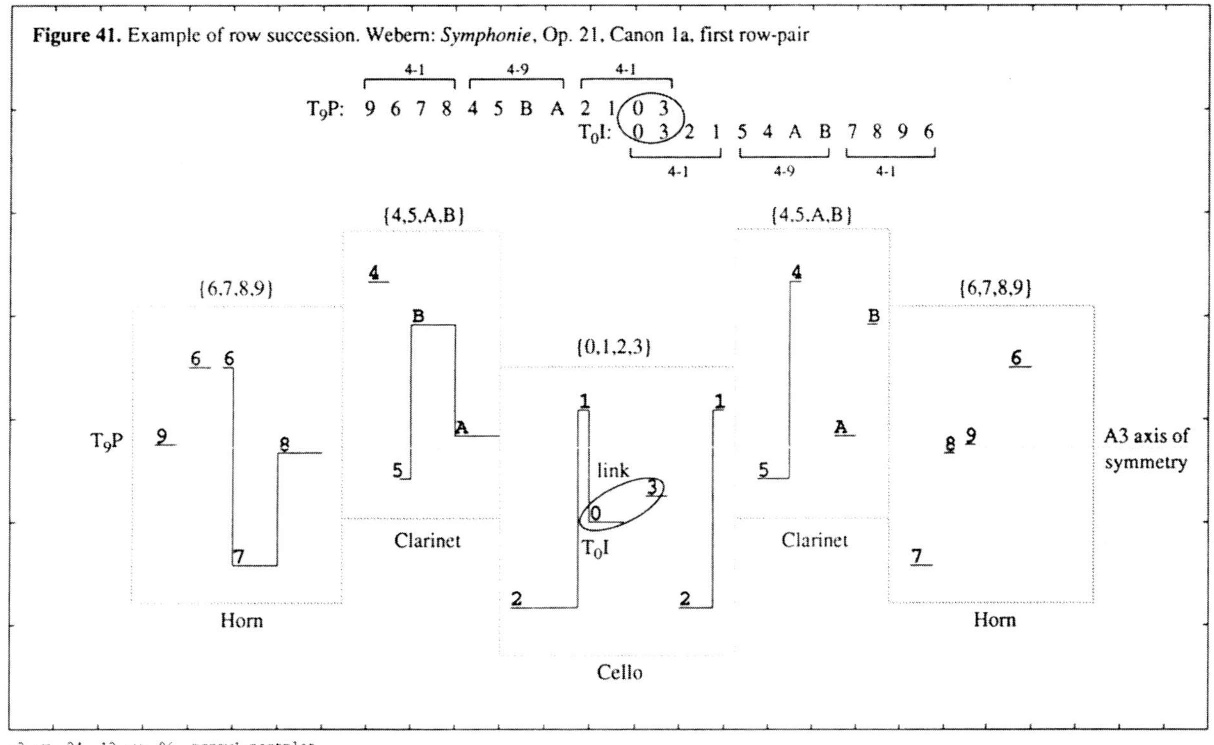

also offers none of the interpretation provided by the author in the text, including the meaning underlying the vertical versus diagonal arrangement of row forms and the fact that the entire piece is a retrograde canon. Adding that information would enhance the image.

Once row forms have been identified, a serial analysis commonly discusses reasons for their selection. One reason sometimes involves invariance, in which small groups of notes appear next to one another within different row forms in the music. The next two examples offer different ways of showing invariance (as did fig. 49.7 above). In figure 49.16, letters *a–f* label the six discrete dyads in the row from Arnold Schoenberg's Third String Quartet. These dyads appear together, though in different positions, in three other forms of the row. To find where a dyad has relocated requires a bit of searching, as the letters labeling them are rather small and their features do not pop out. Placing the letters above the dyads rather than below would help. Online figure 49.17 takes a graphical rather than a textual approach. Three images that explore twelve-tone arrays in various ways appear in online figures 49.18–49.20.

49.12 Alexander R. Brinkman and Martha R. Mesiti, "Computer-Graphic Tools for Music Analysis," *Computers in Music Research* 3 (1991): 34.

Echo Song

I Measures 132-144

O Thrush in the woods I fly among,
RI_{11} P_4

Do you, too, talk with the forest's tongue?
RI_{11} P_4

Stung, Stung

The Sting of Becoming I Sing
RI_{11} P_4

II Measures 146-171

O Hawk in the High and Widening Sky
R_5 I_{10} RI_5

What need I finally do to fly?
R_5 I_{10} RI_5

And see with your unclouded eye?
R_5 I_{10} I_{10}

Die, Die Let the Day
RI_5

Of despairing be done
R_5 I_{10} RI_5

III Measures 180-193

O Owl the wile mirror of the night
P_{11} RI_1 P_{11}

What is the force of the forest's light?
RI_1 P_{11} RI_1

Slight, Slight

With the slipping away of the sun
P_{11} RI_1 P_{11} RI_1 P_{11}

IV Measures 194-204

O Sable Raven call me back!
P_7/R_9 RI_0

What color does my torn robe lack?
I_5 P_9/RI_0

Black, Black

As your blameless and long dried blood
R_2/I_0

V Measures 205-216

O Bright Gull, aid me in my dream
P_1 I_3 P_1 I_3

Above the riddled breaker's cream!
P_1 I_3 I_3

Scream, Scream
P_1

For the shreds of your being
I_3 P_1 I_3

49.14 Emily J. Adamowicz, "Subjectivity and Structure in Milton Babbitt's *Philomel*," *Music Theory Online* 17, no. 2 (2011): ex. 2, http://www.mtosmt.org/issues/mto.11.17.2/mto.11.17.2.adamowicz.html.

SECTION ONE (bars 1-8):

P_4 R_{10} RI_5 I_{11}
P_{10} RI_{11} I_5 R_4

SECTION TWO (bars 9-18):

R_{10} I_{11} I_{11} R_{10} RI_5 P_4 P_4 RI_5
P_{10} RI_{11} RI_{11} I_5 P_{10} R_4 I_5 R_4

SECTION THREE (bars 19-28):

I_{11} R_{10} R_{10}
P_4 RI_5 P_4 I_{11} P_4 P_4 RI_5
R_{10} RI_5 RI_5 I_{11} R_{10} I_{11}

SECTION FOUR (bars 29-38):

RI_{11} P_{10} RI_{11} I_5 I_5 P_{10}
I_5 R_4 R_4 RI_{11} R_4 I_5 R_4
P_{10} P_{10} RI_{11}

SECTION FIVE (bars 39-48):

P_4 RI_5 P_4 R_{10} RI_5 I_{11} I_{11} R_{10}
I_5 R_4 R_4 I_5 P_{10} RI_{11} RI_{11} P_{10}

SECTION SIX (bars 49-56):

P_{10} RI_{11} I_5 R_4
RI_5 I_{11} P_4 R_{10}

49.15 George Perle, *Serial Composition and Atonality: An Introduction to the Music of Schoenberg, Berg, and Webern* (Berkeley: University of California Press, 1991), 128–29.

a.

P_7
a b c d e f

P_1
e b d c a f

I_{10}
e f c d a b

I_4
a f d c e b

b.

P_7
a b c d e f

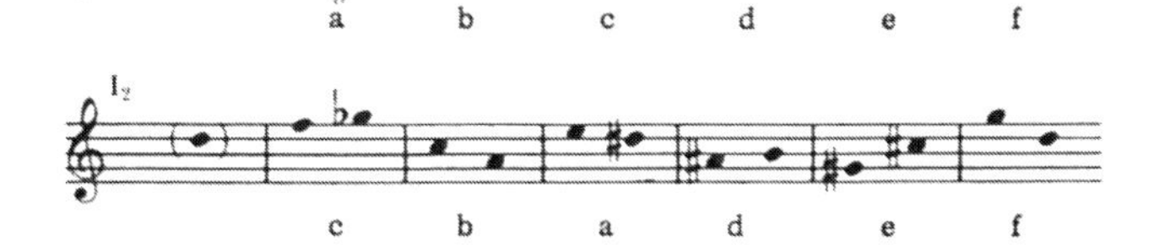

49.16 George Perle, *Serial Composition and Atonality: An Introduction to the Music of Schoenberg, Berg, and Webern* (Berkeley: University of California Press, 1991), 117.

Like analysis of any other kind of music, analysis of serial music can address that music at a minutely detailed level or a more general level. While an accounting of the details is sometimes more important than in analysis of other musics because it allows the viewer to verify the analysis, the offering of a synoptic view is still important here.

CHAPTER 50 Corpus Studies

Corpus studies are explorations of characteristics of large bodies of works. Corpus studies in music have emerged relatively recently. The creation of explorable databases and the development of computer tools for working with them are still in their infancy. Although the scope of their inquiry is different, as with any analytic endeavor, corpus studies should yield insights into the repertoire being studied, and their visualizations should make those insights plain.

The small multiple in figure 50.1 graphs the number of times each key is pressed in each of the op. 10 Etudes of Frédéric Chopin. The design is clever and simple (and you can generate your own at http://joeycloud.net/v/pianogram/). The view conveys clearly the relative frequency of black versus white notes, differences in registral emphasis, and the prevalence of smooth versus gapped distribution of pitches.

The tables in figure 50.2 show how often various phrase structures are used in themes from a corpus of films spanning nine decades, above in raw numbers, below as a percentage within each decade. Besides the now predictable refrain about the overuse of gridlines to structure tabular information, the image reminds us that the meaning of numerical information does not pop out at us. In this case, the all-important third dimension of information is flattened. Viewers must exert effort to track through the 108 pieces of information in each table and develop a conception of their own dimensionality. Summing the values in order to compare totals for

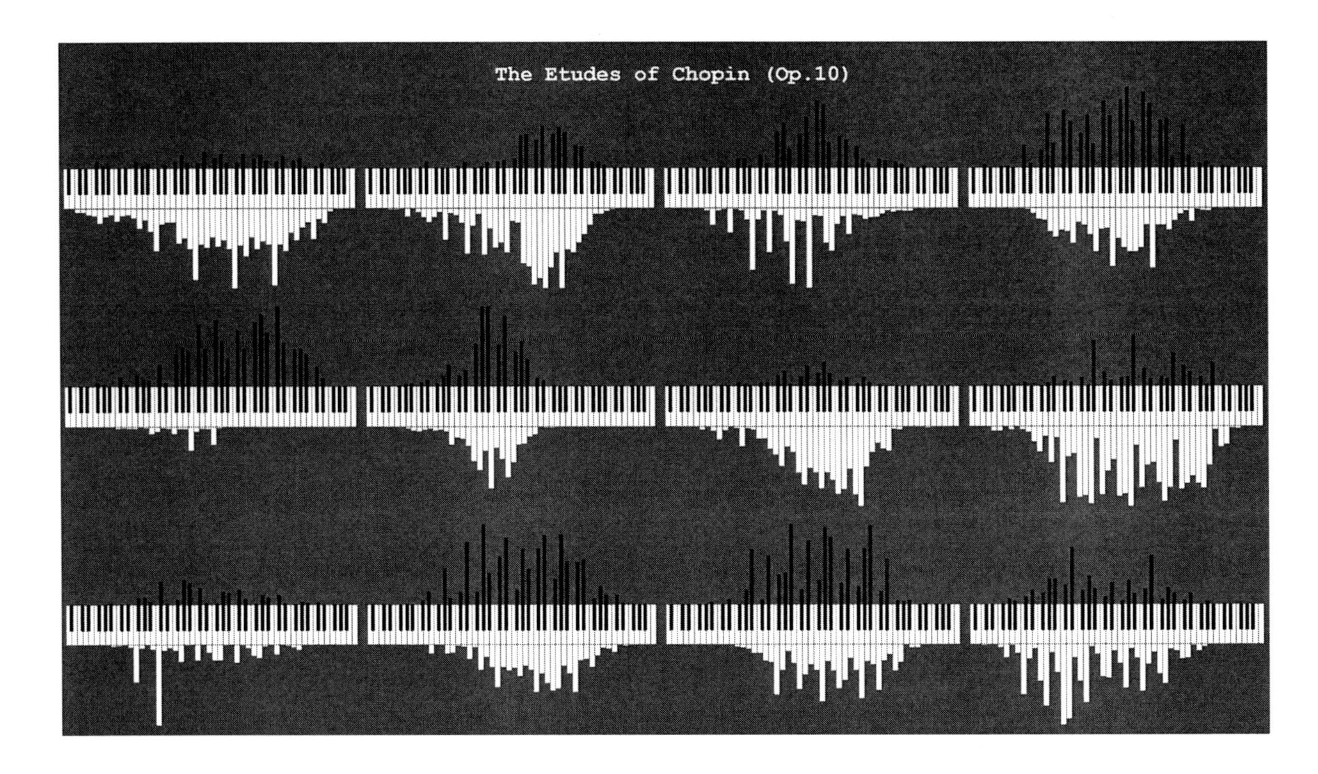

50.1 Joseph Yuen, histogram of keypresses in "The Etudes of Chopin, op. 10," accessed January 10, 2021, https://i.imgur.com/rx54l8Z.png, created at http://joeycloud.net/v/pianogram/.

***Facing top*, 50.2** Mark Richards, "Film Music Themes: Analysis and Corpus Study," *Music Theory Online* 22, no. 1 (2016): ex. 3, https://mtosmt.org/issues/mto.16.22.1/mto.16.22.1.richards.html.

the four broader types is yet more difficult. Differences and changes across these various dimensions are therefore frustratingly unapparent.

See how this third dimension becomes clear in the "skyline" image in figure 50.3. Like figure 50.2, it starts with a two-dimensional grid (let's call them Pitch Class Streets and Temporal Position Avenues). In place of numbers, the image erects buildings whose heights map the frequency with which melodic pitches from a corpus of tonal melodies (transposed to begin on C or A, depending on whether they were in a major or minor key) appear at different metrical positions. There are of course other challenges here. Just as in Manhattan, taller buildings occlude shorter ones, sometimes almost entirely, and without numerical information, one cannot easily make close comparison between "buildings" of similar height. Nevertheless, the data has a stronger impact than a table of building heights would. Different gray shades help mark the buildings in the long Temporal Position Avenues, while gray vertices serve a critical role in defining the extent of each building.

Visualizations of large data sets often employ scatterplots. Figure 50.4 charts changes in the use of chord types across six decades of published jazz fake books. To help maintain visual clarity, the image creates separate graphs for the three chord types that decrease in frequency and the three that increase. Trend lines can be calculated in a number of ways, and it is important to choose wisely. Particularly

Total Number

	Sentence			Clause			Period			Composite		
	Basic	*Developing*	*Periodic*	*Basic*	*Developing*	*Periodic*	*Basic*	*Developing*	*Sentential*	*Basic*	*Developing*	*Sentential*
1930s	11	2	1	5	0	2	8	3	2	1	1	2
1940s	23	1	2	12	3	0	20	6	1	5	4	5
1950s	5	3	4	6	1	1	9	2	0	4	0	4
1960s	7	1	0	9	3	0	5	10	2	0	1	3
1970s	7	2	0	8	1	2	6	10	0	3	0	1
1980s	5	2	0	4	4	1	6	14	3	3	0	1
1990s	5	1	0	5	0	2	9	7	1	1	1	2
2000s	6	0	0	3	1	0	7	4	3	1	0	3
2010s	4	0	0	0	1	0	2	2	2	0	0	0

Percentage (of Total Grammatical Themes per Decade)

	Sentence			Clause			Period			Composite		
	Basic	*Developing*	*Periodic*	*Basic*	*Developing*	*Periodic*	*Basic*	*Developing*	*Sentential*	*Basic*	*Developing*	*Sentential*
1930s	29	5	3	13	0	5	21	8	5	3	3	5
1940s	28	1	2	15	4	0	24	7	1	6	5	6
1950s	13	8	10	15	3	3	23	5	0	10	0	10
1960s	17	2	0	22	7	0	12	24	5	0	2	7
1970s	18	5	0	20	3	5	15	25	0	8	0	3
1980s	12	5	0	9	9	2	14	33	7	7	0	2
1990s	15	3	0	15	0	6	26	21	3	3	3	6
2000s	21	0	0	11	4	0	25	14	11	4	0	11
2010s	36	0	0	0	9	0	18	18	18	0	0	0

in the left image, the linear trend lines might imply that the data is more regular than it actually is (see the wide variations for the Dom7 and Maj chords).

Figures 50.5 and 50.6 both examine the relationship between variation in durational contrast within a spoken language and variation in melodies written by composers who speak that language, using a normalized Pairwise Variability Index (nPVI). Both figures show minimum and maximum values, plus a dot for the

50.3 Jon B. Prince and Mark A. Schmuckler, "A Corpus Analysis: The Tonal-Metric Hierarchy," *Music Perception* 31, no. 3 (2014): 263.

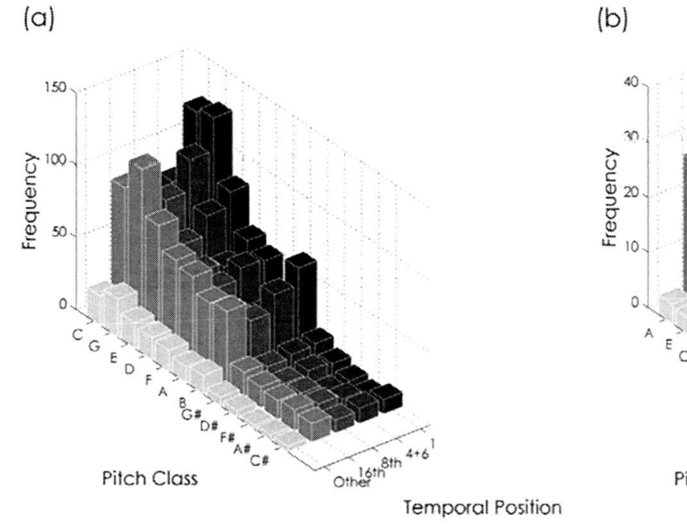

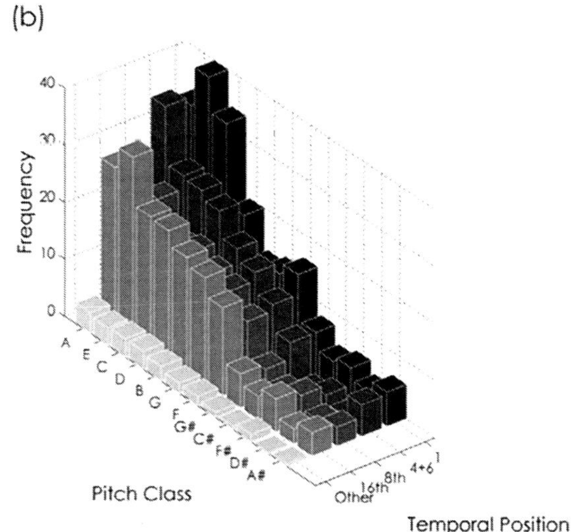

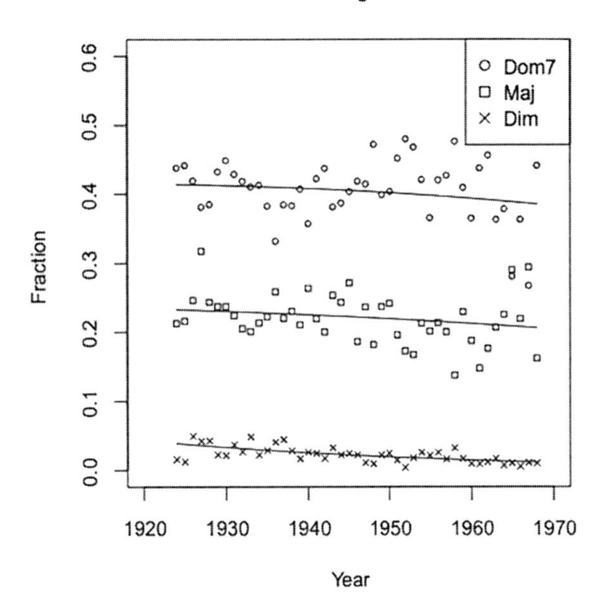

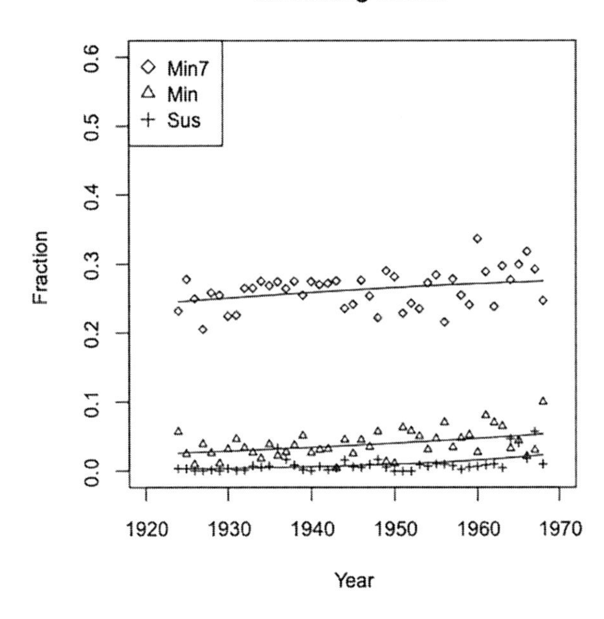

50.4 Yuri Broze and Daniel Shanahan, "A Cognitive Perspective: Diachronic Changes in Jazz Harmony," *Music Perception* 31, no. 1 (2013): 37.

average value, for each composer. Values in figure 50.5 are sorted alphabetically (French-speaking composers above, German below), while those in figure 50.6 are arranged chronologically by each composer's midpoint. While this type of plot generalizes the data, it also hides potentially useful insight by eliminating individual data points. Outliers can affect both the average and the extremes, and that is likely a factor in the data in figure 50.5 for Massé. It is better to both map every point and show the average and extrema.

Trend lines are valuable in generalizing data, particularly where there is variability in that data, as is inevitable in human research. They aim to find a path through a data set that minimizes the amount of variability in the data. Statistics research often faces the challenge of choosing an appropriate number of inflection points in a trend line. Figure 50.6 illustrates this nicely, computing three possible trend lines: first order (linear), second order (with one inflection point), and third order (with two inflection points); a fourth was computed in an accompanying table but omitted from this image. The authors conclude that the second-order trend line best summarizes the changes in values (for an explanation, see Hansen, Sadakata, and Pearce [2016]).

From the same source, figure 50.7 adds information from Austro-German composers to the French-language composers of figure 50.6. As graphics are always more effective when they facilitate comparison, figure 50.7 is ultimately more meaningful. Different text colors represent composers associated with each language. The design decision to center the composer names directly over where their

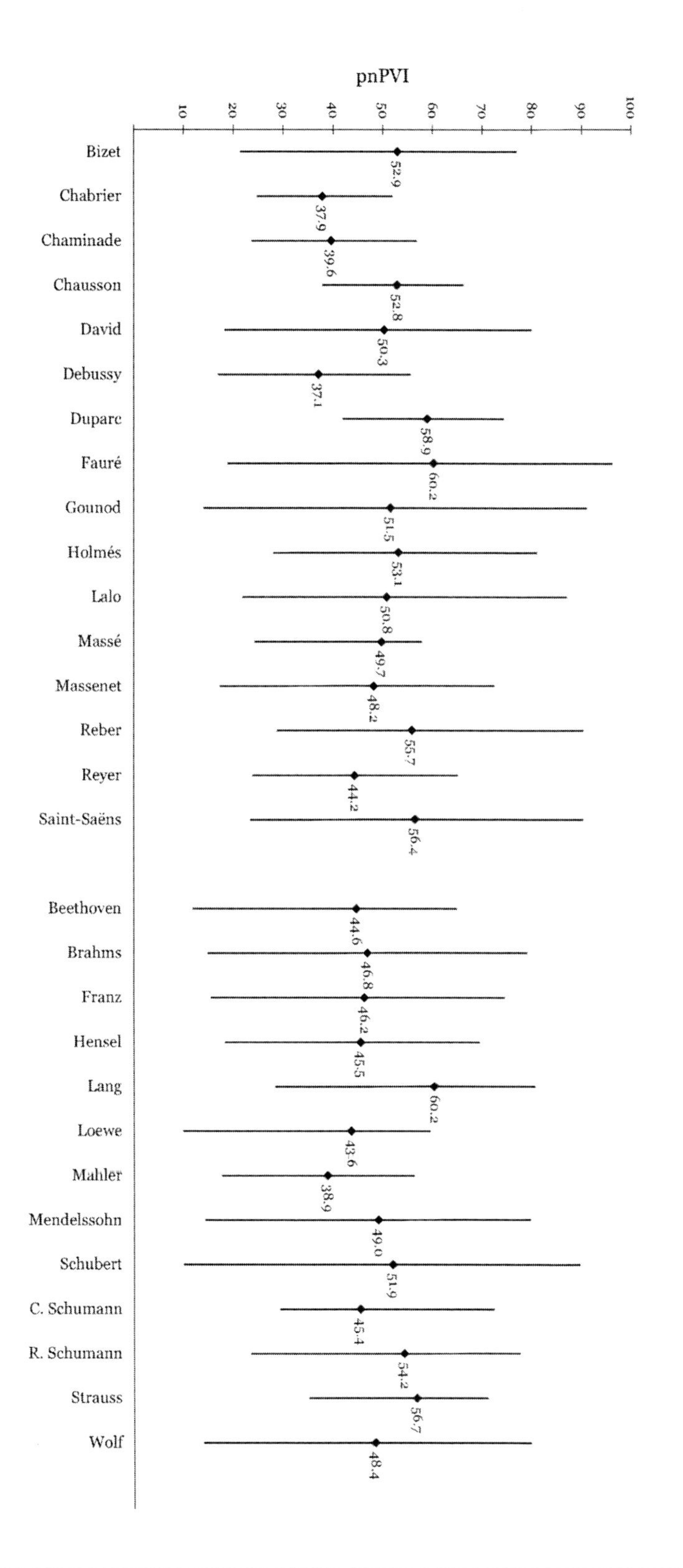

50.5 Leigh VanHandel and Tian Song, "The Role of Meter in Compositional Style in 19th Century French and German Art Song," *Journal of New Music Research* 39, no. 1 (2010): 4.

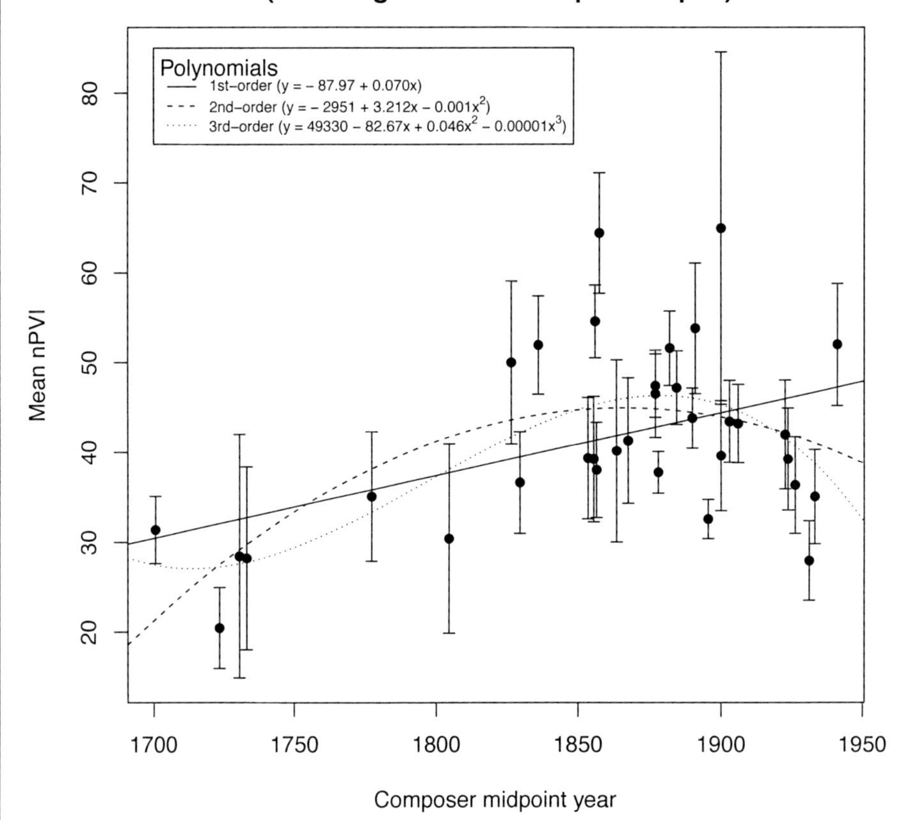

50.6 Niels Chr. Hansen, Makiko Sadakata, and Marcus Pearce, "Nonlinear Changes in the Rhythm of European Art Music: Quantitative Support for Historical Musicology," *Music Perception* 33, no. 4 (2016): 420.

data point should be leads to a number of name collisions, as well as an apparent loss of data precision, which would be restored by the use of a dot for each point. On the other hand, the use of dots implies a precision that is not statistically warranted, as we saw in figure 50.6. The trend lines, colored to match the composer names, tell the image's story clearly.

In figure 50.8, heat maps show the frequency of two-chord successions in two ten-year periods excerpted from a corpus of jazz works, 1924–33 and 1959–68. Lighter shades represent transitions that are more common in the corpus. Because the graphs have the same structure, it is relatively easy to compare corresponding blocks. Unfortunately, because it is more difficult to distinguish between shades of highly saturated colors, such as characterize both graphs, the changes are difficult to tease out here. A scale representing the darkest value with a medium-dark gray would make the images clearer. See online figure 50.9 for an example of a stacked line graph in a study of journal article topics.

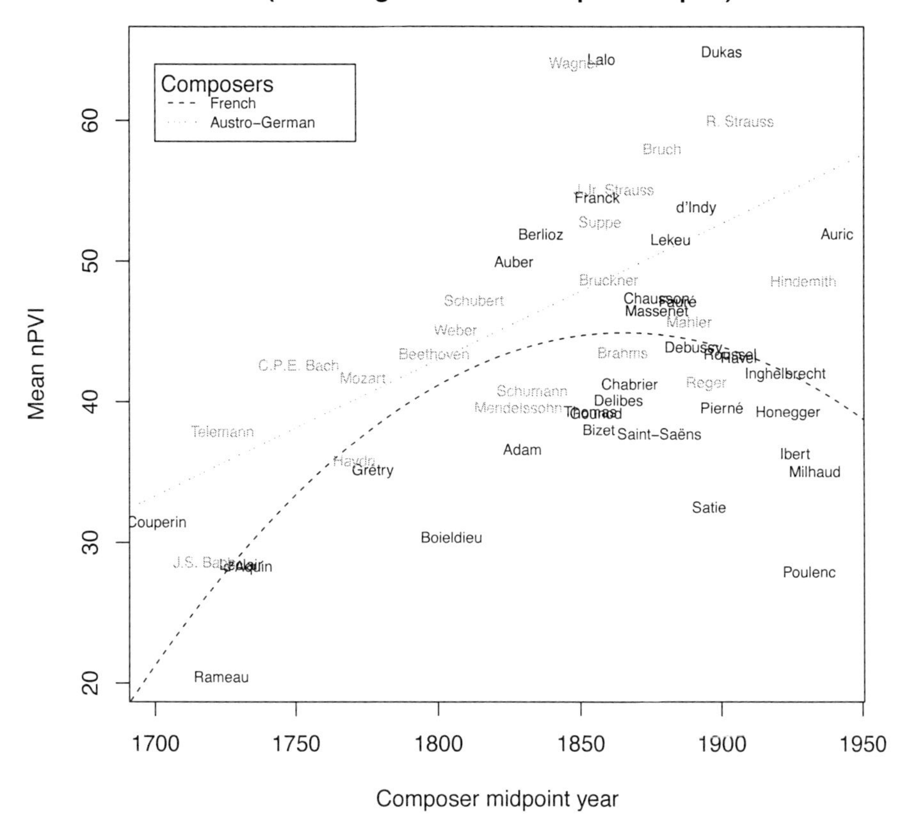

50.7 Niels Chr. Hansen, Makiko Sadakata, and Marcus Pearce, "Nonlinear Changes in the Rhythm of European Art Music: Quantitative Support for Historical Musicology," *Music Perception* 33, no. 4 (2016): 426.

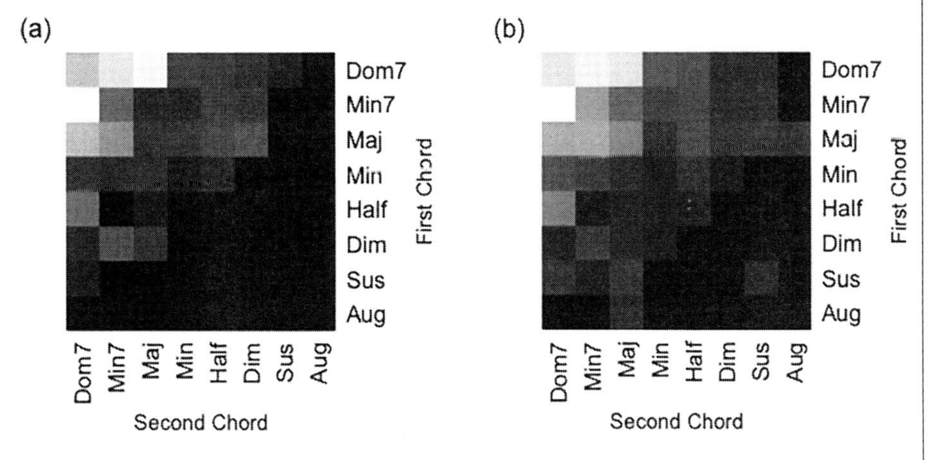

50.8 Yuri Broze and Daniel Shanahan, "A Cognitive Perspective: Diachronic Changes in Jazz Harmony," *Music Perception* 31, no. 1 (2013): 39.

In 2009, a terrific website called An Eroica Project went dark. Built around a database of over four hundred commercial recordings of Beethoven's *Eroica Symphony*, it offered multiple views into the treatment of tempo in performances spanning nearly a century. These include tempo relationships between movements, whether conductors with multiple recordings took tempos faster or slower later in their careers, which conductors had the most flexible tempos, and many more. Unfortunately, the data underlying the site has subsequently been lost (Eric Grunin, personal communication, December 9, 2020). A sampling of screenshots from the site appears in online figure 50.10.

Corpus studies and other representations of large data sets emphasize forest more than trees and above all should find the story that underlies the information. One must take care not to be too reductive, however, as data is often messy, as a result of competing influences and random elements. For this reason, shaping a truthful narrative often requires that the trees be represented as well.

CHAPTER 51

Musical Chronologies, Influences, and Styles

It is common to trace historical succession and influence in visual form. In music, we might be interested in dates of composers, important compositions, or the development of genres, styles, or instruments. Or we might want to know how various musical styles influence or develop into others. While they are less inherently "musical" than many of the book's other images, historical timelines are, after all, topic-neutral; the recurrence of such images in scholarly and pedagogical music literature invites an exploration of best practices. From the design perspective, the most successful images provide a context that facilitates comparisons among the elements. The most compelling lead to fresh understanding and insight. In addition, some subjects lend themselves to the construction of narratives, which requires special consideration. In all such images, as in corpus studies, one should keep in mind the potential for scholarly mischief through the choice of what information to include (and exclude).

The first three images provide variations on the basic timeline in which years are plotted on the horizontal axis. Figure 51.1 neatly separates three categories of information and employs fading lines to reflect the gradual introduction and decline of various forms and styles. The layout is richly comparative. The placement of text and bar on the same line makes it ambiguous whether the name of the genre is part of the time span or merely labels it, however. "Dance Forms" fade in beginning just before 1160, but do "Strophic Forms" appear suddenly with the line, ca. 1050, or with the text, ca. 1000? And are the composer names left-justified with the birth date or centered under the years of their life? Or their active period as a composer? The latter is more explicit in online figure 51.2.

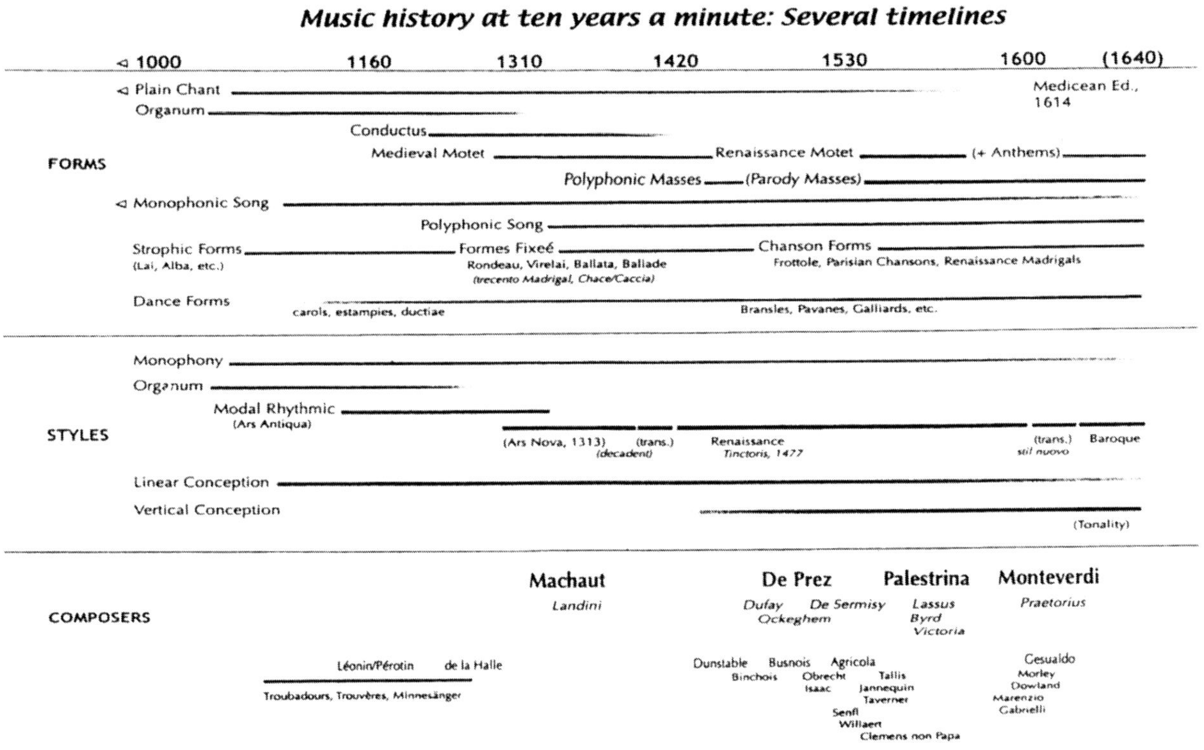

Music history at ten years a minute: Several timelines

◁ 1000 1160 1310 1420 1530 1600 (1640)

FORMS
◁ Plain Chant
Medicean Ed., 1614
Organum
Conductus
Medieval Motet
Renaissance Motet
(+ Anthems)
Polyphonic Masses
(Parody Masses)
◁ Monophonic Song
Polyphonic Song
Strophic Forms (Lai, Alba, etc.)
Formes Fixeé
Rondeau, Virelai, Ballata, Ballade (trecento Madrigal, Chace/Caccia)
Chanson Forms
Frottole, Parisian Chansons, Renaissance Madrigals
Dance Forms
carols, estampies, ductiae
Bransles, Pavanes, Galliards, etc.

STYLES
Monophony
Organum
Modal Rhythmic (Ars Antiqua)
(Ars Nova, 1313)
(trans.) (decadent)
Renaissance Tinctoris, 1477
(trans.) stil nuovo
Baroque
Linear Conception
Vertical Conception
(Tonality)

COMPOSERS
Machaut
Landini
De Prez
Dufay
De Sermisy
Ockeghem
Palestrina
Lassus
Byrd
Victoria
Monteverdi
Praetorius
Léonin/Pérotin
de la Halle
Troubadours, Trouvères, Minnesänger
Dunstable
Binchois
Busnois
Obrecht
Isaac
Agricola
Tallis
Jannequin
Taverner
Senfl
Willaert
Clemens non Papa
Gesualdo
Morley
Dowland
Marenzio
Gabrielli

51.1 Art Samplaski, "Music History at Ten Years a Minute," *College Music Symposium* 44 (2004): 106.

Figure 51.3 combines design elements of both images. It plots both composer dates (*below*) and the dates of some significant twentieth-century works (*above*). Names and titles tussle with the lines and dots, whose orientation to the implied grid of the timeline is strained as a result. The redrawing in online figure 51.4 adopts useful features from all three images in an improved version.

Figures 51.5 and 51.6, which trace the history of jazz, come from editions of the same book published twenty-eight years apart. In addition to content updates, improvements in design are striking. The noisy fills of the long vertical arrows have been replaced with gentler shading. Gray replaces the black text backgrounds that render their text all but illegible, and attractive sideless boxes replace the heavy boxes surrounding other text. In addition, the implied vertical grid is now spaced evenly, while a pair of dashed lines unobtrusively segments the jazz era into three broad eras (traditional, modern, postmodern).

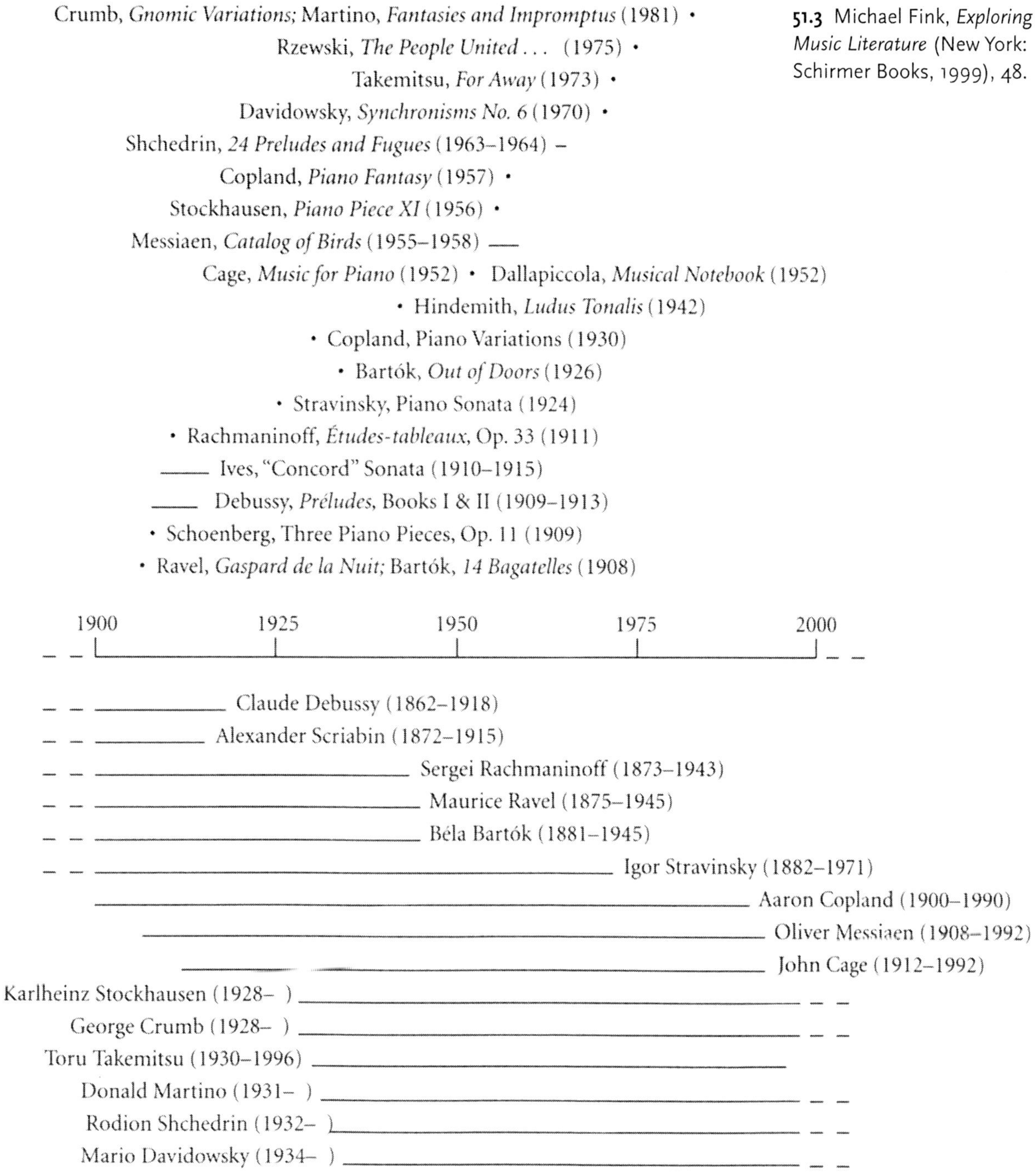

Crumb, *Gnomic Variations;* Martino, *Fantasies and Impromptus* (1981) •
Rzewski, *The People United . . .* (1975) •
Takemitsu, *For Away* (1973) •
Davidowsky, *Synchronisms No. 6* (1970) •
Shchedrin, *24 Preludes and Fugues* (1963–1964) –
Copland, *Piano Fantasy* (1957) •
Stockhausen, *Piano Piece XI* (1956) •
Messiaen, *Catalog of Birds* (1955–1958) —
Cage, *Music for Piano* (1952) • Dallapiccola, *Musical Notebook* (1952)
• Hindemith, *Ludus Tonalis* (1942)
• Copland, Piano Variations (1930)
• Bartók, *Out of Doors* (1926)
• Stravinsky, Piano Sonata (1924)
• Rachmaninoff, *Études-tableaux*, Op. 33 (1911)
— Ives, "Concord" Sonata (1910–1915)
— Debussy, *Préludes*, Books I & II (1909–1913)
• Schoenberg, Three Piano Pieces, Op. 11 (1909)
• Ravel, *Gaspard de la Nuit;* Bartók, *14 Bagatelles* (1908)

1900 1925 1950 1975 2000

Claude Debussy (1862–1918)
Alexander Scriabin (1872–1915)
Sergei Rachmaninoff (1873–1943)
Maurice Ravel (1875–1945)
Béla Bartók (1881–1945)
Igor Stravinsky (1882–1971)
Aaron Copland (1900–1990)
Oliver Messiaen (1908–1992)
John Cage (1912–1992)
Karlheinz Stockhausen (1928–)
George Crumb (1928–)
Toru Takemitsu (1930–1996)
Donald Martino (1931–)
Rodion Shchedrin (1932–)
Mario Davidowsky (1934–)

51.3 Michael Fink, *Exploring Music Literature* (New York: Schirmer Books, 1999), 48.

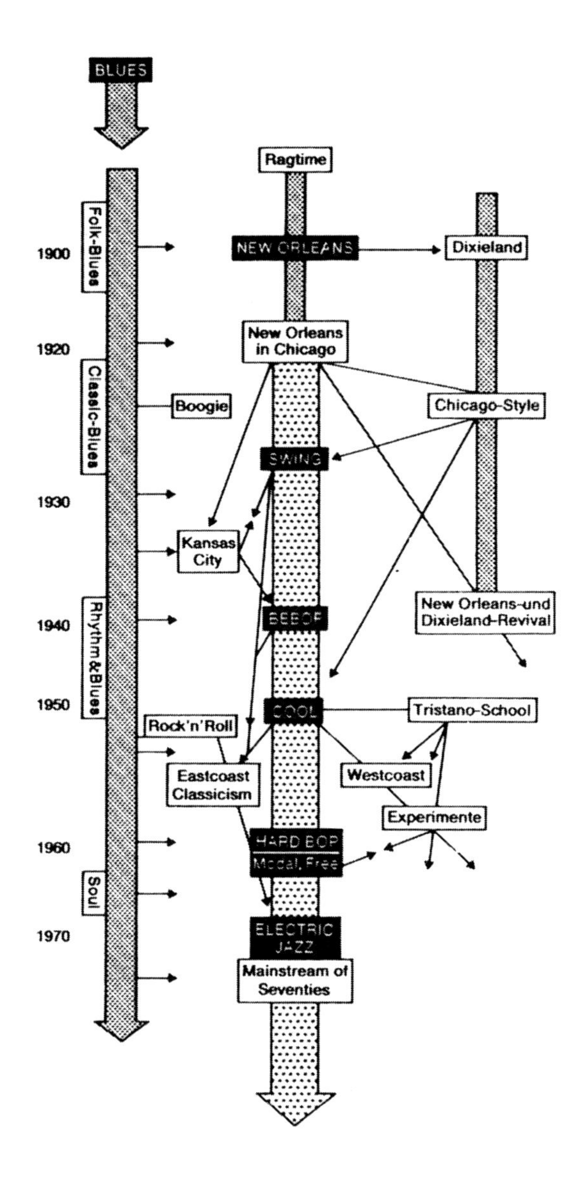

51.5 Joachim-Ernst Berendt, *Das Jazzbuch—Von Rag bis Rock* (Frankfurt am Main, 1973), 14, cited in Werner Breckoff et al., *Musik Aktuell: Informationen, Dokumente, Aufgaben* (Kassel: Bärenreiter, 1971), 231.

Attempting a similar stylistic tracing but for rock, figure 51.7 paints a messier and less successful picture. Musical traditions from African and European origins follow a tangled path leading to a panoply of rock genres in the early 1970s. Except for the primary (solid) and secondary (dashed) influences at the top of the image, and the secondary influence "soul" had on "Liverpool 1958–1962," no distinction is made in the degree or significance of influence among forms. Oddly, the image implies that all stylistic roads lead by the mid-1960s to two styles (white blues, folk rock) and one group (the Beatles). At that point, it throws up its hands. The image manages to both oversimplify and underspecify all at once.

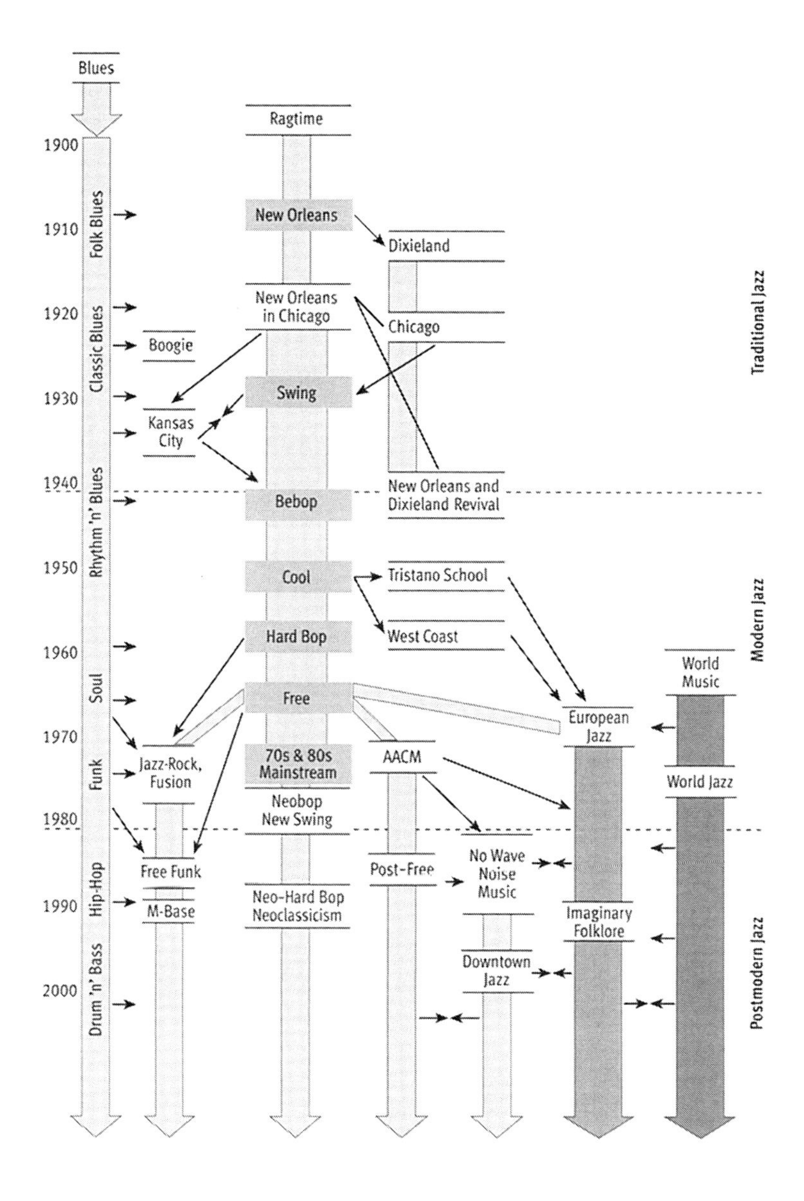

51.6 Joachim-Ernst Berendt, *The Jazz Book from Ragtime to the 21st Century*, 7th ed., ed. Günther Huesmann (Chicago: Lawrence Hill Books, 2009), 3.

In figure 51.8, we find a more credible web of influences, "a visual rendering of the discourse around the concept of statistical form" in the 1950s (Iverson 2014, 343). However, this multifaceted and complex network of contributors (main in bold, secondary in regular typeface) and relationships (also of two weights) would benefit from clearer visual differentiation between the two levels of objects and connectors.

Figure 51.9 depicts the musical styles of the Aka around a circle. It provides three types of information: performing forces, the Aka name for the category, and a description of the category. In addition, shading indicates the types of pieces that

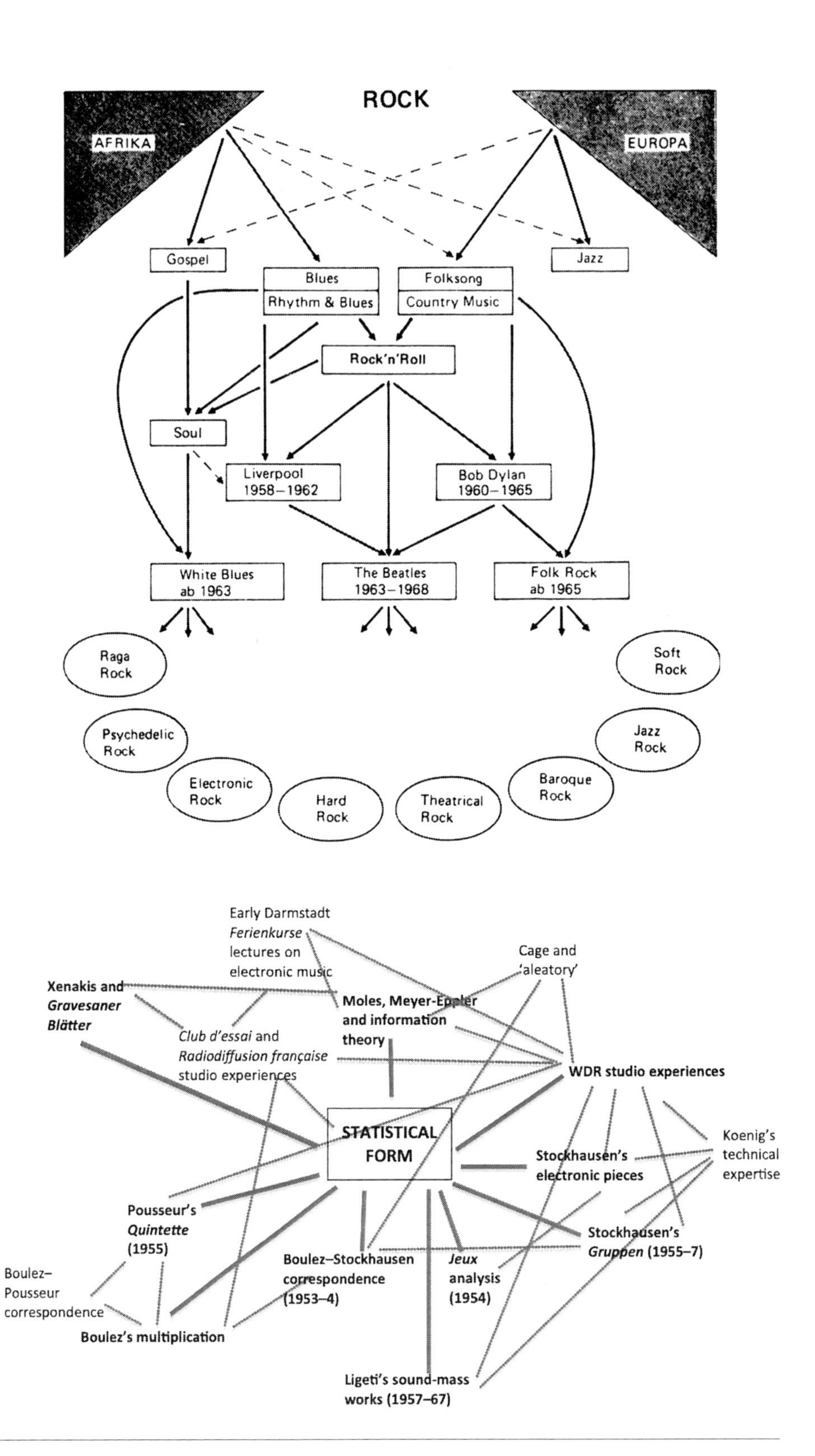

51.7 Werner Breckoff et al., *Musik Aktuell: Informationen, Dokumente, Aufgaben* (Kassel: Bärenreiter, 1971), 242.

51.8 Jennifer Iverson, "Statistical Form amongst the Darmstadt School," *Music Analysis* 33, no. 3 (2014): 344.

do not involve hand clapping. The information is essentially tabular, and while it might effectively be expressed that way, this layout provides a more visceral picture of the various categories of piece that are sung a cappella, with bunches of leaves or with drums, sticks, and machetes. The treble clef in the center is, of course, unnecessary.

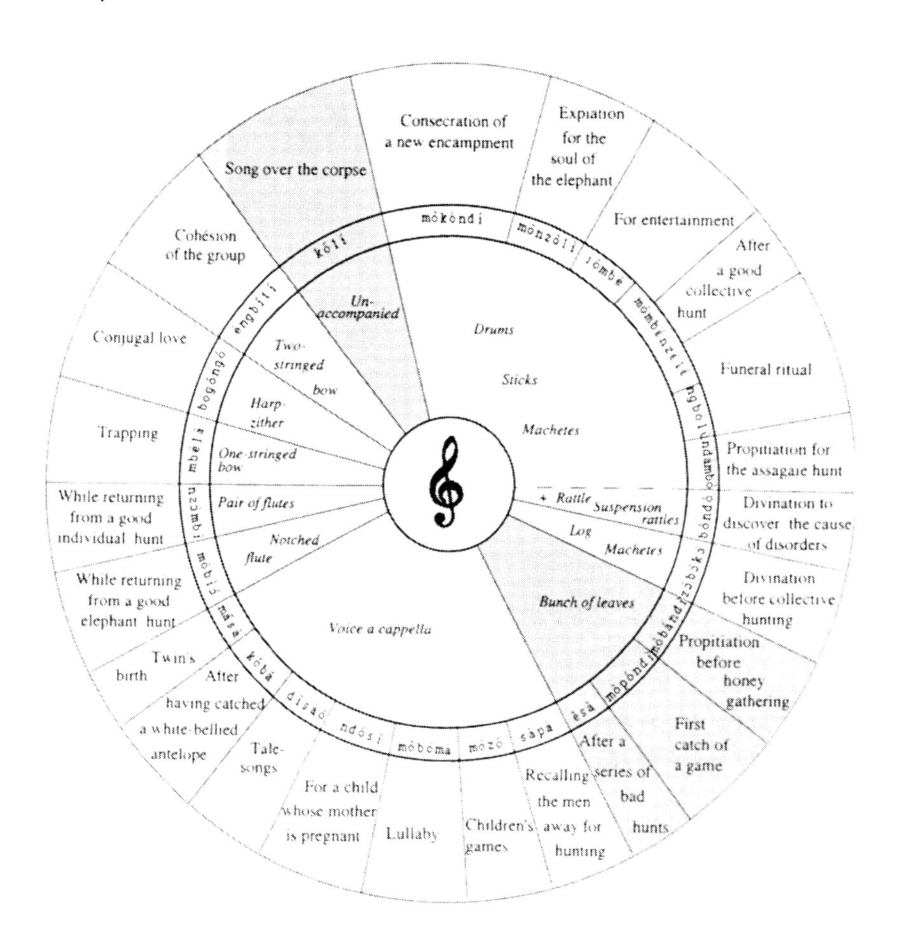

51.9 Susanne Fürniss, "Aka Polyphony: Music, Theory, Back and Forth," in *Analytical Studies in World Music*, ed. Michael Tenzer, 163–204 (New York: Oxford University Press, 2006), 166.

Finally, figures 51.10–51.12 together brilliantly describe the contents of two edited volumes of essays on world music. As summarized in the table in figure 51.10, twenty chapters address music from six continents. The music is geolocated on the map in figure 51.11, which brings meaning back to the table in figure 51.10. That table organizes the musics described in these twenty chapters into ten temporal categories. Figure 51.12 then plots these ten categories onto a two-dimensional grid that maps their temporal organization on a continuum from "ostinato cyclic" to "unmeasured rhythm" and their transformational character on a continuum from "repetition" to "transformation." Readers of the book who reference these

1. Temporal Category	2. Author (volume[a] and chapter)	3. Origin and Genre	4. Selection Title	5. Time Organization	6. Place of Articulation	7. Formal Continuity
A. "Pure" ostinato-cyclic	1. Fürniss (*ASWM* 5)	Central African song	*Dikòbò dámù dá sòmbé*	Ostinato cycle	Cycle boundaries	Cyclic
B. Cyclic—discursive	2. Tenzer (*ASWM* 6)	Balinese gamelan	*Oleg Tumulilingan*	Ostinato cycle and expansions	Metacycle boundaries	Through-composed with cycles
	3. Sutton/Vetter (*ASWM* 7)	Javanese gamelan	*Ladrang Pangkur*			
C. "Pure" hybrid (transformative/ sectional/cyclic)	4. Hesselink (*ACCSWM* 7)	Korean *p'ungmul*	*P'an Kut*	Succession of ostinato cycles	Metacycle boundaries	Sectional
D. Sectional with ostinato cycle basis	5. Manuel (*ASWM* 3)	Spanish *flamenco*	*A Quién le Contaré Yo*	Ostinato cycle layer, metered layer	Metacycle boundaries	Sectional
	6. Terauchi (*ACCSWM* 1)	Japanese *gagaku*	*Etenraku*			
	7. Moore/Sayre (*ASWM* 4)	Cuban *batá*	*Obatalá*			
E. Sectional with nonmetric (pulsed, unmeasured rhythm) basis	8. Blum (*ASWM* 1)	Xorasani *navā'i*	*Sāqi-nāme* of Qomri	Unmetered/ measured/cyclic	Configured group boundaries	Sectional/cyclic
	9. Levine/Nettl (*ACCSWM* 8)	*Arapaho* song	*Wolf Dance Song*			
	10. Barwick (*ACCSWM* 9)	Murriny Patha *djanba*	*Kunyibinyi Tjingarru*			
F. Sectional—cyclic	11. Stanyek/Oliveira (*ACCSWM* 3)	Brazilian *samba pagode*	*Sorriso Aberto*	Metered/cyclic	Configured group boundaries	Cyclic
	12. Ziporyn/Tenzer (*ACCSWM* 4)	American *jazz*	*I Should Care*			
	13. Leach (*ACCSWM* 2)	French medieval *balade*	*De Petit Po*			
G. Sectional—metered	14. Buchanan/Folse (*ASWM* 2)	Bulgarian *horo*	*Georgi, le Lyubile*	Metered/cyclic	Configured group boundaries	Through-composed with sectional articulations
	15. Morris (*ASWM* 9)	S. Indian *varnam*	*Valachi Vacchi*			
H. Transformative—sectional	16. Stock (*ASWM* 8)	Chinese *huju*	*Jin Yuan Seeks Her Son*	Metered	Configured group boundaries	through-composed with sectional articulations
	17. Benjamin (*ASWM* 10)	European *piano concerto*	*Concerto 17 in G Major, K. 453, I*			
I. Open transformative	18. Roeder (*ASWM* 11)	American *chamber music*	*Enchanted Preludes*	Multiply-pulsed free rhythm/ *unmeasured* rhythm	Configured group boundaries	Through-composed with sectional articulations
	19. Widdess (*ACCSWM* 5)	North Indian *ālāp*	*rāg Pūriyā-Kalyān*			
J. "Pure" transformative	20. Bunk (*ACCSWM* 6)	American *"timbre-and-form"*	*Phoneme (3)*	*Unmeasured* rhythm	(Weakly) configured group boundaries	Through-composed with weak sectional articulations

[a] *ASWM* = *Analytical Studies in World Music* (Tenzer 2006); *ACCSWM* = *Analytical and Cross-Cultural Studies in World Music* (current volume)

51.10 Michael Tenzer, "A Cross-Cultural Topology of Musical Time: Afterword to the Present Book and to *Analytical Studies in World Music* (2006)," in *Analytical and Cross-Cultural Studies in World Music*, ed. Michael Tenzer and John Roeder, 415–40 (New York: Oxford University Press, 2011), 420–21.

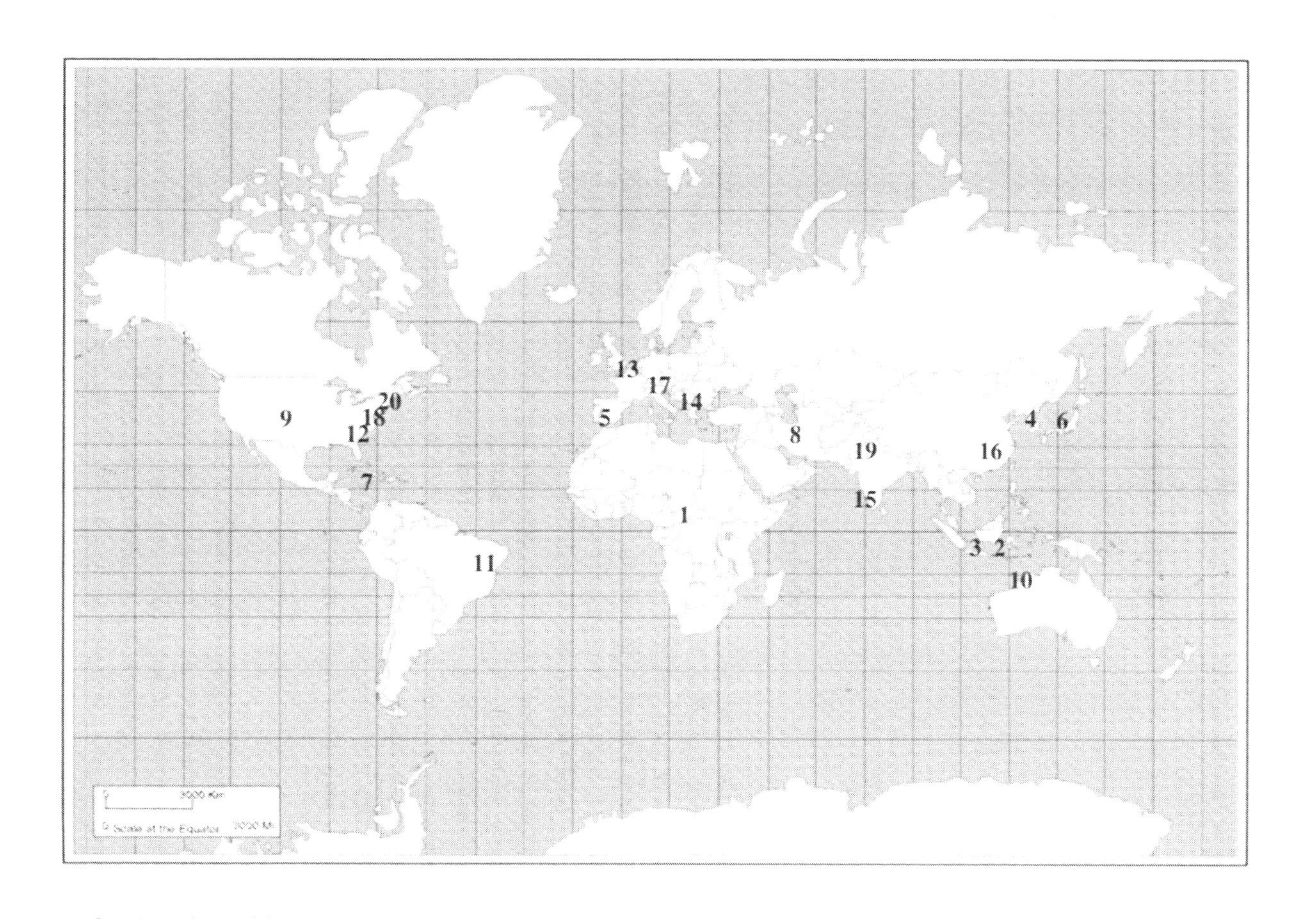

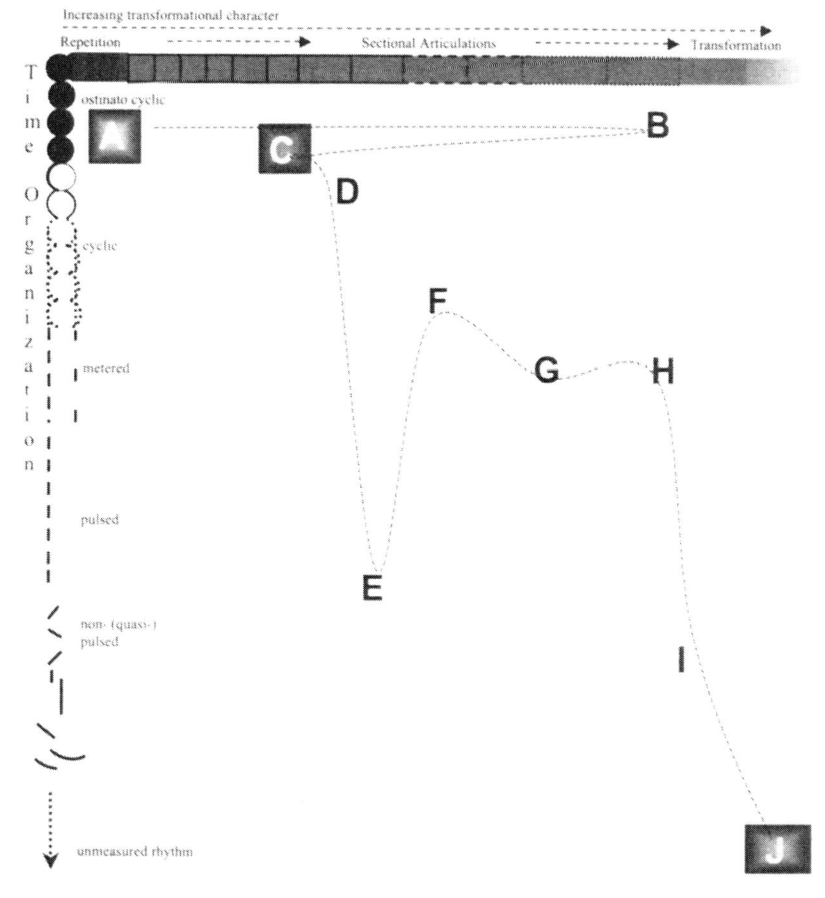

***Above*, 51.11** Michael Tenzer, "A Cross-Cultural Topology of Musical Time: Afterword to the Present Book and to *Analytical Studies in World Music* (2006)," in *Analytical and Cross-Cultural Studies in World Music*, ed. Michael Tenzer and John Roeder, 415–40 (New York: Oxford University Press, 2011), 416.

***Left*, 51.12** Michael Tenzer, "A Cross-Cultural Topology of Musical Time: Afterword to the Present Book and to *Analytical Studies in World Music* (2006)," in *Analytical and Cross-Cultural Studies in World Music*, ed. Michael Tenzer and John Roeder, 415–40 (New York: Oxford University Press, 2011), 422.

figures as they read the chapters will gain a richer understanding of the characteristics of the musics under discussion.

The chapter has examined musical chronologies, influences, and styles. Each image has either facilitated comparisons among the information presented or created a narrative of some kind. Most of them are effective models for the kinds of information they present.

Chapter 52 Animation

This book has taken pains to valorize the still image. Still images can convey a great deal about music, including about its temporal aspects. Many images transform time into a physical dimension, often mapped onto the horizontal axis, or represent snapshots in time through small-multiple formats (see the discussion of fig. 10.3). Nevertheless, since the experience of music unfolds in time, it makes sense that visualizations of music might benefit from doing so as well. For most of time, however, there has been almost no way to make this happen. But with the advent of computing and high-speed networks, the possibility for time-based music visualization has grown dramatically. As we see in this chapter, animated music visualizations often include audio, allowing our eyes and ears to work in synchronization and ideally enhancing the experience of both.

We will exclude from this discussion the kinds of abstract, quasi-stochastic visualizations like figure 52.1, which media players like the MacOS Music program can produce, as well as artistic animations such as in online figure 52.2, which, while they accompany music and even visually respond to musical cues, are not themselves aimed at communicating about music. Since many of the animations discussed in this chapter are in color, still images from most appear in the online supplement, with external links to the animations on the web.

Musical animations are bound by the same principles as any other visualization: clarity of design, meaningful underlying metaphors, effective layering of information, and so on. They have additional challenges, however. With a still image,

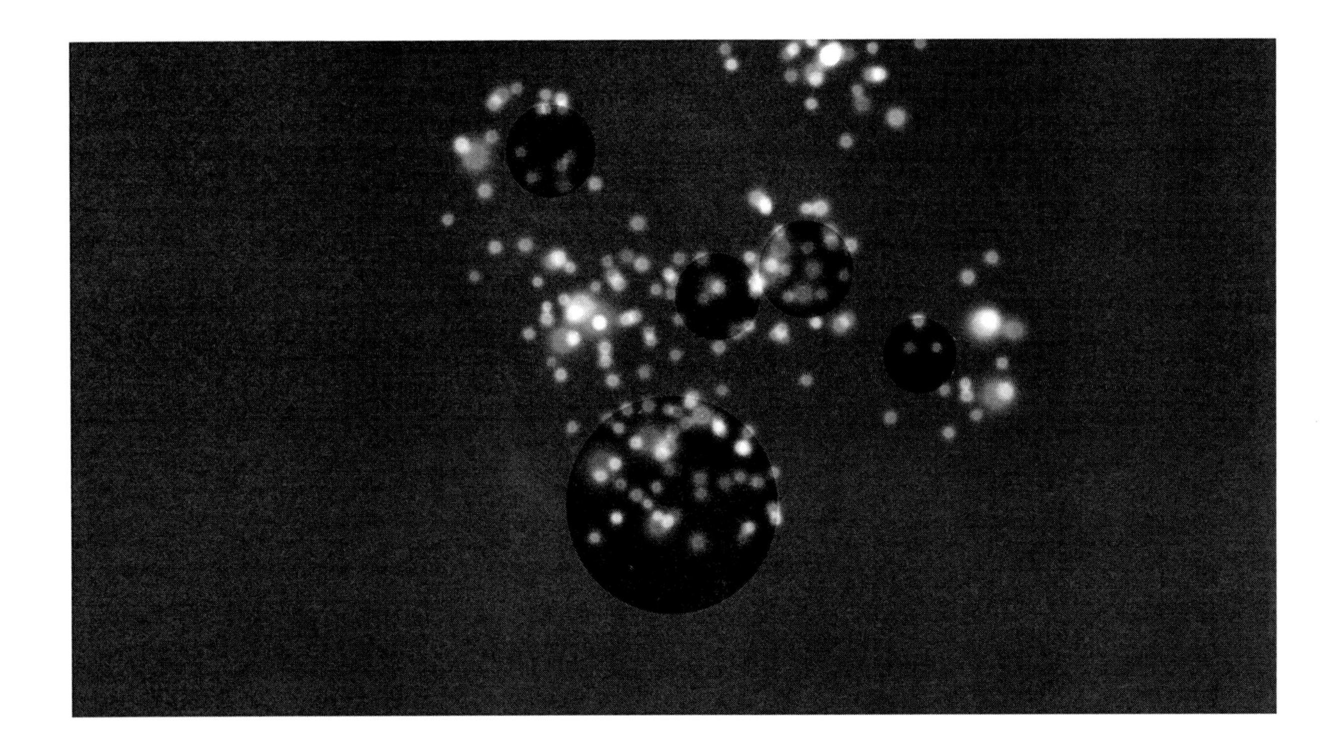

52.1 Screenshot of visualization produced by Music (version 1.2.28, Apple, MacOS) during playback of Stacy Garrop, *In Eleanor's Words: Music of Stacy Garrop*, V, "The Dove of Peace," track 6, Cedilla, 2010, compact disc.

whether or not viewers follow the designer's techniques to guide their viewing, the content of the visualization remains available to them, and initially missed details remain available. The viewer retains ultimate control of the viewing experience. In an animation, control shifts toward the designer, who determines what is on the screen at each moment. If you want someone to focus on something, it's best not to startle them with it. By the time the viewer understands what just appeared, it will be old news and something else will already be vying for attention. Therefore, unless there is a compelling reason to do otherwise, animators should give viewers as much agency as possible over how they experience the animation. This can take one or both of two forms. The first allows them to start and stop the animation and to jump to any point they wish. Commercial platforms such as YouTube or Vimeo automatically provide this functionality for videos uploaded to their sites. But animated GIFs generally do not support this, and videos embedded directly into web pages do not necessarily support this unless programmed explicitly to do so. The second form embeds "now" in a context that helps viewers see both what has recently happened and what is about to occur. This will be illustrated shortly.

We see this control issue play out in figure 52.3, in which a nicely nuanced reading of hypermetrical irregularity in a piece by Béla Bartók is marred by its execution as an animation. (This is not the fault of the article's author. The animation and its well-intentioned design were suggested by the journal's editor—me!) As the music plays, solid arrows representing "present" metrical hearing unfold in sync with the music. So far, so good. Once the animation reaches the arrowhead signaling the next hyperbeat, a dashed arrow representing a continuation that is expected but that will be thwarted suddenly appears, and a new solid arrow representing a new actual in-progress hearing unfurls. While the unfurling of the analysis in sync with the music seemed like a reasonable design, the sudden appearance of each new visual element comes as a surprise. The brain interprets the new element as a (minor!) threat and takes a moment to assess its safety before it can figure out what it means. Meanwhile, the music has continued to play, and additional visual elements have appeared. Even in this simple animation, there is simply too much information to take in.

52.3 Screenshot from Gretchen Horlacher, "Bartók's 'Change of Time': Coming Unfixed," *Music Theory Online* 7, no. 1 (2001): ex. 2, 0:05, https://mtosmt.org/issues/mto.01.7.1/horlacher_examples.php?id=1.

One of the beauties of visualized music is the ability to see how it unfolds outside of the constraints of time, so forcing it back into time constitutes a step backward. In this case, a simple improvement would have helped: to show the entire analysis in gray from the beginning, then to allow the analysis to play out in black just as it does in the unanimated original. In this way, viewers could see what is to come and adjust their expectation so they could listen ahead, guided by the analysis. Listeners would still have the opportunity to listen critically, but their eyes and ears wouldn't *both* be trying to process new information in real time.

We have already discussed Stephen Malinowski's remarkable animated score to Igor Stravinsky's *Rite of Spring* (fig. 28.5, colorplate 9). Malinowski's YouTube channel currently has about a thousand similar animated graphical scores dating back to 2006 (see also https://musanim.com/). While the earliest of these feature now-primitive-sounding synthesized scores and are as visually interesting as a piano roll (see chap. 28), more recent ones are often tied to recorded performances or high-quality synthesized renderings and feature both visual beauty and insightful musical interpretations. The animations enhance the already-effective metaphors that shape traditional Western notation in rich ways. They often literally draw the lines we hear, and at least some animations also reinforce characteristics of the music by mapping visual features such as size, shape, color, highlighting, and connective lines to musical features such as pitch, instrumentation, duration, articulation, and line, drawing attention to features that are particularly pertinent to the work.

The collection is full of delights and well worth visiting. Online figure 52.4 provides screenshots from four contrasting videos by Malinowski. (The fourth of those videos is from a fascinating, immersive iPad app that accompanies the album *Biophilia* by Björk. For each of the album's ten songs, it offers a game that allows the viewer to interact with the song's musical materials, a scrolling score, lyrics, some analytic insights by music theorist Nicola Dibben, and an animated graphical score by Malinowski.)

Besides their high aesthetic qualities, the animations adhere to the two principles outlined above: as YouTube videos, they allow the viewer to pause and rewind, and they both show what has already happened and foreshadow what is to come. The scrolling feature means that the currently sounding music is always in the center. After that moment passes, it remains visible for several seconds as it scrolls toward the left edge. This provides an opportunity to establish a visual memory corresponding to the audio memory we form when listening. Even more importantly, it provides the opportunity to see several seconds ahead. This permits one to see how long an existing pattern will continue or to prepare to attend to something new about to happen. This allows for an enhanced emotional response to the music. To oversimplify Huron (2006), as we listen to music, we experience two kinds of expectations: one based on prior specific knowledge (of a piece, of motives or themes we've already heard, etc.) and one based on our grammatical expectations of the style. Those expectations are continually validated or thwarted, triggering different localized emotional responses. Adding scrolling animations to the music enhances this effect because it provides another source of expectation. We can *see* what we are about to hear, so when we hear it, the confirmation of our expectation provides an extra element of delight that isn't present when we're merely listening.

These visualizations are both more interesting and more informative about the music than videos featuring simply a recording and a scrolling score, countless examples of which can be found online. Sometimes, however, a score synchronized to music is exactly what one is looking for, and this approach can be quite helpful. Online figure 52.5 offers a screenshot from a video in which a 1970 graphical transcription of György Ligeti's 1956 electronic music piece *Artikulation* accompanies the original recording. Rather than scrolling, the transcription appears in pages, while a vertical line shows the current location in the audio. The transcription is already an especially effective listening guide for many of the same reasons the Malinowski videos and indeed all good music visualizations are: it reveals structure, patterns, and difference. Tying the transcription to the recording makes both all the more accessible. The combination of fixed image and scrubber works almost as well as the Malinowski visualizations, except that when the page turns, we lose what has just occurred, and just before the end of the page, we do not know what is about to happen. In both cases, however, even seeing that something new is about to happen typically does not undermine the sense of delight we experience when we hear the new material.

One can also use the approach of online figure 52.5 in presenting an analysis, as in figure 52.6. This is a screenshot from a simple animation that shows the structure of Wilco's song "Poor Places" and provides a brief description of each section as the music plays. The screenshot captures the point when the section marker glides from verse 3 to chorus and the descriptive text cross-fades. This approach can work with text or images and might involve revealing elements one by one. When taking this approach, one must remember how the viewer's visual system works, as described in chapter 1. In addition to the pop-out effects that we have seen at work in all of the still images (color, shape, and so on), animations can also take advantage of *appearance* and *relative motion*, which are among the most elemental pop-out characteristics of our lowest-level visual processing system.

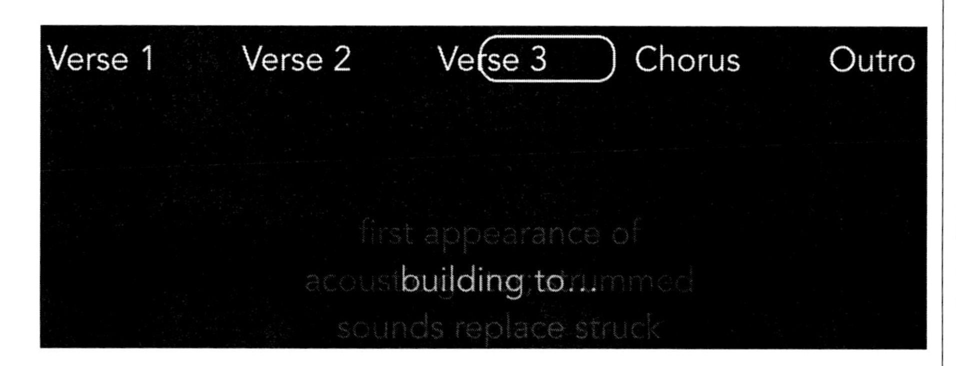

52.6 Screenshot from Steven Rings, "Music's Stubborn Enchantments (and Music Theory's)," *Music Theory Online* 24, no. 1 (2018): slide 1, 0:26, https://mtosmt.org/issues/mto.18.24.1/rings_examples.php?id=0.

The animation in online figure 52.7 illustrates performance timing in a Maria Callas performance of "O Mio Babbino Caro," from Giacomo Puccini's opera *Gianni Schicchi*. The vertical bar scrolls across the image as the performance plays. It does not diminish the listening experience in any way to know in advance that Callas is going to slow dramatically at three moments. The image also benefits from clean design: the grid is gray and sparse. The smoothed tempo line matches the gray gridlines in intensity, but its rounded contour and contrasting color (light red in the original) make it distinct. Points indicating beats appear in the same color but with high saturation so they pop off the screen.

The ability to look ahead may be less important in animations that seek to convey a process that is easy to comprehend. The images in online figure 52.8 come from two animations that trace voice leading in Frédéric Chopin's *Prelude in E*

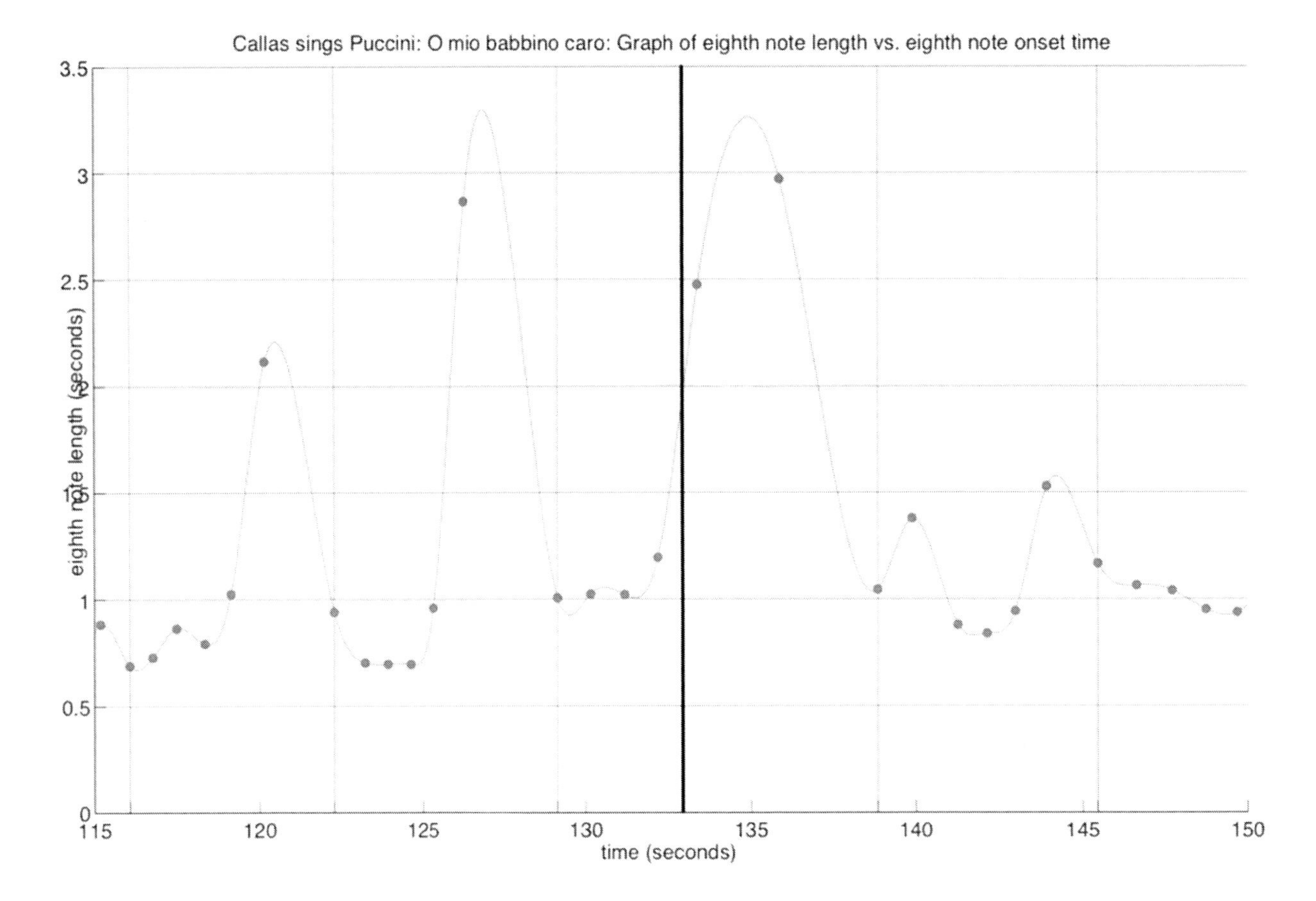

52.7 Screenshot from Elaine Chew, "Tipping Point Example 1—Maria Callas—O Mio Babbino Caro—Giacomo Puccini part 2b," Vimeo, 0:18, accessed November 1, 2020, https://vimeo.com/70618222.

Minor, a piece that famously features almost entirely half-step motion among its voices. The first image tracks the four pitches of each chord one by one on a pitch wheel as the music plays. The smooth voice leading is quite obvious during listening. The second image divides each chord into two pairs of pitches and traces the smooth voice leading through a pitch space consisting of dyads in which the first pitch of each pair changes by semitone when moving southwest to northeast and the other pitch changes by semitone when moving northeast to southwest. In these animations, the primary point is not to track the moment-to-moment changes but rather to experience the overall effect of parsimonious voice leading, which becomes clear during listening. On the other hand, a clearer picture of the piece as a whole would emerge if the animation maintained a trace of the path the two dyad markers (located at (9 3) and (6 11) in the screenshot) followed.

The animation in online figure 52.9 is similar, tracing musical activity through an abstract pitch space. The space involves perfect fifths that are linked into three color-coded squares related by minor third, forming the three (0134679T) octatonic collections. Triangles at each corner are related by major third, forming the four (014589) hexatonic collections. (The combinations of minor thirds, major thirds, and perfect fifths recalls the Tonnetz; see chap. 17.) Highly contrasting visual elements represent three textural elements: Sustained perfect fifth dyads are represented by large stars whose colors are a blend of those of the two octatonic collections they operate within. Descending pizzicato lines, which remain within an octatonic collection, are represented by solid balls whose color matches that of the corresponding octatonic square. Finally, an arco violin melody is represented by a black circle. As the music plays, the pitches move about the pitch space, entering and leaving the space as the phrases they belong to begin and end.

The animation effectively represents a theory of the musical activity. Nevertheless, the application of some design principles found in this book would improve it. For example, the animation would benefit from additional information labeling the elements of pitch space itself. Additionally, it would be enhanced if information from the accompanying narrative made its way onto the screen. Adding the music notation as an additional layer in the animation would link two visualizations. Finally, the layout of the wire-frame representation does not make it easy to follow the animation's primary story, which involves motion within and between the three octatonic collections. As the balls leap from node to node, they become disconnected not just from the octatonic square they are associated with but from the entire conceptual space. The color-coding is not sufficient to counter the effects of the motion, so the events become unanchored from their pitch space. A design in which the three colored squares did not overlap on the two-dimensional field of vision would allow the animation's story to speak with greater clarity.

The technical challenges in creating animations prevent many from even considering using them in the first place. But even those with the necessary skill set will benefit from knowing when the payoff is worth the effort. The chief benefit of animation is the ability to show change over time. As we have seen throughout the book, however, still images can also effectively show change over time, through mapping along the horizontal axis, through small multiples, and by other means. And viewers have more autonomy in how they take in still musical images. An animation potentially takes some of that control away from them. Nevertheless, there are situations in which an animation provides value that justifies this loss of control. The most obvious is when a visualization is tied to a performance of the music. Another is when the activity mapped by the image is more fine-grained than can be captured in a small-multiple format, which cannot contain more than a large handful of snapshots before becoming unwieldy. A third category comprises visualizations that want to convey more information than can be represented on an *x-y* grid in which time occupies the horizontal axis. If an animated format proves the best, one must keep the viewer experience at the forefront, not just in the visual design but also as one empowered to control the viewing experience.

* * *

Part 5, the book's largest, has focused on visualization as a tool for communicating music analysis. After looking at the common task of score annotation, we considered several common music analytic tasks. Many of these built on types of visualization tasks covered in more abstract ways earlier in the book, with the added imperative of illuminating the structure of musical works (or collections of works). Whether an image is still or moving, we have seen that a failure to leverage principles of effective visualization can mask otherwise keen analytic insights. We have also seen that no amount of design skill will make weak content come to life.

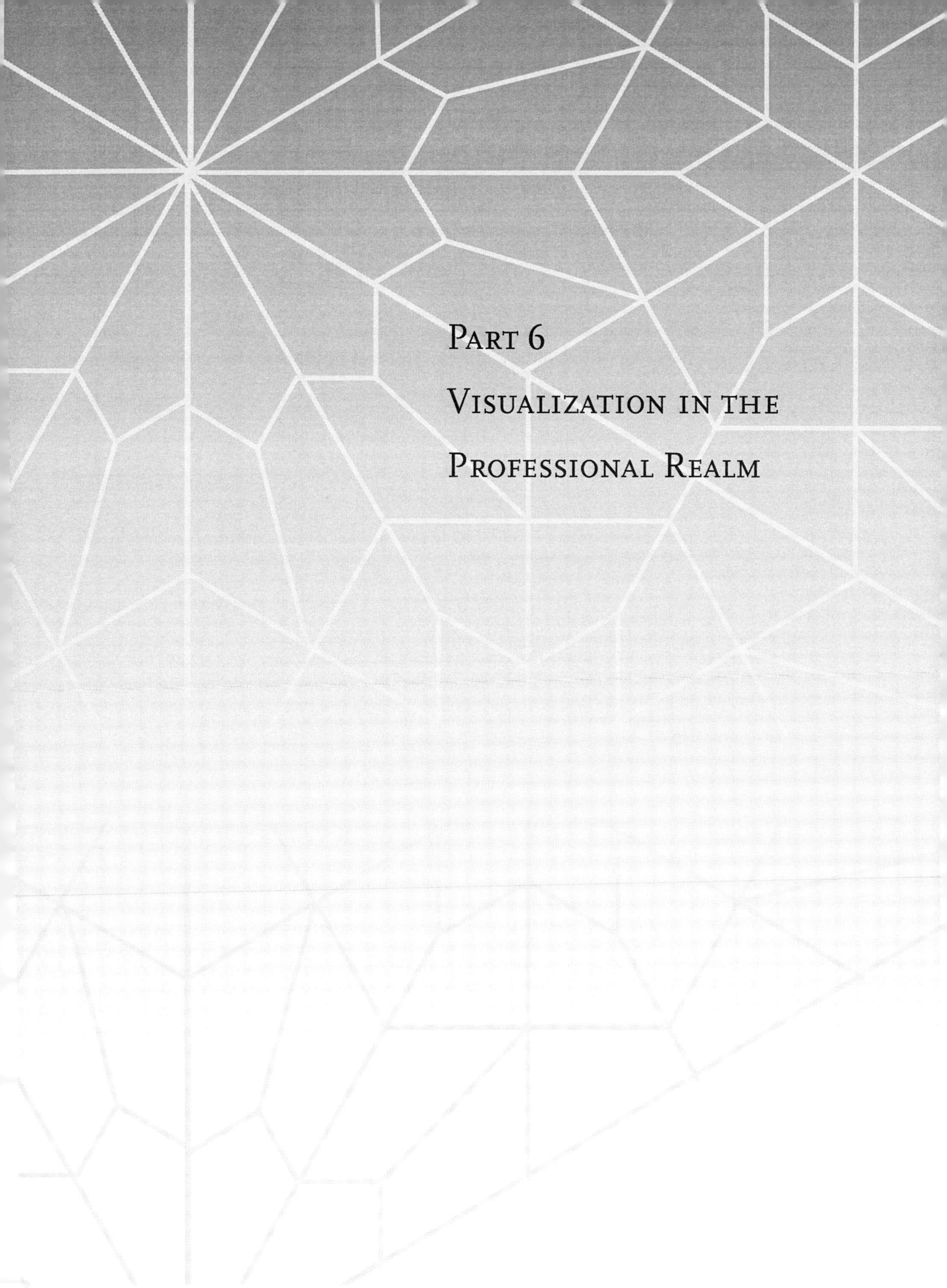

Part 6

Visualization in the Professional Realm

Now that we have looked at ways to create effective music visualizations, we conclude by addressing their professional use. This final part of the book is aimed at the academic world of professional presentations and publication. The first three chapters (53–55) address professional conference presentations. Chapter 56 addresses professional publication, a topic that will be of at least as much interest to editors as it is to authors. The final chapter (57) discusses software (and a little hardware) that supports the creation of beautiful musical visualizations. I have styled these chapters more as checklists than as prose narrative. My intention is to make it easier to refer to the principles each time a presentation of each type is designed.

Presenting a paper at a professional conference is a common way to disseminate one's research. These conference papers commonly include visual aids in some form to help those in attendance understand the topic of the paper. While historically these aids have been provided in the form of a printed handout that includes figures and musical examples similar to those one expects to present in a published article, increasingly, presentations are accompanied by a projected slideshow using PowerPoint, Keynote, or other software (see chap. 57). The use of presentation software has a unique set of considerations. A third type of presentation, the poster session, is an entirely different animal. While the principles of effective visualization apply to the design of all three, each format presents special challenges, which should be taken into consideration.

There are pros and cons to all three types of presentation. As the table below shows, a one-page handout can show 8.6 times as much information as a projected slide, while a typical poster has the same resolution as over three hundred projected screens of information. Of course, resolution also needs to take into account the distance of the viewer. The handout will be viewed at arm's length, while information on a projector screen needs to be visible from the back of a conference room, perhaps while projected on a too-small screen that may be viewed at an angle. A poster will also be viewed from several feet away, and its content cannot be spread across multiple pages or screens, so its resolution advantage is not as great as the figure might imply. Each chapter will address the pros and cons of the relevant format.

Format	Size	Total Bits	Scale
Typical projectors	1024 × 768	0.78M	1.0
Common laptops	1366 × 768	1.05M	1.3
MacBook Pro	1920 × 1600	3.07M	3.9
Paper (300 dpi)	8½" × 11" (½" margins)	6.75M	8.6
Poster (300 dpi)	4' × 5' (½' margins)	240.12M	305.3

The next three chapters each begin with a brief summary of the general principles involved with the presentation type. A set of numbered points addressing the special design considerations for that format follows. The chapters deal with overall layout considerations. It is assumed that the content is already beautiful, in keeping with the practices outlined to this point in the book. Whatever the format, the first consideration in design must always be the viewer's experience.

Chapter 53 Conference Handouts

While the conference handout is quickly giving way to the presentation slideshow, it remains an important way of providing information that supplements your oral presentation. It should flow seamlessly with your presentation. This chapter is relevant both to printed handouts at an in-person conference and those shared online in PDF form as part of a live or online conference.

Preparing Your Handout

1. Legal-sized paper (8.5" × 14") has 30 percent more capacity than letter-sized paper, and sometimes this extra space is helpful, particularly for very long examples. It is more awkward for the user, however, and is better avoided unless you have a compelling reason to use it.
2. At the top of the first page, provide the presentation title, your name, your institutional affiliation, the conference name, and the presentation date. This is all important bibliographic information should someone be interested in following up with you later or citing your presentation years from now. Including your email address is also a good idea.
3. Use a running header with your name; the paper's title, perhaps in shortened form; and, above all, page numbers. That way, if your pages get separated, it will still be clear where they came from.

4. Keep the complexity of each image in line with the time you plan to spend discussing it. Your viewers will be looking at the handout while they listen, so if you provide a highly detailed Schenkerian diagram and then only say, "See figure 3 where there is a five-line hidden in the inner voice," you will be moving on to the next thing before your audience finds figure 3. In general, however, dense examples are better avoided because they present special accessibility difficulties.
5. For images with any complexity, leverage pop-out effects (see chap. 1) to allow the viewer to find what you are talking about without having to hunt for the information.
6. The point above is especially true when you include score excerpts. Annotate these to show your main points. Your annotations will guide your audience's attention as you describe it, reinforce the point of your talk, and make it more likely that they will remember what was important to you when they refer back to it later. See chapter 41.
7. Be mindful of the size of the figures. Shrinking figures so they fit on fewer pages is eco-friendly and cost-effective, but if people can't follow figures when the handout is held at a comfortable height, you're doing a disservice to your talk and to those attending it.
8. Keep the figures in order. It is sometimes tempting to rearrange things to save printing costs, but doing so makes it harder on the audience, not to mention distracting during the talk, because of the need to flip back and forth to figure out where the errant figure is and then get back to the flow. Avoid this if at all possible.
9. Number your items sequentially with a single prefix throughout (*figure* is a good, generic option) or, better, without any prefix at all.
10. Give each figure a clear and concise caption and make the figure reference bold so it pops off the paper. See the figures in this book for examples.
11. If the example comes from another source, provide a full bibliographic citation.
12. Print double-sided to save paper.
13. Staple in the top-left corner according to the orientation of the pages.
14. Before you make copies, look through your handout, imagining it as your audience will experience it. This perspective may lead you to make changes.

During the Presentation

1. References to figures can be brief; "please see figure 2 on page 4 of your handout" is too much. "See figure 2" is plenty (or "see 5," if you've dispensed with the word *figure* altogether). Since you have already put the figures in order, your audience is aware that the thing in their hands is probably your handout, and they know how to turn pages.
2. Ask your audience to look at a figure *before* you start talking about it. Don't go on at length about something and then tag on "see figure 4."
3. Even if your images are designed effectively, make sure that your viewers always know where to look.
4. If you include a long quote that you want people to actually think about, either read it to them or give them time to read it. People can process symbols and words in parallel to some extent, but the language portion of the brain cannot multitask, so if you're asking your audience to read something and listen to it to you talk about something else at the same time, one of these modalities will lose out.

A well-designed and effectively laid-out handout will not make your talk better. That's for the content. But a poorly designed handout can undermine the effectiveness of even the most brilliant conference paper.

CHAPTER 54 Presentation Slideshows

A presentation slideshow serves precisely the same purpose as a conference handout: it supplements your spoken word with images that illustrate and illuminate your points. The potential benefits of using projected slides rather than a handout are clear: a slideshow saves paper, gives you control over the sequence and timing of your audience's attention, and allows for easy playing of audio and video. It can introduce a more performative quality to your presentation and (let's admit it) can be tweaked right up to the time of presentation.

The downsides include the fact that the audience cannot take the slides home with them (unless they are printed out and shared, a dreadful way to review a presentation, or posted online). Most significant, however, is the fact that a screen projection has about the same information capacity as a 3"×5" index card. In addition, you cannot know in advance how big the screen will be or how far it will be from the people at the back of the conference room, who have the same expectation of visibility as those in the front row.

A projected slideshow must therefore be thought of entirely differently than a handout. Not knowing how to design an effective slideshow leads to all kinds of intellectual woes (and others, including the *Challenger* space shuttle disaster, as is summarized in Edward Tufte's [2006, 156] chapter "The Cognitive Style of PowerPoint: Pitching Out Corrupts Within," which is also available for individual purchase at https://www.edwardtufte.com/tufte/books_pp).

A list of best practices and design considerations for creating a slideshow follows.

Preparing Your Slides

1. Avoid using both a projected presentation and a handout. This is especially true if they have similar content. People will look at the handout because it's easier to see. If you must do this, carefully script how you will direct your audience's attention between them.
2. Choose a theme with a plain background. To maximize visibility, ensure that the text and background colors are highly contrasting. Against a white background, black, dark gray, deep blue, or deep green are good choices for text. Against a black background, white or bright yellow are good options.
3. Use the same font set throughout the presentation. Avoid decorative fonts, script fonts, and monospaced fonts. Sans serif fonts that are not "narrow" or "light" will be most legible.
4. To maximize the amount of space available, avoid headers and footers, particularly running headers and footers that credit your institution.
5. Do not fill your slides with text parceled out bullet point by bullet point and then proceed to read it. Even worse is putting up page upon page of bullet points and then narrating a heterophonic gloss over them. Your audience will be trying to simultaneously read and listen to text that is similar but not the same, getting the gist of neither.
6. If you do have a text-heavy screen, for an extended quote, for example, either read it for your audience or plan to stop talking to give them time to do so. Don't display a block of text and then talk over it. See the previous point.
7. Here is the main point: slides should be primarily graphical. They can feature musical examples, photographs, charts, diagrams, or videos. Slides should be used to show things that cannot be as easily said.
8. If displaying music notation, be mindful that typically only four to six measures and no more than about four staves will be legible, and even that is pushing it. If you have a larger musical example, you may need to break it into pieces.
9. Think carefully about how your audience will best understand your images. Sometimes the point of a slide will be self-evident, but sometimes, particularly in musical examples or detailed charts, you may need to take extra action. Rather than simply describing where to look ("see the second violin part in measure 4"), use the presentation software's annotation tools to highlight where you want people to focus. Review the points of chapter 41

and take particular advantage of the pop-out effects provided by color and shape. Gentle motion effects can help direct attention, with *gentle* being the operative word. This is better than keeping a laser pointer and shooting a squiggling point at the screen for a second. Anyone whose attention is momentarily diverted may miss the point(er) altogether.

10. If you need to highlight multiple features, do so one feature at a time. For instance, allow previously highlighted items to recede by changing them to a more neutral color on a subsequent slide and introducing the new item in a bold color.
11. Avoid animations whose only value is amusement, unless amusement serves a point.
12. When your scintillating narrative is sufficient, insert a blank slide to avoid distraction.

Preparing for and Delivering the Presentation

1. Test your presentation by standing several feet from your computer screen. Make sure you can see all the detail you want your audience to see. Better, find a classroom with a projector, sit at the back of the room while someone advances your slides (or use a remote), and see what you can see.
2. If you are reading from a script, make sure you insert cues precisely when you want to advance the slide or trigger a programmed animation.
3. If you plan to read your script from the notes area of the presentation software, be sure you know well in advance that the physical arrangement of the conference site will allow for this.
4. Rehearse the presentation while imagining that you are in the audience. Insert pauses in your script at moments when the audience might need a little time to process what they are seeing.

The conference slideshow has increasingly become the norm. While it has many of the same aims as the handout in playing a supporting role for your spoken word, from a design perspective it is a whole other medium.

CHAPTER 55 Conference Posters

An excellent conference poster is fundamentally similar to its more popular cousin, the conference paper: it grabs its audience's attention, concisely makes a main point, provides clear supporting evidence for that claim in a well-organized narrative, inspires thoughtful discussion, and leaves a lasting, positive impression. A poster, however, must convey these elements visually rather than verbally.

When preparing to participate in a poster session, do not try to produce a *poster*; make a *presentation*. A poster may be a physical realization of your presentation, but your purpose is to present your ideas, both in visual form and in your conversations with those who stop to discuss your work. You must therefore get your ideas into shape before you start designing the poster itself.

Your presentation needs to convey three things: What question did you set out to answer? What did you learn that you want your audience to understand? And what information will demonstrate how you got from your research question to your conclusions? These three pieces of information need to appear on the poster.

The relationship between presenter and audience is different for a poster presentation than for a traditional paper. For traditional papers, the session title, paper title, abstract, and sometimes the presenter's name are the main draws. By the time you stand up, those present have already decided to engage with your paper. At poster sessions, attendees more often simply wander in and wander around. Some will have particular posters in mind, and some will purposefully stop by every poster, but the "glance and go" is also common. Increasing traffic to your poster thus requires a bit of a sales job. Below, I outline four stages in getting to the poster session: preparing the content, designing the poster, producing the poster, and finally presenting the poster.

Preparing Your Content

1. The content of your poster should contain three main elements: its purpose, the conclusions, and the intellectual path that leads from one to the other. Your *purpose* is a research question that conveys concisely what your presentation is about. It should grab attention and inspire visitors to linger. Your *conclusion* should summarize the points you most want visitors to take away from your presentation. The supporting information should focus only on what is most important. Emphasize your findings and the big picture of how you got there. Focus on key processes or methodological points, illustrative examples, and high-level summary data. Your poster should be self-sufficient, but it doesn't need to be comprehensive.
2. To the extent possible, plan to tell your story through pictures rather than text. Graphical images, which can include brief text annotations, will almost always be more powerful than text paragraphs or sets of bullet points in a poster presentation. These images might take many forms, including musical examples, charts and tables, photographs, data representations, schematic diagrams, and so on. Images should be rich with information and designed according to best practices outlined in this book.

Designing Your Poster

1. Determine the permitted poster size, orientation, and mounting plan. Heights of three to four feet and widths of four to six feet are typical, and you may be required to orient your poster vertically or (more commonly) horizontally. Conference sites may provide easels (requiring you to have your poster mounted on a rigid surface), large corkboards with pushpins, or tape or adhesive putty that can attach your poster to a wall or other hard surface.
2. Think carefully about your layout. Your poster design will undergo several stages of successive refinement and experimentation with alternatives. To a large extent, the amount, shape, and representation of your information will determine its optimal arrangement on the poster. Most posters have discrete components. It can be helpful to experiment with layouts using pencil and paper or slips of paper you move around on the table, though

software like Adobe Illustrator or Adobe XD can serve much the same purpose.

3. Some content invites (or rewards) a more radical layout. Don't be afraid to do something bold, so long as effective visual communication drives your decision.
4. In most cases, you can express the information sequentially. In planning the content, remember that people's tendency when reading large, printed items is to go top to bottom, then left to right, as with columns in a newspaper. Typical arrangements involve two to four columns, with your purpose at the upper left and conclusions at lower right. Other layouts are possible. For example, if you are looking at the same piece of information from different perspectives, you can place that information in the center and reference it with commentary around the periphery.
5. Sketch out your layout.
6. Review your title to make sure it is short and compelling. It should be only as long as necessary to make the topic of your presentation clear. Finding the sweet spot between brief and informative is tricky, with informative being more important.

Producing Your Poster

1. It is most convenient to create your poster on the computer and print it using a large-format printer. Most commercial printing companies can print posters at a somewhat hefty cost. Many campuses have plotters on which you can create a high-quality, full-color poster more economically. A printed poster can be easily transported in a cardboard tube.
2. Ensure that there is sufficient separation between pieces of content to make the poster's information structure clear. You can do this purely through white space that separates columns and rows. You can also use a dark background for the poster, then put the content in white boxes with rounded corners and a thin gray border. In any case, the content should always feature dark text on a light background. This is generally more legible than light text on a dark background and will use far less ink.
3. Fonts should be simple. Avoid script or highly characteristic fonts that only draw attention to themselves, except in those rare cases in which the font choice truly enhances your presentation. An ordinary sans serif font is best.

4. The title font should be legible from at least twenty feet away. A good starting place is a 144-point font size, though this will depend on the length of your title and the characteristics of the chosen font. Include a subtitle only if it helps clarify your focus, since it will encroach on your content area. Make it visually inferior by using a smaller font size and possibly a lighter color for the text. Include your name and school affiliation as well. These can go either below your title or, if your title is short enough, in a box in the upper right corner.
5. If the rest of your content divides into clear sections, consider using a background of a different color for each. Bold colors will overwhelm the content, so instead use light, complementary colors or two shades of the same color.
6. If the ordering is not obvious, draw attention to the desired flow. Common options include nesting regions into larger groups, color-coding rows or columns, using arrows, or numbering boxes.
7. Structure your story with information-rich headers that harmonize with your title and regular text fonts, using either a smaller version of your title font or a larger version of your main text font.
8. For text longer than a header or short phrases, choose an open font with nondramatic serifs that is easily legible from at least six feet. Some visitors will come closer, but you don't want to exclude the shy. Something in the 24- to 30-point range generally works well.

Presenting Your Poster

1. Don't make your poster do all your work for you. Make eye contact with and smile at passersby so they feel welcome to engage with you.
2. Conferences sometimes schedule time for each presenter to make a short pitch of one to two minutes at the start of the poster session. Given the short amount of time, this pitch should be carefully planned and either read or largely memorized. It should introduce the general topic, describe the content area (for example, the particular repertoire you are dealing with), and end with your central finding. If there is additional time, you can insert a sentence of two about your general methodology. Details are not important. Your principal aim is to convince those listening to come hear more.

3. During the rest of the session, have a five- to ten-second pitch ready to draw people in and a more robust thirty-second "elevator pitch" to convey the essence of your presentation.
4. Be prepared to elaborate with supporting narrative during more extended discussions with those who are particularly drawn to your topic.
5. Don't be afraid to ask those who stop to chat for their reactions. One of the advantages of the poster presentation is that it is interactive, making it possible to get more engaging feedback than in a paper presentation.
6. Consider making a small handout available (even a half or quarter sheet) with a link to your poster or associated website, perhaps with a QR code for easy access.

The poster presentation demands the same level of scholarly rigor as the conference paper. The time it takes to design an exceptional poster is not all that different from the time it takes to write a twenty-minute paper. Don't settle for your first conception for the poster. Each topic invites its own best solution, and experimenting until you find that solution often brings rewards.

Chapter 55 Print Publication

Authors generally have only minimal control over how their articles or books will appear when published. Publishers appropriately have house styles that ensure a level of uniformity among their publications and also simplify the production process. Nevertheless, some aspects of publication detract from effective visual communication and are done simply out of habit or because alternatives haven't been considered. In this chapter, I advocate for three changes to common practice. These recommendations pertain to varying extents to presses and editors, as well as to authors deciding where to submit work for publication. In writing this, my publisher, editor, and I are all mindful that this book violates each of these principles!

First, do everything possible to keep images and the text that discusses them visible at the same time. This is an extension of the principles discussed in chapter 7, which advocates for richly integrating text and images. It is also a direct application of the general design principle to put information at the point of need. For print publications, if readers need to continually flip back and forth to relate text to image, there is a good chance they will stop flipping and lose the impact of both text and image. Take considerable care in page layout to ensure that an image and the text discussing it are on the same or facing pages. For image-heavy books, this might mean choosing a larger page size. It might also mean repeating all or part of an image so it will still be visible after a page turn. (Tufte's books have elevated this principle to high art.) For electronic publications, it means ensuring that readers can keep an image visible the entire time they are reading the associated text.

Because screen resolutions vary and are often considerably lower than print resolution, this task requires thoughtful web design.

Second, provide full support for color. Electronic formats have a built-in advantage since adding color is free. For print media, while you can design effective images in grayscale, color adds a dimension that sometimes makes the effort worth the cost. Color printing is roughly quadruple the cost of grayscale printing, but other costs (editing, layout, binding, marketing) remain unchanged. Relegating color images to dedicated colorplates or to an online supplement as in this book means the images are not adjacent to their discussion (see the first point). As authors come to insist on including color in their images, print-oriented presses will feel increasing pressure to support them or risk losing their pipeline of manuscripts.

Third, simplify image referencing. You do not need separate numbering schemes for figures, examples, tables, and equations. While it is probably impractical to eliminate numbers altogether (as Tufte does in his books), make the referencing system as concise as possible. Replacing fussy references like "figure 45.1" (or worse, "figure 4.6.2") with simple sequential numbering and using boldface as a marker for "figure reference" ("**223**") would save ink, space, and time with no loss of clarity. This is a good time to remind authors and editors that references to images should come at the beginning of the discussion of that image, not at the end ("Oh, by the way, all the things I've just spent four paragraphs discussing are illustrated in figure 5").

These changes will showcase your exceptional images to full advantage.

Chapter 55 The Essential Visualization Toolbox

As we reach this final chapter, some readers may be wondering what resources are required for creating beautiful music visualizations. While some of the kinds of visualization included in this book require more specialized software, the vast majority could be created with some combination of a drawing program, music notation software, and an image-editing program. This chapter summarizes the essential contents of a visualization toolbox. There are countless programs in each of the categories listed below, and I make no attempt to be comprehensive. I also do not cover certain additional, more specialized categories of tools that can be used to create visualizations (including digital audio workstations).

Drawing

The workhorse of the visualization toolbox is the drawing program. Such programs can create standalone visualizations, annotate (or even edit) files created by a music notation program, and annotate scanned images (such as scores).

Much standard office software includes rudimentary drawing tools. This software includes word processing programs such as Word (Microsoft) and Pages (Apple), whose drawing capabilities are quite limited and unwieldy. Word processors are generally effective at creating tables, however.

Presentation programs such as PowerPoint (Microsoft), Keynote (Apple), and Slides (Google) are somewhat better suited to the task and can be used for simple drawing tasks. Preview (Mac) also has some rudimentary markup tools that can directly edit images.

For data-derived visualizations, spreadsheet applications have built-in tools to generate charts of many kinds. The most common are Excel (Microsoft), Numbers (Apple), and Sheets (Google). You can customize these charts in sophisticated ways. In general, the default settings for a chart are rarely the most effective.

For images of any complexity, however, a dedicated drawing program works better. Among the best options in this category are Adobe Illustrator (used for all but a handful of the redrawings in this book), CorelDRAW, and the more affordable Affinity Designer. In addition to providing precise control over the appearance and positioning of objects, their support of layers allows them to manage large numbers of drawing objects. You can import a scanned image and add annotations. Additionally, you can import a PDF generated by music notation software (see below) and directly manipulate the objects, changing features such as color and positioning. You can also produce basic music notation manually, provided you have a music font installed. (See the following section.) These programs are also effective at producing posters (see chap. 55).

These programs come with a learning curve, and most tasks require only a fraction of their capabilities, but the effort is often rewarded.

Music Notation

If you just need to annotate an existing score, you can scan a published version and manipulate it as described above. If you need more customized music notation, and if that notation is more complex than a drawing program can handle (for instance, if an image needs beams, ties, slurs, and other complex shapes), use a dedicated music notation program. Most notation programs are designed with composers and arrangers in mind, and they are less amenable to the kinds of things music scholars often do. And they all come with substantial learning curves. The venerable programs Finale (MakeMusic) and Sibelius (Avid) are a bit overpowered and expensive if you are paying for them out of pocket. The newer programs Notion 6 and Dorico are getting positive reviews. The open-source MuseScore is free and produces very good notation. Finally, LilyPond is another free program, though it uses a text rather than a graphical interface. It is more like creating an HTML or XML file by hand.

From any of these programs, you can print to a PDF file, which can be opened and edited in a vector drawing program such as Illustrator. Note that for that to work, the music font (or fonts) used by the notation program needs to be installed on the computer with the drawing program. Many of the font libraries are available for free download on the developers' websites.

Image Editing (plus Scanners)

Some visualizations require reproduction of existing images, often score excerpts. This requires the ability to (1) scan an image, (2) rotate it so it is straight, (3) crop the scan so it contains only the information desired, and (4) adjust the colors to reduce or remove artifacts, including bleed-through from the opposite page, stray blemishes, and the general faint grayness that many printed pages have.

For step 1, many offices and libraries have good-quality scanners. For home use, for under $100 you can purchase a high-resolution scanner powered only by a USB cable. Look for resolutions of at least 300 dots per inch and scanning that supports at least 256 shades of gray and millions of colors. It is best to scan in TIF format, a lossless format that can always be downscaled later.

Step 2 is not always necessary, but it is difficult to precisely align a document on a scanner bed, and even seemingly perfect scans may be off by a tenth of a degree, which becomes noticeable after annotations are added. Unfortunately, the Photos apps that come standard on Macs and Windows computers do not allow precise rotation, nor does the Mac Preview app, which handles steps 3 and 4 reasonably well. Photoshop, Adobe's high-end photo editor, allows for arbitrary rotation, and since it also allows the creation of horizontal or vertical guidelines to measure the scan against, you can use it to perfectly deskew an image.

Step 3 (cropping) can be done in Photoshop, as well as Mac Preview. The former offers better control. For step 4 (eliminating background grays), Mac Preview offers basic global adjustments, but Photoshop allows for greater precision. And its brush tool allows you to paint over blemishes on an image.

Presentations

For creating presentation slideshows, you can use the desktop programs PowerPoint (Microsoft) and Keynote (Apple). Cloud-based applications include Google Slides, Canva, and the funky Prezi. All of these programs are capable of producing both excellent and terrible slideshows. See chapter 54 for important design considerations in designing slideshows.

Animation

A detailed exploration of this topic is beyond the scope of this book. Some basic animation can be done in Keynote or PowerPoint. At the upper end, various tools from Adobe can create animations of great sophistication.

* * *

Part 6 of the book has provided some basic guidelines for employing music visualizations in various professional settings: the conference handout, the conference slideshow, the presentation poster, and the publication. There remains ample room for creativity in response to the nature of the task at hand, the target audience, and your own personality. As with the rest of the book, the key lesson to remember is that you should always design your presentations with the viewer's perspective foremost in your mind and reduce as many barriers to seeing, processing, and understanding as possible.

Epilogue

Music's power is a beautiful thing. Music is rich with meaning on deeply personal levels, and at the same time, it plays a central societal role in human cultures the world over. How important it is that communication *about* music try to capture and reflect that beauty. A reverence for music and its impact should suffuse any writing about music, whether in program notes, concert reviews, or scholarship about music. And, as I hope I have made clear in this book, it should also be reflected in musical images. Beautiful musical visualizations cannot replace the direct experience of music, any more than a professional photo of an amazing sunset can ever replace the experience of seeing the sky blaze across a rocky shoreline looking out over the ocean. The primary value of the lived experience does not undermine the power of an image, however. We spend more of our lives using our eyes, after all, than we do listening to music. In the introduction, I discussed the many kinds of information visualization that can play a role when deciding to buy a house. These images provide valuable enhancements to the experience of being physically in a house. Images about music play the same role.

This book has promoted viewer-centered design. An image that does not communicate to the viewer is not successful, no matter how insightful its content. Successful images leverage both humans' low-level perceptual and their higher-level cognitive apparatus. They recognize that the same automatic mechanisms that evolved to enable our prehistoric ancestors to survive a dangerous world also govern how we look at musical images. A blue circle drawn on a score may not be life-threatening, but the most primitive part of our visual system doesn't know that; the circle pops out of the visual field just as if it were a snake dropping from a jungle tree. Designing an image without keeping pop-out characteristics such as color, shape, orientation, intensity, and motion in mind is effectively like dressing up ideas in camouflage and sending them out to *not* be seen. Our higher-level cognitive apparatus recognizes patterns, and patterns of patterns, and it understands through analogy, which can range from side-by-side comparison to metaphors, including those relating to our physical experiences. This is what makes visualization such a powerful tool for communicating understanding.

Compelling images are rich with information, don't waste ink, are self-contained, and don't rely entirely on surrounding prose to explain them. They put different information layers in different visual layers, employ color effectively, and, above all, tell a story. These principles, drawn from Edward Tufte and vision science, are explored in detail in part 1. Employ them, and an image can come alive with fresh insights.

In part 2, we saw how these principles can play out in images depicting various musical spaces. There we saw the power of comparison. Placing a diatonic collection on a staff does not visually show its irregularity the way mapping it onto a chromatic collection does. We saw the benefits of information richness, along with the challenges that added information causes for design clarity. We saw the pros and cons of depicting the circularity of a modular space, which requires essentially an extra dimension. This is not so hard when the space itself is one-dimensional. The one-dimensional circle of fifths requires two dimensions to draw as a circle. But a two-dimensional modular Tonnetz needs a third dimension to show its circularity.

Part 3 explored the differences between measuring musical time internally (in terms of musical markers such as beats or measures) or externally (by the clock). There can be good reasons for doing it either way, but the decision has implications for the viewer's understanding. This issue comes to the fore when picturing musical proportion, though as we saw, images that seem to be about proportion don't always take advantage of the opportunity to actually depict musical proportions graphically.

Part 4, which focuses on nontemporal aspects of music, demonstrates that there aren't many areas untouched by the basic design principles outlined in part 1. The pedagogy of voice leading, schematic images (including of form), visualizations of procedures, pitch-class set tables, instrument ranges, and translations each came with special problems, with solutions readily informed by the principles articulated throughout the book. In exploring the strengths and weaknesses of various alternatives to common Western notation, we found that the notation system discussed favorably in chapter 13 sometimes works quite effectively and sometimes does not. When looking at the visualization of tuning and temperament, comparisons to the familiar equal temperament proved helpful. Likewise, notations for microtonal music benefited from modifying familiar notation rather creating entirely new symbols. And we found that graphical representations of timbre are challenging, perhaps owing to a lack of spatial metaphors, while representations of texture are easy and effective, perhaps owing to a surfeit of metaphors.

We turned the lens of those principles on music analytic images in part 5. Comparisons played an important role in thematic and contour analysis and in form. Metaphors played a key role in chronologies, rhythmic analysis, and symmetry. The layering of information proved important in depiction of tonal plans, serialism, and musical hierarchies. Storytelling was central in the discussion of tonal plans and formal analysis. And we found excellent uses for tabular design, small multiples, and especially color. In short, everything discussed in part 1 came back in part 5.

As we saw in part 6, even when we design for professional settings, such as conference handouts, slideshows, posters, or publications, we design with the viewer in mind.

The number of beautifully conceived images has increased each year over the nearly twenty years since I started thinking about how music *looks*. It is my hope that this book has inspired you to think more deeply about how you craft your own musical pictures. I look forward to seeing more clearly what you hear.

Bibliography

Adamowicz, Emily J. 2011. "Subjectivity and Structure in Milton Babbitt's *Philomel*." *Music Theory Online* 17, no. 2 (July). http://www.mtosmt.org/issues/mto.11.17.2/mto.11.17.2.adamowicz.html.

Adams, Kyle. 2009. "On the Metrical Techniques of Flow in Rap Music." *Music Theory Online* 15, no. 5 (October). http://www.mtosmt.org/issues/mto.09.15.5/mto.09.15.5.adams.html.

Albrechtsberger, Johann Georg. 1837. *J. G. Albrechtsberger's sämmtliche Schriften über Generalbaß, Harmonie-Lehre, und Tonsetzkunst, zum Selbstunterrichte*. Edited by Ignaz Seyfried. 2nd ed. Vol. 2. Vienna: Tobias Haslinger.

Aldwell, Edward, and Carl Schachter. 2003. *Harmony and Voice Leading*. Belmont, CA: Thomson/Schirmer.

Aron, Pietro. 1539. *Toscanello in musica*. Venice: M. Sessa.

Ashton, Anthony. 2003. *Harmonograph: A Visual Guide to the Mathematics of Music*. New York: Walker.

Bailey, Kathryn, ed. 1996. *Webern Studies*. New York: Cambridge University Press.

Baker, James M. 1993. "Chromaticism in Classical Music." In *Music Theory and the Exploration of the Past*, edited by Christopher Hatch and David W. Bernstein, 233–308. Chicago: University of Chicago Press.

Ballière de Laisement, Denis. 1764. *Thèorie de la musique*. Paris: Chez P. F. Didot le jeune.

Barwick, Linda. 2011. "Musical Form and Style in Murriny Patha *Djanba* Songs at Wadeye (Northern Territory, Australia)." In *Analytical and Cross-Cultural Studies in World Music*, edited by Michael Tenzer and John Roeder, 316–54. New York: Oxford University Press.

Bauyn, II, f. 12v, reduced. Prelude de Monseiur Couperin. In Paris, Bibliothèque nationale de France, Rés. Vm7 674–675, the Bauyn Manuscript, edited by Bruce Gustafson. New York: Broude Trust, 2014, 12v.

Bazayev, Inessa. 2018. "Scriabin's Atonal Problem." *Music Theory Online* 24, no. 1 (March). https://mtosmt.org/issues/mto.18.24.1/mto.18.24.1.bazayev.html.

Beach, David. 1979. "Pitch Structure and the Analytic Process in Atonal Music: An Interpretation of the Theory of Sets." *Music Theory Spectrum* 1:7–22.

Benadon, Fernando. 2007. "A Circular Plot for Rhythm Visualization and Analysis." *Music Theory Online* 13, no. 3 (September). http://www.mtosmt.org/issues/mto.07.13.3/mto.07.13.3.benadon.html.

Benjamin, William. 2006. "Mozart: Piano Concerto no. 17 in G Major, K. 453, Movement 1." In *Analytical Studies in World Music*, edited by Michael Tenzer, 332–76. New York: Oxford University Press.

Berendt, Joachim-Ernst. 1973. *Das Jazzbuch: von Rag bis Rock; Entwicklung, Elemente, Defenition des Jazz, Musiker, Sänger, Combos, Big Bands, Electric Jazz, Jazz-Rock der siebziger Jahre*. Frankfurt am Main: Fischer

———. 2009. *The Jazz Book from Ragtime to the 21st Century*. Edited by Günther Huesmann. 7th ed. Chicago: Lawrence Hill Books.

Bernard, Jonathan W. 1981. "Pitch/Register in the Music of Edgard Varèse." *Music Theory Spectrum* 3:1–25.

Bernick, Thomas. 1982. "Modal Digressions in the *Musicalische Exequien* of Heinrich Schütz." *Music Theory Spectrum* 4:51–65.

Bernstein, Zachary. 2018. "The Seam in Babbitt's Compositional Development: *Composition for Tenor and Six Instruments*, Its Precedents, and Its Consequences." *Perspectives of New Music* 56, no. 1 (Winter): 191–244.

Berry, Wallace. 1980. "On Structural Levels in Music." *Music Theory Spectrum* 2:19–45.

———. 1987. *Structural Functions in Music*. New York: Dover.

Besson, Mireille, and Daniele Schön. 2003. "Comparison between Language and Music." In *The Cognitive Neuroscience of Music*, edited by Isabelle Peretz and Robert J. Zatorre, 269–93. New York: Oxford University Press.

Blood, Brian. 2018. "Sounding Range of Orchestral Instruments." Music Theory Online: Score Formats. Last modified November 8, 2018. http://www.dolmetsch.com/musictheory26.htm.

Breckoff, Werner, Günter Kleinen, Werner Krützfeld, Werner S. Nicklis, Lutz Rössner, Wolfgang Rogge, and Helmut Segler, eds. 1971. *Musik Aktuell: Informationen, Dokumente, Aufgaben: Ein Musikbuch für die Sekundar- und Studienstufe*. Kassel: Bärenreiter.

Brinkman, Alexander R. 1980. "The Melodic Process in Johann Sebastian Bach's *Orgelbüchlein*." *Music Theory Spectrum* 2:46–73.

Brinkman, Alexander, and Martha Mesiti. 1991. "Graphic Modeling of Musical Structure." *International Computer Music Association Proceedings*: 53–56. http://hdl.handle.net/2027/spo.bbp2372.1991.009.

Broze, Yuri, and Daniel Shanahan. 2013. "A Cognitive Perspective: Diachronic Changes in Jazz Harmony." *Music Perception* 31, no. 1 (September): 32–45.

Brunner, David L. 1994. "Choral Program Design Structure and Symmetry." *Music Educators Journal* 80, no. 6 (May): 46–49. https://doi.org/10.2307/3398713.

Buchler, Michael. 2008. "Every Love but True Love: Unstable Relationships in Cole Porter's 'Love for Sale.'" In *PopMusicology*, edited by Christian Bielefeldt and Rolf Grossman, 184–200. Luneburg, Germany: Transcript.

———. 2016. "Licentious Harmony and Counterpoint in Porter's 'Love for Sale.'" In *A Cole Porter Companion*, edited by Don M. Randel, Matthew Shaftel, and Susan Forscher Weiss, 207–21. Urbana: University of Illinois Press.

Burkhart, Charles. 1978. "Schenker's 'Motivic Parallelisms.'" *Journal of Music Theory* 22, no. 2 (Autumn): 145–175.

———. 1980. "The Symmetrical Source of Webern's Opus 5, no. 4." In *The Music Forum*, vol. 5, edited by Felix Salzer and Carl Schachter, 317–34. New York: Columbia University Press.

Burstein, L. Poundie, and Joseph N. Straus. 2016. *Concise Introduction to Tonal Harmony*. New York: W. W. Norton.

Busby, Thomas. 1818. *A Grammar of Music*. London: J. Walker.

Butler, Mark J. 2005. *Unlocking the Groove: Rhythm, Meter, and Musical Design in Electronic Dance Music*. Bloomington: Indiana University Press.

Byrnes, Jason T. 2005. "Pedagogical Applications of the Spectrogram in the Low Brass Studio." DM document, Indiana University.

Caplin, William E. 2013. *Analyzing Classical Form: An Approach for the Classroom*. New York: Oxford University Press.

Carpenter, Patricia. 1983. "*Grundgestalt* as Tonal Function." *Music Theory Spectrum* 5:15–38.

Castine, Peter. 1994. *Set Theory Objects: Abstractions for Computer-Aided Analysis and Composition of Serial and Atonal Music*. New York: P. Lang.

Chan-Hartley, Hannah. 2018. "Listening Guide to Antonín Dvořák, Symphony no. 9, mvt. 1." Unpublished manuscript.

Chew, Elaine. 2000. "Towards a Mathematical Model of Tonality." PhD diss., Massachusetts Institute of Technology.

Chew, Elaine, and Alexandre R. J. Francois. 2005. "Interactive Multi-scale Visualizations of Tonal Evolution in MuSA.RT Opus 2." *ACM Computers in Entertainment* 3, no. 4 (October): 1–16.

Choquel, Henri-Louis. (1762) 1972. *La musique rendue sensible par la mécanique*. Nouv. éd. Geneva: Minkoff Reprints.

Christensen, Erik. 1996. *The Musical Timespace: A Theory of Music Listening*. Aalborg: Aalborg University Press.

Cleveland, William S. 1993. "A Model for Studying Display Methods of Statistical Graphics." *Journal of Computational and Graphical Statistics* 2, no. 4 (December): 323–43.

Cogan, Robert. 1984. *New Images of Musical Sound*. Cambridge, MA: Harvard University Press.

———. 1998. *Music Seen, Music Heard: A Picture Book of Musical Design*. Cambridge, MA: Publication Contact International.

Cogan, Robert, and Pozzi Escot. 1976. *Sonic Design: The Nature of Sound and Music*. Englewood Cliffs, NJ: Prentice-Hall.

Cohn, Richard. 2012. *Audacious Euphony: Chromaticism and the Triad's Second Nature*. New York: Oxford University Press.

Cone, Edward T. 1962. "The Uses of Convention: Stravinsky and His Models." *Musical Quarterly* 48, no. 3 (July): 287–99.

Cooper, Grosvenor, and Leonard B. Meyer. 1960. *The Rhythmic Structure of Music*. Chicago: University of Chicago Press.

Cox, Frank. 2004. "Rhythmic Morphology and Temporal Experience: *Doubles*, for Piano and Taped Synthesizers (1990–1993)." In *Musical Morphology: New Music and Aesthetics in the 21st Century*, edited by Claus Steffen Mahnkopf, Frank Cox, and Wolfram Schurig, 86–122. Hofheim: Wolke.

Curwen, John. 1875. *The Teacher's Manual of the Tonic Sol-Fa Method*. London: Tonic Sol-Fa Agency.

Dart, Thurston, John Morehen, and Richard Rastall. 2001. "Tablature." *Grove Music Online*. January 20, 2001. https://www.oxfordmusiconline.com/grovemusic/view/10.1093/gmo/9781561592630.001.0001/omo-9781561592630-e-0000027338.

Davis, Andrew. 2014. "Chopin and the Romantic Sonata: The First Movement of op. 58." *Music Theory Spectrum* 36, no. 2 (Fall): 270–94.

Delaere, Mark. 1993. *Funktionelle Atonalität: Analytische Strategien für die frei-atonale Musik der Wiener Schule*. Wilhelmshaven: F. Noetzel.

de la Motte, Diether. 2005. "Theory—Lehre, Wagnis Analyse." In *Musiktheorie*, edited by Helga De la Motte-Haber and Oliver Schwab-Felisch, 489–98. Laaber: Laaber.

de la Motte-Haber, Helga, and Oliver Schwab-Felisch, eds. 2005. *Musiktheorie*. Laaber: Laaber.

Derr, Ellwood. 1981. "The Two-Part Inventions: Bach's Composers' Vademecum." *Music Theory Spectrum* 3:26–48.

Dibben, Nicola. 1999. "The Perception of Structural Stability in Atonal Music: The Influence of Salience, Stability, Horizontal Motion, Pitch Commonality, and Dissonance." *Music Perception* 16, no. 3 (April): 265–94.

———. 2006. "Subjectivity and the Construction of Emotion in the Music of Björk." *Music Analysis* 25, no. 1/2 (March): 171–97.

Douthett, Jack, and Peter Steinbach. 1998. "Parsimonious Graphs: A Study in Parsimony, Contextual Transformations, and Modes of Limited Transposition." *Journal of Music Theory* 42, no. 2 (Autumn): 241–63.

Dowling, W. Jay. 1991. "Pitch Structure." In *Representing Musical Structure*, edited by Peter Howell, Robert West, and Ian Cross, 33–58. London: Academic.

Duinker, Ben, and Hubert Léveillé Gauvin. 2017. "Changing Content in Flagship Music Theory Journals, 1979–2014." *Music Theory Online* 17, no. 4 (December). https://mtosmt.org/issues/mto.17.23.4/mto.17.23.4.duinker.html.

Dunnick, D. Kim. 1980. "A Physical Comparison of the Tone Qualities of Four Different Brands of B-flat Trumpets with Regard to the Presence and Relative Strengths of Their Respective Partials." DM document, Indiana University.

Encyclopaedia Britannica. 2016. S.v. "church mode." Last modified February 15, 2016. https://www.britannica.com/art/church-mode.

Epstein, David. 1995. *Shaping Time: Music, the Brain, and Performance*. New York: Schirmer Books.

Erpf, Hermann. 1927. *Studien zur Harmonie- und Klangtechnik der nueueren Musik*. Leipzig: Breitkopf und Härtel.

———. 1959. *Lehrbuch der Instrumentation und Instrumentenkunde*. Mainz: Schott.

Euler, Leonhard. 1739. *Tentamen novae theoriae musicae: Ex certissismis harmoniae principiis dilucide expositae*. Petropoli: Ex typographia Academiae scientiarum.

Everett, Yayoi Uno. 2009. "Signification of Parody and the Grotesque in György Ligeti's *Le Grand Macabre*." *Music Theory Spectrum* 31, no. 1 (Spring): 26–56.

Ewell, Philip A. 2020. "Music Theory and the White Racial Frame." *Music Theory Online* 26, no. 2 (June). https://mtosmt.org/issues/mto.20.26.2/mto.20.26.2.ewell.html.

Fauconnier, Gilles, and Mark Turner. 2002. *The Way We Think: Conceptual Blending and the Mind's Hidden Complexities*. New York: BasicBooks.

Feldman, Morton. 2000. "Crippled Symmetry (1981)." In *Give My Regards to Eighth Street: Collected Writings of Morton Feldman*, edited by B. H. Friedman. Cambridge, MA: Exact Change.

Finale. n.d. "Finale 2012 Instrument Ranges." Accessed June 7, 2017. https://usermanuals.finalemusic.com/Finale2012Win/Content/Finale/Instrument_Ranges.htm.

Fink, Michael. 1999. *Exploring Music Literature*. New York: Schirmer Books.

Ford, Lysbeth. 2011. "Marri Ngarr Lirrga Songs: A Linguistic Analysis." *Musicology Australia* 28, no. 1: 26–58.

Forte, Allen. 1964. "A Theory of Set-Complexes for Music." *Journal of Music Theory* 8, no. 2 (Autumn): 136–83.

———. 1973. *The Structure of Atonal Music*. New Haven, CT: Yale University Press.

———. 1980. "Aspects of Rhythm in Webern's Atonal Music." *Music Theory Spectrum* 2:90–109.

———. 1998. *The Atonal Music of Anton Webern*. New Haven, CT: Yale University Press.

Friedmann, Michael L. 1985. "A Methodology for the Discussion of Contour: Its Application to Schoenberg's Music." *Journal of Music Theory* 29, no. 2 (Autumn): 223–48.

Fürniss, Susanne. 2006. "Aka Polyphony: Music, Theory, Back and Forth." In *Analytical Studies in World Music*, edited by Michael Tenzer, 163–204. New York: Oxford University Press.

Garza, Loida Raquel. 2018. "Circle of Fifths Color Palette." Adobe Illustrator. Unpublished.

Gauldin, Robert. 1983. "The Cycle-7 Complex: Relations of Diatonic Set Theory to the Evolution of Ancient Tonal Systems." *Music Theory Spectrum* 5:39–55.

———. 2004. *Harmonic Practice in Tonal Music*. 2nd ed. New York: W. W. Norton.

Gjerdingen, Robert O. 1991. "Using Connectionist Models to Explore Complex Musical Patterns." In *Music and Connectionism*, edited by Peter M. Todd and D. Gareth Loy, 138–49. Cambridge, MA: MIT Press.

Godøy, Rolf Inge, and Harald Jørgensen, eds. 2001. *Musical Imagery*. Lisse, Netherlands: Swets and Zeitlinger.

Goetschius, Percy. 1915. *The Larger Forms of Musical Composition: An Exhaustive Explanation of the Variations, Rondos, and Sonata Designs, for the General Student of Musical Analysis, and for the Special Student of Structural Composition*. New York: G. Schirmer.

Gould, Elaine. 2011. *Behind Bars: The Definitive Guide to Music Notation*. London: Faber.

Grunin, Eric. n.d. "An Eroica Project." Accessed December 27, 2009. http://www.grunin.com/eroica/index.htm. (Site discontinued.)

Hanninen, Dora A. 2012. *A Theory of Music Analysis: On Segmentation and Associative Organization*. Rochester, NY: University of Rochester Press.

———. 2019. "Images, Visualization, and Representation." In *The Oxford Handbook of Critical Concepts in Music Theory*, edited by Alexander Rehding and Steven Rings, 699–741. New York: Oxford University Press.

Hansen, Niels Chr., Makiko Sadakata, and Marcus Pearce. 2016. "Nonlinear Changes in the Rhythm of European Art Music: Quantitative Support for Historical Musicology." *Music Perception* 33, no. 4 (April): 414–31.

Hanson, Howard. 1960. *Harmonic Materials of Modern Music: Resources of the Tempered Scale*. New York: Appleton-Century-Crofts.

Hasty, Christopher. 1981. "Segmentation and Process in Post-tonal Music." *Music Theory Spectrum* 3:54–73.

———. 1997. *Meter as Rhythm*. New York: Oxford University Press.

Hepokoski, James, and Warren Darcy. 2011. *Elements of Sonata Theory: Norms, Types, and Deformations in the Late-Eighteenth-Century Sonata*. New York: Oxford University Press.

Hesselink, Nathan. 2011. "Rhythm and Folk Drumming (P'ungmul) as the Musical Embodiment of Communal Consciousness in South Korean Village Society." In *Analytical and Cross-Cultural Studies in World Music*, edited by Michael Tenzer and John Roeder, 263–87. New York: Oxford University Press.

Hindemith, Paul. 1945. *The Craft of Musical Composition*. Edited by Arthur Mendel and Otto Ortmann. New York: Associated Music.

Holzer, Andreas. 2005. "Das Wiederaufleben pythagoreischer Tradicionen im 20. Jahrhundert." In *Musiktheorie*, edited by Helga De la Motte-Haber and Oliver Schwab-Felisch, 73–90. Laaber: Laaber.

Hook, Julian L. 2002. "Hearing with Our Eyes: The Geometry of Tonal Space." In *Bridges: Mathematical Connections in Art, Music, and Science*, edited by Reza Sarhangi, 123–34. Winfield, KS: Southwestern College.

———. 2006. "Exploring Musical Space." *Science* 313 (July 7): 49–50.

———. 2023. *Exploring Musical Spaces: A Synthesis of Mathematical Approaches*. New York: Oxford University Press.

Horlacher, Gretchen. 2001. "Bartók's 'Change of Time': Coming Unfixed." *Music Theory Online* 7, no. 1 (January). http://www.mtosmt.org/issues/mto.01.7.1/mto.01.7.1.horlacher.html.

Houghton-Webb, Charles. 2001. "Tableau des Tessitures." BW Music. Accessed June 7, 2017. https://bwmusic.com/range.php.

Huovinen, Erkki. 2002. *Pitch-Class Constellations: Studies in the Perception of Tonal Centricity*. Turku: Finnish Musicological Society.

Huron, David. 2001. "Tone and Voice: A Derivation of the Rules of Voice-Leading from Perceptual Principles." *Music Perception* 19, no. 1 (September): 1–64.

———. 2006. *Sweet Anticipation: Music and the Psychology of Expectation*. Cambridge, MA: MIT Press.

Hyer, Brian. 1995. "Reimag(in)ing Riemann." *Journal of Music Theory* 39, no. 1 (Spring): 101–38.

Isaacson, Eric J. 1990. "Similarity of Interval-Class Content between Pitch-Class Sets: The IcVSIM Relation." *Journal of Music Theory* 34, no. 1 (Spring): 1–28.

Iverson, Jennifer. 2014. "Statistical Form amongst the Darmstadt School." *Music Analysis* 33, no. 3 (February): 341–87.

Jaffe, Andrew. 1983. *Jazz Harmony*. 2nd ed. Dubuque, IA: Wm. C. Brown.

Jeong, Hye Min. 2007. *Das musikalische Material und seine Behandlung im Frühwerk von Krzysztof Penderecki: Eine Studie zum Cluster und zur Klangfarbe*. Frankfurt am Main: P. Lang.

Johnson, Jeffrey. 1997. *Graph Theoretical Models of Abstract Musical Transformation: An Introduction and Compendium for Composers and Theorists*. Westport, CT: Greenwood.

Johnston, Blair. 2012. "Between Romanticism and Modernism and Postmodernism: George Crumb's *Black Angels*." *Music Theory Online* 18, no. 2 (June). http://www.mtosmt.org/issues/mto.12.18.2/mto.12.18.2.johnston.html.

———. 2014. "Modal Idioms and Their Rhetorical Associations in Rachmaninoff's Works." *Music Theory Online* 20, no. 4

(December). http://www.mtosmt.org/issues/mto.14.20.4/mto.14.20.4.johnston.html.

Joseph, Charles M. 1982. "Structural Coherence in Stravinsky's *Piano-Rag-Music*." *Music Theory Spectrum* 4:76–91.

Judd, Cristle Collins. 2006. *Reading Renaissance Music Theory: Hearing with the Eyes*. Cambridge: Cambridge University Press.

Kelly, Thomas Forrest. 2015. *Capturing Music: The Story of Notation*. New York: W. W. Norton.

Kerman, Joseph. 1996. *Listen*. 3rd brief ed. New York: Worth.

Kielian-Gilbert, Marianne. 2006. "Inventing a Melody with Harmony: Tonal Potential and Bach's 'Das alte Jahr vergangen ist.'" *Journal of Music Theory* 50, no. 1 (Spring): 77–101.

Klavarskribo Institute. [1950s]. *What Is Klavarskribo?* 2nd ed. Slikkerveer, Netherlands: Klavarskribo.

Klein, Michael. 1999. "Texture, Register, and Their Formal Roles in the Music of Witold Lutosławski." *Indiana Theory Review* 20, no. 1 (Spring): 37–70.

Koch, Heinrich Christoph. 1782. *Versuch einer Anleitung zur Composition*. Leipzig: Bey A. F. Böhme.

Kojs, Juraj. 2011. "Notating Action-Based Music." *Leonardo Music Journal* 21 (December): 65–72.

Kokoras, Panayiotis A. 2017. "Towards a Holophonic Musical Texture." *JMM: The Journal of Music and Meaning* 4 (Winter). http://www.musicandmeaning.net/issues/showArticle.php?artID=4.5.

Koozin, Timothy. 1999. "On Metaphor, Technology, and Schenkerian Analysis." *Music Theory Online* 5, no. 3 (May). http://www.mtosmt.org/issues/mto.99.5.3/mto.99.5.3.koozin.html.

Kostka, Stefan M., and Dorothy Payne. 2009. *Tonal Harmony: With an Introduction to Twentieth-Century Music*. 6th ed. New York: McGraw-Hill.

Kramer, Jonathan D. 1988. *The Time of Music: New Meanings, New Temporalities, New Listening Strategies*. New York: Schirmer Books.

Krebs, Harald. 1999. *Fantasy Pieces: Metrical Dissonance in the Music of Robert Schumann*. New York: Oxford University Press.

Krumhansl, Carol L. 1990. *Cognitive Foundations of Musical Pitch*. New York: Oxford University Press.

Krumhansl, Carol L., and Petri Toiviainen. 2003. "Tonal Cognition." In *The Cognitive Neuroscience of Music*, edited by Isabelle Peretz and Robert J. Zatorre, 95–108. New York: Oxford University Press.

Krützfeldt, Werner. 2005. "Polyphonie in der Musik des 20. Jarhhunderts: Die Logik der Linie." In *Musiktheorie*, edited by Helga De la Motte-Haber and Oliver Schwab-Felisch, 311–34. Laaber: Laaber.

Kurth, Richard. 1999. "Partition Lattices in Twelve-Tone Music: An Introduction." *Journal of Music Theory* 43, no. 1 (Spring): 21–82.

Kuusi, Tuire. 2001. *Set-Class and Chord: Examining Connection between Theoretical Resemblance and Perceived Closeness*. Studia Musica. Helsinki: Sibelius Academy.

Lakoff, George, and Mark Johnson. 1980. *Metaphors We Live By*. Chicago: University of Chicago Press.

Larson, Steve. 2012. *Musical Forces: Motion, Metaphor, and Meaning in Music*. Bloomington: Indiana University Press.

LaRue, Jan. 2011. *Guidelines for Style Analysis*. Expanded 2nd ed. Warren, MI: Harmonie Park.

Leichtentritt, Hugo. 1951. *Musical Form*. Cambridge, MA: Harvard University Press.

Lendvai, Ernő. 1971. *Béla Bartók: An Analysis of His Music*. London: Kahn and Averill.

Lendvai, Ernő, Mikløs Szabø, and Mikløs Mohay. 1993. *Symmetries of Music: An Introduction to Semantics of Music*. Kecskemét: Kodály Institute.

Lerdahl, Fred. 1989. "Atonal Prolongational Structure." *Contemporary Music Review* 4:64–87.

———. 2001. *Tonal Pitch Space*. New York: Oxford University Press.

———. 2003. "The Sounds of Poetry Viewed as Music." In *The Cognitive Neuroscience of Music*, edited by Isabelle Peretz and Robert J. Zatorre, 413–29. New York: Oxford University Press.

Lerdahl, Fred, and Ray Jackendoff. 1983. *A Generative Theory of Tonal Music*. Cambridge, MA: MIT Press.

Lester, Joel. 1979. "Articulation of Tonal Structures as a Criterion for Analytic Choices." *Music Theory Spectrum* 1:67–79.

Lewin, David. 1987. *Generalized Musical Intervals and Transformations*. New Haven, CT: Yale University Press.

———. 1993. *Musical Form and Transformation: 4 Analytic Essays*. New Haven, CT: Yale University Press.

Lewis, Christopher. 1981. "Tonal Focus in Atonal Music: Berg's op. 5/3." *Music Theory Spectrum* 3:84–97.

Ligeti, György. 1970. *Artikulation (an Aural Score)*. Edited by Rainer Wehinger. Mainz: B. Schott's Söhne.

Lochhead, Judith. 2016. "'Difference Inhabits Repetition': Sofia Gubaidulina's String Quartet no. 2." In *Analytical Essays on Music by Women Composers: Concert Music, 1960–2000*, edited by Laurel Parsons and Brenda Ravenscroft, 102–26. New York: Oxford University Press.

Longy-Miquelle, Renée. 1925. *Principles of Musical Theory*. Boston: E. C. Schirmer Music.

Losada, C. Catherine. 2009. "Between Modernism and Postmodernism: Strands of Continuity in Collage Compositions by Rochberg, Berio, and Zimmermann." *Music Theory Spectrum* 31, no. 1 (Spring): 57–100.

MacPherson, Stewart. 1930. *Form in Music with Special Reference to the Designs of Instrumental Music*. New and rev. ed. London: J. Williams.

Madden, Charles. 1999. *Fractals in Music: Introductory Mathematics for Musical Analysis*. Salt Lake City: High Art.

Malawey, Victoria. 2020. *A Blaze of Light in Every Word: Analyzing the Popular Singing Voice*. New York: Oxford University Press.

Malinowski, Stephen, and Jay Baca. 2013. "Animated Graphical Score of Stravinsky, *Rite of Spring*." YouTube. Accessed February 10, 2015. https://www.youtube.com/watch?v=5IXMpUhuBMs.

Maniates, Maria Rika. 1993. "The Cavalier Ercole Bottrigari and His Brickbats: Prolegomena to the Defense of Don Nicola Vicentino against Messer Gandolfo Sigonia." In *Music Theory and the Exploration of the Past*, edited by Christopher Hatch and David W. Bernstein, 137–88. Chicago: University of Chicago Press.

Margulis, Elizabeth Hellmuth. 2003. "Melodic Expectation: A Discussion and Model." PhD diss., Columbia University.

———. 2017. "An Exploratory Study of Narrative Experiences of Music." *Music Perception* 35, no. 2 (December): 235–48.

Marvin, Elizabeth West. 1983. "The Structural Role of Complementation in Webern's *Orchestra Pieces (1913)*." *Music Theory Spectrum* 5:76–88.

Marvin, Elizabeth West, and Paul A. Laprade. 1987. "Relating Musical Contours: Extensions of a Theory for Contour." *Journal of Music Theory* 31, no. 2 (Autumn): 225–67.

Mazzola, Guerino, Stefan Göller, and Stefan Müller. 2002. *The Topos of Music: Geometric Logic of Concepts, Theory, and Performance*. Basel: Birkhauser.

McClimon, Michael. 2017. "Transformations in Tonal Jazz: ii–V Space." *Music Theory Online* 23, no. 1 (March). http://mtosmt.org/issues/mto.17.23.1/mto.17.23.1.mcclimon.html.

Mead, Andrew W. 1983. "Detail and the Array in Milton Babbitt's *My Complements to Roger*." *Music Theory Spectrum* 5:89–109.

———. n.d. "Expanded Phrase Model." Graphical design by John Heilig (Abobe Illustrator), 2016. Unpublished.

Mecklenburg, Carl Gregor, and Waldemar Scheck. 1963. *Die Theorie des Blues im modernen Jazz*. Strasbourg: Heitz.

Meyer, Felix, and Anne Shreffler. 1996. "Performance and Revision: The Early History of Webern's Four Pieces for Violin and Piano, op. 7." In *Webern Studies*, edited by Kathryn Bailey, 135–69. Cambridge: Cambridge University Press.

Middleton, Richard. 1993. "Popular Music Analysis and Musicology: Bridging the Gap." *Popular Music* 12, no. 2 (May): 177–90.

Monzo, Joe. 2000. "Partch 43-Tone JI Scale: Shaded Inverted Monzo Lattice." Accessed October 25, 2018. http://www.tonalsoft.com/monzo/partch/scale/partch43-lattice.aspx.

Morgan, Robert P. 1993. "Coda as Culmination: The First Movement of the *Eroica* Symphony." In *Music Theory and the Exploration of the Past*, edited by Christopher Hatch and David W. Bernstein, 357–76. Chicago: University of Chicago Press.

Morley, Thomas. 1597. *A Plaine and Easie Introduction to Practicall Musicke*. London: Peter Short.

Morris, Robert. 1987. *Composition with Pitch-classes: A Theory of Compositional Design*. New Haven, CT: Yale University Press.

———. 2001. *Class Notes for Advanced Atonal Music Theory*. Lebanon, NH: Frog Peak.

Ms. 241. [1150–1199]. Angers. Bibliotheque Municipale. Laon, France.

Neidhöfer, Christoph. 2016. "Improvvisazioni Concertanti no. 1 by Norma Beecroft: Serialism, Improvisatory Discourse, and the Musical Avant-Garde." In *Analytical Essays on Music by Women Composers: Concert Music, 1960–2000*, edited by Laurel Parsons and Brenda Ravenscroft, 33–66. New York: Oxford University Press.

Neumeyer, David. 1982. "Organic Structure and the Song Cycle: Another Look at Schumann's *Dicterliebe*." *Music Theory Spectrum* 4:92–105.

Neuwirth, Markus. 2013. "Surprise without a Cause?: 'False Recapitulations' in the Classical Repertoire and the Modern Paradigm of Sonata Form." *Zeitschrift der Gesellschaft für Musiktheorie* 10, no. 2. http://www.gmth.de/zeitschrift/artikel/722.aspx.

Ockelford, Adam. 2005. *Repetition in Music: Theoretical and Metatheoretical Perspectives*. Burlington, VT: Ashgate.

Oettingen, Arthur von. 1866. *Harmoniesystem in dualer Entwicklung*. Leipzig: Verlag von W. Glaser.

———. 1916. *Die Grundlage der Musikwissenschaft und das duale Reininstrument*. Leipzig: B. G. Teubner.

Ohriner, Mitchell S. 2007. "Instrument Ranges." Adobe Illustrator. Unpublished.

———. 2012. "Grouping Hierarchy and Trajectories of Pacing in Performances of Chopin's Mazurkas." *Music Theory Online* 18, no. 1 (April). http://mtosmt.org/issues/mto.12.18.1/mto.12.18.1.ohriner.php.

Osmond-Smith, David. 1985. *Playing on Words: A Guide to Luciano Berio's "Sinfonia."* London: Royal Musical Association.

Pantev, C., A. Engelien, V. Candia, and T. Elbert. 2003. "Representational Cortex in Musicians." In *The Cognitive Neuroscience of Music*, edited by Isabelle Peretz and Robert J. Zatorre, 382–95. New York: Oxford University Press.

Parks, Richard S. 1980. "Pitch Organization in Debussy: Unordered Sets in 'Brouillards.'" *Music Theory Spectrum* 2:119–34.

Partch, Harry. 1974. *Genesis of a Music: An Account of a Creative Work, Its Roots, and Its Fulfillments*. 2nd ed. Madison: University of Wisconsin Press.

Peretz, Isabelle. 2003. "Brain Specialization for Music: New Evidence from Congenital Amusia." In *The Cognitive Neuroscience of Music*, edited by Isabelle Peretz and Robert J. Zatorre, 192–203. New York: Oxford University Press.

Perle, George. 1991. *Serial Composition and Atonality: An Introduction to the Music of Schoenberg, Berg, and Webern*. 6th ed. Berkeley: University of California Press.

Pierce, Alexandra. 2007. *Deepening Musical Performance: The Theory and Practice of Embodied Interpretation*. Bloomington: Indiana University Press.

Piston, Walter. 1962. *Harmony*. 3rd ed. New York: W. W. Norton.

Polth, Michael. 2005. "Dodekaphonie und Serialismus." In *Musiktheorie*, edited by Helga De la Motte-Haber and Oliver Schwab-Felisch, 421–40. Laaber: Laaber.

Prince, Jon B., and Mark A. Schmuckler. 2014. "A Corpus Analysis: The Tonal-Metric Hierarchy." *Music Perception* 31, no. 3 (February): 254–70.

Probst, Stephanie. 2021. "Music Appreciation through Animation: Percy Scholes's 'AudioGraphic' Piano Rolls." *SMT-V: The Society for Music Theory Videocast Journal* 7, no. 1 (January). https://www.smt-v.org/archives/volume7.html#music-appreciation-through-animation-percy-scholess-audiographic-piano-rolls.

Protopopov, Sergei. 1930. *Элементы строения музыкальной речи* [Elements of the structure of musical speech]. 2 vols. Moscow: Izdatel'stvo Muzykal'nyĭ Sektor.

Puffett, Derrick. 1996. "Gone with the Summer Wind; or, What Webern Lost." In *Webern Studies*, edited by Kathryn Bailey, 32–73. Cambridge: Cambridge University Press.

Quinn, Ian. 1997. "Fuzzy Extensions to the Theory of Contour." *Music Theory Spectrum* 19, no. 2 (Fall): 232–63.

———. 1999. "The Combinatorial Model of Pitch Contour." *Music Perception* 16, no. 4 (July): 439–56.

Rahn, John. 1987. *Basic Atonal Theory*. New York: Longman.

Read, Gardner. 1990. *20th-Century Microtonal Notation*. New York: Greenwood.

Reck, David B. 1996. "India/South India." In *Worlds of Music*, edited by Jeff Todd Titon, 252–315. New York: Schirmer.

Reef, John S. 2019. "Subjects and Phrase Boundaries in Two Keyboard Fugues by J. S. Bach." *Music Theory Spectrum* 41, no. 1 (Spring): 48–73.

Reicha, Anton. 1824. *Traité de haute composition musicale*. Vol. 2. Paris: Zetter.

Reuter, Christoph. 2002. *Klangfarbe und Instrumentation: Geschichte, Ursachen, Wirkung*. Frankfurt am Main: P. Lang.

Richards, Mark. 2016. "Film Music Themes: Analysis and Corpus Study." *Music Theory Online* 22, no. 1 (March). https://www.mtosmt.org/issues/mto.16.22.1/mto.16.22.1.richards.html.

Riemann, Hugo. 1914–15. "Ideen zu einer 'Lehre von den Tonvorstellungen.'" *Jahrbuch der Musikbibliothek Peters*, vol. 21–22, 1–26. Trans. by Robert W. Wason and Elizabeth W. Marvin, *Journal of Music Theory* 36, no. 1 (Spring 1992): 81–117.

———. 1919. *L. van Beethoven sämtliche Klavier-solosonaten; Ästhetische und formal-technische Analyse mit historischen Notizen*. Berlin: M. Hesse.

Rings, Steven. 2013. "A Foreign Sound to Your Ear: Bob Dylan Performs 'It's Alright, Ma (I'm Only Bleeding),' 1964–2009." *Music Theory Online* 19, no. 4 (December). https://mtosmt.org/issues/mto.13.19.4/mto.13.19.4.rings.html.

———. 2018. "Music's Stubborn Enchantments (and Music Theory's)." *Music Theory Online* 24, no. 1 (March).

https://mtosmt.org/issues/mto.18.24.1/mto.18.24.1.rings.html.

Risset, Jean-Claude, and David L. Wessel. 1982. "Exploration of Timbre by Analysis and Synthesis." In *The Psychology of Music*, edited by Diana Deutsch, 25–58. San Diego: Academic. https://doi.org/10.1016/B978-0-12-213562-0.50006-1.

Rode-Breymann, Susanne. 1996. "'. . . Gathering the Divine from the Earthly . . .': Ferdinand Avenariou and His Significance for Anton Webern's Early Settings of Lyris Poetry." In *Webern Studies*, edited by Kathryn Bailey, 1–31. Cambridge: Cambridge University Press.

Roeder, John. 1994. "Interacting Pulse Streams in Schoenberg's Atonal Polyphony." *Music Theory Spectrum* 16, no. 2 (Fall): 231–49.

———. 2006. "Autonomy and Dialogue in Elliott Carter's *Enchanted Preludes*." In *Analytical Studies in World Music*, edited by Michael Tenzer, 377–414. New York: Oxford University Press.

———. 2009a. "Constructing Transformational Signification: Gesture and Agency in Bartók's Scherzo, op. 14, no. 2, measures 1–32." *Music Theory Online* 15, no. 1 (March). http://www.mtosmt.org/issues/mto.09.15.1/mto.09.15.1.roeder_signification.html.

———. 2009b. "A Transformational Space Structuring the Counterpoint in Adès's 'Auf dem Wasser zu singen.'" *Music Theory Online* 15, no. 1 (March). https://mtosmt.org/issues/mto.09.15.1/mto.09.15.1.roeder_space.html.

———. 2016. "Superposition in Kaija Saariaho's 'The claw of the magnolia . . .'" In *Analytical Essays on Music by Women Composers: Concert Music, 1960–2000*, edited by Laurel Parsons and Brenda Ravenscroft, 156–75. New York: Oxford University Press.

Rogers, Nancy M., and Michael H. Buchler. 2003. "Square Dance Moves and Twelve-Tone Operators: Isomorphisms and New Transformational Models." *Music Theory Online* 9, no. 4 (December). https://mtosmt.org/issues/mto.03.9.4/mto.03.9.4.rogers_buchler.html.

Roig-Francolí, Miguel. 2007. *Understanding Post-tonal Music*. New York: McGraw-Hill.

———. 2018. "From Renaissance to Baroque: Tonal Structures in Tomás Luis de Victoria's Masses." *Music Theory Spectrum* 40, no. 1 (Spring): 27–51.

Rusch, René. 2013. "Crossing Over with Brad Mehldau's Cover of Radiohead's 'Paranoid Android': The Role of Jazz Improvisation in the Transformation of an Intertext." *Music Theory Online* 19, no. 4 (December). http://www.mtosmt.org/issues/mto.13.19.4/mto.13.19.4.rusch.html.

Samarotto, Frank. 2004. "Sublimating Sharp $\hat{4}$: An Exercise in Schenkerian Energetics." *Music Theory Online* 10, no. 3 (September). https://mtosmt.org/issues/mto.04.10.3/mto.04.10.3.samarotto.html.

Samplaski, Art. 2004. "Music History at Ten Years a Minute." *College Music Symposium* 44. https://www.jstor.org/stable/40374493.

Sapp, Craig Stuart. 2011. "Computational Methods for the Analysis of Musical Structure." PhD diss., Stanford University.

Schachter, Carl. 1987. "Rhythm and Linear Analysis: Aspects of Meter." In *The Music Forum*, edited by Felix Salzer and Carl Schachter, 1–60. New York: Columbia University Press.

Scheideler, Ullrich. 2005. "Analyse von Tonhöhenordnungen: Allen Fortes pitch-class-set System." In *Musiktheorie*, edited by Helga De la Motte-Haber and Oliver Schwab-Felisch, 391–408. Laaber: Laaber.

Schenker, Heinrich. 1969. *Five Graphic Music Analyses*. New York: Dover.

Schmalfeldt, Janet. 2017. "Beethoven's 'Violation': His Cadenza for the First Movement of Mozart's Concerto in D Minor, K. 466." *Music Theory Spectrum* 39, no. 1 (Spring): 1–17.

Schmidt-Jones, Siegfried, and Barry Graves. 1973. *Rock-Lexikon*. Hamburg: Reinbek.

Schøyen Collection. [ca. 1950]. MS 1275/19. Metz linear staffless neumes. France [or Germany?].

Schröder, Hermann. 1902. *Die symmetrische Umkehrung in der Musik*. Leipzig: Breitkopf und Härtel.

Schuller, Gunther. 1998. *The Compleat Conductor*. New York: Oxford University Press.

Secor, George D., and David C. Keenan. 2012. "Sagittal: A Microtonal Notation System." Accessed February 22, 2019. http://sagittal.org/sagittal.pdf.

Senior, Mike, and Paul White. 2001. "Understanding Instrument & Voice Frequency Ranges." Sound on Sound. Accessed June 1, 2022. https://www.soundonsound.com/techniques/using-eq.

Shepard, Roger. 1982. "Geometrical Approximations to the Structure of Musical Pitch." *Psychological Review* 89:305–33.

Simms, Bryan R. 1996. *Music of the Twentieth Century: Style and Structure*. 2nd ed. New York: Schirmer Books.

———. 2000. *The Atonal Music of Arnold Schoenberg, 1908–1923*. New York: Oxford University Press.

Slawson, Wayne. 1981. "The Color of Sound: A Theoretical Study in Musical Timbre." *Music Theory Spectrum* 3:132–41.

Slonimsky, Nicolas. 1960. *The Road to Music*. Rev. ed. New York: Dodd, Mead.

Smith, Tim. 1997. "Animated Analysis of Schoenberg, *Klavierstück*, op. 11, no. 1." Accessed June 22, 2012. URL lost.

Spitzer, Michael. 2004. *Metaphor and Musical Thought*. Chicago: University of Chicago Press.

Stanyek, Jason, and Fabio Oliveira. 2011. "Nuances of Continual Variation in the Brazilian *Pagode* Song 'Sorriso Aberto.'" In *Analytical and Cross-Cultural Studies in World Music*, edited by Michael Tenzer and John Roeder, 98–146. New York: Oxford University Press.

Straus, Joseph N. 1997. "Voice Leading in Atonal Music." In *Music Theory in Concept and Practice*, edited by James M. Baker, David Beach, and Jonathan W. Bernard, 237–74. Rochester, NY: University of Rochester Press.

———. 2005. *Introduction to Post-tonal Theory*. 3rd ed. Upper Saddle River, NJ: Prentice Hall.

———. 2016a. *Introduction to Post-tonal Theory*. 4th ed. New York: W. W. Norton.

———. 2016b. "'Twelve Tone in My Own Way': An Analytical Study of Ursula Mamlok's *Panta Rhei*, Third Movement, with Some Reflections on Twelve-Tone Music in America." In *Analytical Essays on Music by Women Composers: Concert Music, 1960–2000*, edited by Laurel Parsons and Brenda Ravenscroft, 18–31. New York: Oxford University Press.

Tarasti, Eero. 1994. *A Theory of Musical Semiotics*. Bloomington: Indiana University Press.

Taruskin, Richard. 1993. "The Traditional Revisited: Stravinsky's Requiem Canticles as Russian Music." In *Music Theory and the Exploration of the Past*, edited by Christopher Hatch and David W. Bernstein, 525–50. Chicago: University of Chicago Press.

Temperley, David. 2001. *The Cognition of Basic Musical Structures*. Cambridge, MA: MIT Press.

Tenzer, Michael. 2011. "A Cross-Cultural Topology of Musical Time: Afterword to the Present Book and to *Analytical Studies in World Music* (2006)." In *Analytical and Cross-Cultural Studies in World Music*, edited by Michael Tenzer and John Roeder, 415–40. New York: Oxford University Press.

Terauchi, Naoko. 2011. "Surface and Deep Structure in the *Tôgaku* Ensemble of Japanese Course Music (*Gagaku*)." In *Analytical and Cross-Cultural Studies in World Music*, edited by Michael Tenzer and John Roeder, 19–55. New York: Oxford University Press.

Tramo, Mark Jude, Peter A. Cariani, Bertrund Delgutte, and Louis D. Braida. 2003. "Neurobiology of Harmony Perception." In *The Cognitive Neuroscience of Music*, edited by Isabelle Peretz and Robert J. Zatorre, 127–51. New York: Oxford University Press.

Tufte, Edward R. 1995. *Envisioning Information*. Cheshire, CT: Graphics.

———. 2001. *The Visual Display of Quantitative Information*. 2nd ed. Cheshire, CT: Graphics.

———. 2003. *Visual Explanations: Images and Quantities, Evidence and Narrative*. Cheshire, CT: Graphics.

———. 2006. *Beautiful Evidence*. Cheshire, CT: Graphics.

———. 2020. *Seeing with Fresh Eyes: Meaning, Space, Data, Truth*. Cheshire, CT: Graphics.

Turek, Ralph. 1996. *The Elements of Music: Concepts and Applications*. 2nd ed. Vol. 1. New York: McGraw-Hill.

Turner, William. 1724. *Sound Anatomiz'd*. London: William Pearson.

Tymoczko, Dmitri. 2012. "The Generalized Tonnetz." *Journal of Music Theory* 56, no. 1 (Spring): 1–51.

VanHandel, Leigh, and Tian Song. 2010. "The Role of Meter in Compositional Style in 19th Century French and German Art Song." *Journal of New Music Research* 39, no. 1 (April): 1–11.

Ware, Colin. 2008. *Visual Thinking for Design*. Burlington, MA: Morgan Kaufmann.

Wason, Robert W. 1996. "A Pitch-Class Motive in Webern's George-Lieder, op. 3." In *Webern Studies*, edited by Kathryn Bailey, 111–34. Cambridge: Cambridge University Press.

Wattenberg, Martin. n.d. "The Shape of Song." Turbulence. Accessed October 12, 2009. http://turbulence.org/Works/song/gallery/gallery.html.

Weather Underground. 2015. "Forecast for Bloomington, Indiana." Accessed February 17, 2015. http://www.wunderground.com/cgi-bin/findweather/getForecast?query=Bloomington%2C+IN.

Weber, Gottfried. 1851. *The Theory of Musical Composition: Treated with a View to a Naturally Consecutive Arrangement of Topics*. Vol. 1. Translated by James Warner. London: Messrs. Robert Cocks.

Wennerstrom, Mary. 1975. "Form in Twentieth-Century Music." In *Aspects of Twentieth-Century Music*, edited by Gary E. Wittlich, 1–65. Englewood-Cliffs, NJ: Prentice-Hall.

Westergaard, Peter. 1996. “Geometries of Sounds in Time.” *Music Theory Spectrum* 18, no. 1 (Spring): 1–21.

Widdess, Richard. 2011. “Dynamics of Melodic Discourse in Indian Music: Budhaditya Mukherjee’s *Ālāp* in *Rāg* Pūriyā-Kalyān.” In *Analytical and Cross-Cultural Studies in World Music*, edited by Michael Tenzer and John Roeder, 187–224. New York: Oxford University Press.

Wikipedia. 2021a. S.v. “circle of fifths.” Last modified September 12, 2021. https://en.wikipedia.org/wiki/Circle_of_fifths.

Wikipedia. 2021b. S.v. “mensuration notation.” Last modified May 12, 2021. https://en.wikipedia.org/wiki/Mensural_notation.

Williams, Chas. 2001. *The Nashville Number System*. 6th ed. N.p.: Big Timbre Music.

Williams, J. Kent. 1997. *Theories and Analyses of Twentieth-Century Music*. Fort Worth: Harcourt Brace.

Winold, Allen. 1986. *Harmony: Patterns and Principles*. Englewood Cliffs, NJ: Prentice-Hall.

Wintle, Christopher. 1976. “Milton Babbitt’s *Semi-simple Variations*.” *Perspectives of New Music* 14, no. 2 (Spring–Summer)/15, no. 1 (Fall–Winter): 111–54.

———. 1996. “Webern’s Lyric Character.” In *Webern Studies*, edited by Kathryn Bailey, 229–63. Cambridge: Cambridge University Press.

Wittlich, Gary E. 1975. “Sets and Ordering Procedures in Twentieth-Century Music.” In *Aspects of Twentieth-Century Music*, edited by Gary E. Wittlich, 388–476. Englewood-Cliffs, NJ: Prentice-Hall.

Wolff, C. A. Herm. [1894]. *Kurzgefasste Allgemeine Musiklehre*. Leipzig: P. Reclam.

Zarlino, Gioseffo. 1571. *Dimostrationi Harmoniche*. Venice.

Zbikowski, Lawrence. 2002. *Conceptualizing Music: Cognitive Structure, Theory, and Analysis*. Oxford: Oxford University Press.

Ziporyn, Evan, and Michael Tenzer. 2011. “Thelonious Monk’s Harmony, Rhythm, and Pianism.” In *Analytical and Cross-Cultural Studies in World Music*, edited by Michael Tenzer and John Roeder, 147–86. New York: Oxford University Press.

Zorzini, Catalin. 2010. “How the Web Has Changed the Way We Buy Music [Infographic].” Last modified September 27, 2010. http://inspiredm.com/how-the-web-has-changed-the-way-we-buy-music-infographic/. (Site discontinued.)

Index

Page numbers in *italics* indicate a figure; numbers in **bold** indicate a figure in the online supplement; CP indicates a colorplate.

ERIC ISAACSON
is Associate Professor of Music Theory at the Indiana University Jacobs School of Music and a faculty member in the Cognitive Science Program at Indiana University.